Magnificent Menagerie

LUCINDA LAMBTON'S

Magnificent Menagerie

— OR —

Queer Pets and Their Goings-On

HarperCollins*Publishers*

HarperCollins*Publishers*
77–85 Fulham Palace Road,
Hammersmith, London w6 8jb

Published by HarperCollins*Publishers* 1992
9 8 7 6 5 4 3 2 1

A catalogue record for this book is
available from the British Library

ISBN 0 00 217723 4

Set in Linotron Fournier by
Rowland Phototypesetting Limited
Bury St Edmunds, Suffolk

Printed in Hong Kong

To my sweetheart husband, Perry

❧ CONTENTS ❧

Acknowledgements		viii
Introduction		xi
1	Wakenings of Life	1
2	The Nature of the Beast	19
3	Admirable Antics	37
4	Our Dumb Companions	51
5	Outrage, Revenge and Betrayal	73
6	Vim and Verve	95
7	Service to Man	113
8	Saved from Pot and Grave	129
9	Strange As Can Be	143
10	Out of the Run of Things	161
11	Tricks and Treats	179
12	Improbabilities	195
13	Of Beasts and Men	215
14	Sentiments of Beauty, Wonder and Passion	233
15	Fables and Fancies, Myths and Magic	253
16	Have a Good Laugh	271
17	A Cautionary Tale	289
18	Tears and Farewell	307
	List of Illustrations	333
	Index of Subjects	335
	Index of Authors and Sources	341

❧ Acknowledgements ❧

I want to thank all the following people from the bottom of my heart for their varied and many kindnesses:

Lancelot Allgood; Pauline Allwright, Imperial War Museum; Ian Anstruther; Nicholas Bagnall; George Bailey, Bates; Hugh Belsey; Carrie and Glynn Boyd Harte; Julia Brightwell, Reflections; Alan Broomfield; Frances Broomfield; Thomas Chitty; David E. Coke; Neil Crombie; Duncan Dallas; Warren Davies; Ansie de Swardt; Paul Dove, British Museum; Sue Dudley, London Library; Barbara Fulford; Hazel Garlick, Royal Naval Museum; Jason Goodwin; John Gross; Huckleberry Harrod; Sharon Haskell; Michael Higgins, London Library; Myles Hildyard; Pamela Hoare; Jennifer Leggatt; Bill and Shirley Letwin; Marina Majdalany; Douglas Matthews; my mother and father; David and Jeannie Oliver; Lynn Osborne, British Library; Graham Ovenden; Hugh Pawsey; Lucienne Phillips; Nicholas David Phillips; Michael and Elizabeth Sandford; John Saumerez Smith; Mary Sheehan, *Nature* magazine; Susan Small, *Daily Telegraph*; Graham Snow; Donald Steven; Sir Ralph Verney; Richard Wheeler, National Trust; Gerald L. Wood.

Special thanks must be showered on to: Dan Franklin for commissioning the book, which gave me the incentive to wade through so many literary delights; Ariane Goodman for encouraging it on its way; Kate Harris for her spot-on sympathy in nosing out so many gems; Michael Fishwick for cheerfully helping to battle through the mountain of files that eventually accumulated; Esther Jagger for her expert efficiency in sorting through the subsequent mass of papers; Francis Graham for both nosing out and finding out so many of the final details; Juliet Van Oss for days of calming support as we struggled through the final order; Perry, my husband, for listening with ever smiling patience to tales of operatic mice, a trained oyster and a bejewelled tortoise; and lastly my assistant, Justine Oliver, without whose patience, inspired initiative, never-failing forbearance, good humour and friendship this book would never have seen the light of day.

The author and publishers would like to thank the following copyright holders for their permission to use material:

Ernie Bayly (19 Glendale Road, Bournemouth BH6 4JA) for extract from *The Story of Nipper and the 'His Master's Voice' picture painted by Francis Barraud* by Leonard Petts, revised edition 1990;

A & C Black for extract from *The Animal World of Albert Schweitzer*, ed. Charles Joy;

Curtis Brown Group Ltd on behalf of the Executors of Sir Patrick Coghill, Bt for extracts from *Happy Days* and *Some Experiences of an Irish RM*, © E. OE. Somerville and Martin Ross;

Curtis Brown Ltd, London for extract from *Drawn from Memory*, © E. H. Shepard 1956;

Andre Deutsch Ltd for extracts from *Journals of Canetti*, Elias Canetti, 1942;

Florence Feiler for extracts from *Out of Africa* by Karen Blixen, © 1937, 1938 by Random House, Inc., copyright renewed 1956 by Rungstedlund;

Mrs M. Forge for extract from *The Dogs' Cemetery, Oatlands* by J. W. Lindus Forge;

Victor Gollancz Ltd for extract from *Courtiers* by Thomas Hinde;

Daphne Eden Blanche Halstead for extract from *Sold for a Farthing* by Clare Kipps;

Hamish Hamilton Ltd for extract from *Time Was* by W. Graham Robertson, pub. 1931;

HarperCollins for extract from *My Family and Other Animals* by Gerald Durrell, pub. Rupert Hart-Davies;

Harvard University Press for extracts from *Natural History* by Pliny, trans. H. Rackham, pub. Heinemann 1940;

A. M. Heath for extract from *Swans Reflecting Elephants* by Edward James, ed. George Melly, pub. Weidenfeld & Nicolson;

David Higham Assocs for extracts from *English Eccentrics* by Edith Sitwell, pub. Penguin;

Little, Brown & Co. Ltd for extract from *Life with Rossetti* by Gale Pedrick, pub. Macdonald;

Ewan MacNaughton Assocs for extracts from the *Daily Telegraph*: letter, November 1985, © Daily Telegraph plc, 1985; obituary of Charles Addams, © Daily Telegraph plc, 1992, Peregrine Worsthorne, 24 April 1991, © Daily Telegraph plc 1991;

John Murray (Publishers) Ltd for extract from Lytton Strachey, *Letters 1892–1956*, ed. Rupert Hart-Davis; and for extract from *The Story of San Michele* by Axel Munthe;

Penguin Books Ltd for extract from *Germinal* by Emile Zola, trans. L. W. Tancock (Penguin Classics, 1954), © Leonard Tancock 1954; and for extract from *Against Nature* by Huysman, trans. Robert Baldrick (Penguin Classics, 1959), © the Estate of Robert Baldrick 1959; and for extract from 'A Simple Heart' from *Three Tales* by Gustave Flaubert, trans. Robert Baldrick (Penguin Classics, 1961), © Robert Baldrick 1961; and for extracts from *Other People's Trades* by Primo Levi, trans. Raymond Rosenthal (Michael Joseph, 1989), © Giulio Einaudi editore s p a, 1985, translation © Summit Books, a division of Simon & Schuster Inc, 1989; and for extract from *Ring of Bright Water* by Gavin Maxwell (Penguin Books, 1960), © Gavin Maxwell 1960;

Random Century Group on behalf of the Estate of J. R. Ackerley for extract from *My Dog Tulip*, pub. Bodley Head; and on behalf of Mrs Laura Huxley for extract from 'Sermons in Cats' by Aldous Huxley, pub. Chatto & Windus; and on behalf of C. Day Lewis for lines from *Georgics* by Virgil, trans. C. Day Lewis, pub. Jonathan Cape;

Raymond Savage Ltd for extract from *Hippy, The Story of a Dog* by Sir Nevile Henderson;

Martin Secker and Warburg Ltd for lines from 'The Horse' from *Collected Poems of Ronald Duncan*, 1962, and for extract from *If I Die (Sie le Grain Ne Meurt)* by Andre Gide, trans. Dorothy Bussey, 1926, and for extracts from *My Mother's House* by Colette, trans. Una Vicenzo Troubridge and Enid McLeod, 1923;

Sheil Land Associates Ltd for extract from *Irish Eccentrics* by Peter Somerville-Large, pub. Hamish Hamilton, © Peter Somerville-Large 1975;

The Society of Authors on behalf of the Estate of John Masefield for lines from *Right Royal* and *Reynard the Fox*;

The *Tablet* for extracts from 'An Animal Alphabet' by Neville Braybrooke, 21/28 December 1991;

University of Reading on behalf of Hubert Nicholson for part of 'Cats' from *The Collected Poems of A. S. J. Tessimond, with translations from Jacques Prevert*, ed. Hubert Nicholson (Whiteknights Press, 1985);

A. P. Watt on behalf of The Executors of the Estate of David Garnett for extract from *Lady into Fox* by David Garnett; and on behalf of Michael Simone O'Reilly and Sylvia O'Reilly for extract from *The Pre-Raphaelite Tragedy* by William Gaunt;

George Weidenfeld & Nicolson Ltd for extract from *Wellington: The Years of the Sword* by Elizabeth Longford;

Grateful acknowledgement is made to the following copyright holders for their permission to reproduce illustrations:

Divers Exercises des Éléphants (p. 29), The dexterity of deer (p. 38), Fairground tricks (p. 172), Les Serins Savants (p. 184), 'He bid his horse . . .' (p. 220), Horse with plumes (p. 239) by permission of the British Library; 'I never reared a young wombat . . .' (p. 331) reproduced by courtesy of the Trustees of the British Museum; Frances Broomfield for: Sherry (p. 168); Mary Evans Picture Library for: A new musical instrument (p. 85); Robert Hale for: Toby the Sapient Pig (p. 187); Hulton Deutsch Library for: Two maiden ladies and an alligator (p. 7); Imperial War Museum for: A noble sacrifice (p. 100), Help the Horse (p. 111); Methuen for: 'Comfortably tucked up in bed . . .' (p. 65); Daphne Halstead for: 'I can turn round ten times' (p. 46); Royal Naval Museum, Portsmouth for: Her Master's Breath (p. 119), 'It was positively affecting . . .' (p. 139); D. C. Thomson for: Gnasher (p. 288); Sir Ralph Verney for: Florence Nightingale (p. 3); Gerald L. Wood for: Railway Jack (p. 42);

Every effort has been made to contact the copyright owners of material included in this anthology. In the instances where this has not proved possible, the author and publishers offer their apologies to and would be pleased to hear from those concerned.

❦ INTRODUCTION ❦

Why are we moved to tears of laughter, delight, wonder and sentimentality by such tales as a chicken's friendship with a horse, a monkey manning the signals of the Natal railway or a sparrow's war effort of cheering card tricks in the air-raid shelters? The tears well up, the heart is pierced and the soul soothed by the artlessness of animals. With no ulterior motive to poison their every action and with an innocence that is unsullied by either cynicism, scepticism or guile, they are a source of constant and consoling comfort in our often contorted lives. We may have crashed from grace but animals have not, and with them, as if with a magic machine, one can tune into a world of perfect innocence. (The jungle of course is riddled with ruthlessness, but the animals who charge through the pages of this book have left the jungle behind.)

In our ever changing culture animals alone have remained constant. From Mary Queen of Scots' dog, who refused to move from his mournful position between his mistress's shoulders and severed head, to Wellington's charger Copenhagen, of whom the Iron Duke said 'for bottom and endurance I have never known his fellow', such mutual loyalty and devotion has been like a backbone of sobering support through the ages. Admittedly, according to Aristotle, it was this very understanding with beasts that brought about a great Sybarite downfall: having taught their horses to dance to the sound of pipes, they were to be defeated in battle by a cunning enemy who, playing away on their own instruments, whipped the Sybarite chargers into an equine ballroom of confusion. Hundreds of horses were then slain as they cavorted away. It was also the love of animals that put the destiny of the British Isles into the paws of a single monkey in 1599, when the baby Oliver Cromwell was snatched from the cradle by his uncle's pet at Hinchinbrooke. Scampering out of the window and up on to the roof, the creature dangled the 'Fortune of England' in his arms, to the terror of his family, who surrounded the house with mattresses. In the end, the monkey trotted back on his own, bearing his burden to the cradle, thereby determining the course of history.

Great Britain's is a history of a country bristling with unashamed ardour for animals. Such giants as Walter Scott, Matthew Arnold and Thomas Hardy write as if beasts were the very bastion of their lives, and such giantesses as Florence Nightingale and Elizabeth Barrett Browning as if their hearts would break at the loss of their furred and feathered friends. Animals have always taken a prominent place. The first balloon flight over Britain was enjoyed by a cat, a dog and two pigeons in 1789, and the earliest extant Cabinet of Curiosities grew from a pet chameleon stuffed by Canon Bargrave in 1690 (which can still be seen in a permanently frolicsome position in Canterbury today). Many of the first recordings by a gramophone were popularized by 'Nipper' listening to 'His Master's Voice', and Sir Edward Elgar broadcast a concert on the wireless to his dog Marco, knowing how much his canine friend would love the shound of *his* master's voice! I could go on until the cows come home but hope instead that these tempting titbits will make you yearn to feast on the fancies that lie ahead!

Choosing the pieces for this book was like nibbling mere morsels from the richest of banquets. In what other field and in what other country would one be faced with the opportunity to revel in such anecdotes as Jemmy Hirst riding a bull out hunting in the 1840s with a pack of 'vivacious and sagacious pigs', or Philip Thicknesse's European travels in the 1790s with Jacko the monkey dressed as a footman and riding postillion to a carriage, in which Mrs Thicknesse sat with a parrot either on her bosom or 'hanging on her tippet'. What more noble words could be found than in Byron's 'Mazeppa', 'Away, away, my steed and I,/Upon the pinions of the wind', and could you be moved more than by the verse on a pug's memorial stone in Northamptonshire: 'Besprinkled here this tribute lies/With heavenly tears from angels' eyes'.

The issue of whether beasts have souls has been decided by most Englishmen, whose great love for their animals compels them to insist on their immortality. 'Admitted to that equal sky, his faithful dog will bear him company', wrote Pope in the 1700s. His view was supported by the great and the good: Kipling, Southey, Disraeli, Cowper and Horace Walpole all send our spirits soaring with the souls of the animals that they have commemorated. Byron left in his will that he should be buried beside Boatswain, his Newfoundland dog, for whom he built a magnificent memorial on the site of the high altar at Newstead Abbey in Nottingham. His words, once again, give greatest delight: 'To mark a friend's remains these stones arise/I never knew but one and here he lies'. Three words alone, 'Gone for Walkies', on a dog's grave in Cobham, deserve a modern mention, as does the memorial to our own dogs in the front hall at Hedgerley in Buckinghamshire, a soaring Gothic structure on which the dead dogs' names are emblazoned under the words, 'Joyfully barking in the Heavenly Chorus'.

If you think that Christian teaching questions the immortality of animals, then seek your sanction with the highest authority, in *Ecclesiastes*,

3:19: 'For that which befalleth the sons of man befalleth beasts; even one thing befalleth them: As the one dieth, so dieth the other; yea they have all one breath; so that a man hath no pre-eminence above a beast: for all is vanity.' No-one in fact has championed the spirituality of animals more vigorously than certain English clergymen. The vicar of St Mary Redcliffe in Bristol organized a funeral with a proper burial for the church cat Thomas, named after Thomas Chatterton, and as The Rev. Francis Rosslyn Courtenay Bruce lay dying, he saw his pet mouse's ears, illuminated by the sun, as the Gates of Heaven. At Morwenstow in Cornwall The Rev. Steven Hawker would preach a daily sermon in church to his nine cats, with a lone dog on the altar steps in place of a server. 'Turn a dog out of the ark!' he exclaimed to objectors. 'All animals, clean and unclean, should find there a refuge.' (Hawker had a pet pig which he would take '. . . into ladies drawing-rooms, not always to their satisfaction'.) I came upon only one dissenting cleric in this field: an unnamed nineteenth century Divine who allowed vermin to bite him without hindrance, saying 'We shall have Heaven to reward us for all our sufferings, but these poor creatures will have nothing but the enjoyment of this present life'.

Between 1850 and 1890 The Rev. J. G. Wood wrote some fifty-nine tomes on animals, many of them arguing for their immortality. One particularly enraged critic was moved to write that 'he would never condescend to share immortality with a cheesemite'. Wood's account of his reply never tires with the telling: 'In the first place, I told him that it is not likely that he would be consulted on the subject; and in the second place, as he did condescend to share mortality with a good many cheesemites, there could be no harm in extending his condensation a step further'. Such gems pack the pages of The Rev. J. G. Wood. He recommends rats as both pets and pie, he writes of the 'deep contralto' tones of an operatic mouse, and he tells the tale of a frog who was permanently attached to his mistress's nose. He wasted no words, and neither have I in airing so many of them once again. He is one of my great heroes; he and Frank Buckland both soar above all others in their writings of the animal kingdom. I have quoted them over and over again, to my and, I am sure, your delight. With tales of tamed butterflies and a 'coati-mundi' whose greatest pleasure was to rub a scented handkerchief down his tail, and with advice such as 'I doubt very much whether any Brighton lassie could find a prettier object to put in her hat than a live herring', their books make the richest of reading.

Buckland was a naturalist who sampled the meat of every animal at his table – his 'viands were varied', with potted ostrich, panther chops, pickled horse's tongue and roast alligator, much to the misery of one guest, who noted 'tripe for dinner; don't like crocodile for breakfast'. Ruskin regretted missing 'a delicate toast of mice' but immensely enjoyed being waited on at table by 'two polite and graceful Carolina lizards'. Buckland's pets were as mixed as his menu. He had a bear that he took out riding round the Oxfordshire countryside, as well as monkeys galore, an enchanting suricate, an eagle, chameleons, a hyena and ten score more, all running free in the house. I make no apologies, but rather am as proud as Punch, that both Frank Buckland and tales of Frank Buckland have been quoted at such length. Like a great pied piper, with The Rev. J. G. Wood in attendance, he leads the pack through the pages of this book.

Starting with Rabelais' 'barking of Currs . . . tattling of Jackdaws . . . nuzzing of Camels and buzzing of dromedaries' we are off to the merriest party. Festivities abound throughout these pages. There are tales of a clergyman's wife riding a camel out hunting, and of a guinea fowl, as well as a fox and pigs, being trained to join the hounds. Fleas play whist and rats sneak up bustles. A live chameleon is decoratively attached to his mistress's hat and a cow plays 'the games of a puppy . . . with his young biped friends'. Albert Schweitzer writes lovingly of his wild boar attending church services and in *Out of Africa* Karen Blixen describes her baby gazelle's hooves as being like the 'laced feet . . . of a young Chinese lady of the old school'. In Nathaniel Wanley's *The Wonders of the Little World*, a parrot recites the Apostles' Creed, verbatim, in Latin. Prince Pückler-Muskau, who journeyed through the British Isles in the 1820s, observes a myriad of delights, most especially the 'right English religion' of foxhunting, with a hunting parson who kept a tame fox under the altar to be let out occasionally for a good chase. There is a laugh or two to be had at some fantastically bad poetry – 'The Dachshund (with Queen Anne her legs)/Your Sympathy enchains or begs' – while Rossetti pens a cheerful farewell to his wombat and Kipling brings tears to the eyes with the tale of 'Dinah in Heaven', a dog who 'saved a fool from drunkenness/And a coward from his fear'.

Several other books are ragingly represented. One is the nineteenth-century *Parlour Menagerie*, whose cover alone would send you scrambling to hold it in your hands. With an embossed and gilded, kilted Scotsman flying along on an alligator, it promises princely pickings, and you are not disappointed. It holds stories of a lion aboard ship caressing and licking his master's cat-o'-nine-tailed back, and of a crane who took on the daily duty of rounding up cattle. This book was an irresistible source, as was the entrancingly entitled *Queer Pets and Their Doings* (shamelessly fleeced!) of the 1880s, by Olive Thorne Miller – one of the many delectable American offerings. Two other noble tomes of the 1860s, *Our Dumb Neighbours* and *Our Dumb Companions*, produce lively contributions. Both are stuffed with anecdotes for children, both are penned in the most pious and poignant humour, and they are by yet another clergyman, The Rev. T. Jackson.

It was John Bunyan who said that an Englishman would rather walk with a dog than with a fellow Christian. The Englishman, with his characteristics of restraint and reserve among his own kind, would seem to have specifically nurtured this passion for animals as the perfect outlet for all his unexpressed emotions. Certainly words that he would not dream of confiding to his best friend can be poured unreservedly into the ears of his dog, cat or horse. Stoicism is swept aside by swirling sentiment as the floodgates fly open. Pity the poor naval and military mascots, the pigs, the parrots, the bears, monkeys and cats, as well as the goat who was given a funeral with full military honours by the R.A.F. at Ilford Pet Cemetery. What a burden of emotions they must all have had to bear! The composer Dame Ethel Smythe described the liberating joy of being able to bury her face into her dog's neck, crying 'I do love you so' when the sheer gratitude for the boon of being alive came over her. How unwelcome such attentions would have been if she had bestowed them on any of her friends. 'But how natural, how soul-satisfying to both parties is such an action in the case of one's dog!'

This book is set solid with such sentiment. Whether it be George III's giraffe with 'the eyes of the loveliest Southern girl' and the 'beaming expression of tenderness and softness, united in volcanic fire', or Pliny's account of a raven's funeral bier being borne along by Ethiopian warriors, almost every entry in some way links animals with man. Some poor creatures have had to be eaten to be included, others merely imitated, as with Robert Watson who, thinking that he was a fox, lived underground for twenty years in a Gothic entranced 'earth' and a classical temple he built to mark his lair.

The poison of cruelty has seeped into these pages, in the hunting of deer, fox and stag, in coursing and shooting, in 'ratting' in nineteenth-century London and in the training to do tricks – and all the other torments man feels licensed to inflict on so-called lowly creatures. Apart from being very much part of the great order of things between man and beast, such unkindnesses have been teeth-grittingly included as a most necessary antidote to the sugar-sweet sentimentality that is the main bill of fare. As I sink to my knees awash with good will at the mere sight of a dog's wagging tail and feel all the magic of watching 'God's Spies' at work when sitting alone with a cat, and as I am soothingly sucked into a vacuum of quiet and ennobling reflection when nuzzling up to a horse, sentiment is my daily diet; it would have been most unnatural to change it completely, especially when planning such a feast as this.

Domestic animals undoubtedly fulfil a deep spiritual need in man; they are their cosiest confidants in childhood and their faithful companions in death. In 'Fidele's Grassy Tomb' by Sir Henry Newbolt, the Squire's last wish is that his hound be buried at his feet:

> The last that his heart could understand
> Was the touch of the tongue that licked his hand;
> 'Bury the dog at my feet,' he said,
> And his voice dropped and the squire was dead.

There are those who question the reasoning behind the devotion of and to the domesticated animal. Lady Byron encapsulated her husband and his beasts' mutual adoration with supreme cynicism: 'It is because the creatures have no exercise of reason and could not condemn the wickedness of their master'. A sneering solution for many, and one that I would attack to the last, having often and successfully used dogs as Geiger counters as to the character of a caller. However, it is certainly true that Hitler's dog Fuchsl was unquestioning in his devotion. So too was Hippy the Bavarian Bloodhound who belonged to Sir Neville Henderson, that great advocate of appeasement and our ambassador in Berlin during Hitler's day. It is pleasing to think of the two dogs having a good jaw together, no doubt being able to nose each other out so much better than could their masters.

Dogs live in a great well of silence filled to the brim with sensitivity, into which you are forever lured through their loyal and loving eyes. It is a fear that we do use these untainted spirits to our own ends, and that the pleasure that animals, most particularly dogs, give to us is largely the pleasure of having our vanity preened. Many is the time that I have shrieked 'Mirror, mirror on the Wall' to my canine companions, and never once have I been disabused of the illusion that I am 'The Fairest of them All'. As my mother says, 'They do not answer back'. This is the one cloud in the otherwise brilliantly sunny sky of man's relations with the dog.

If this book were not dedicated to my sweetheart husband it would be celebrating the glory of our four dogs: two mongrels, Clover and her son Thistle, and two dachshunds, Violet and Florence. They are the bane and the blessing of every day. Many hours are spent engaging them in one-way conversations, marvelling all the while at their carefree lives. However, the boot could well be on the other foot (which makes me think of the tiny leather boots issued to the rescue digger dogs in the Blitz), for as Elias Canetti says in his journals, 'Whenever you observe an animal closely you feel as if a human being inside were making fun of you'. No further word need be written.

. . . the barking of Currs, bawling of Mastiffs, bleating of Sheep, prating of Parrots, tattling of Jackdaws, grunting of Swine, girning of Boars, yelping of Foxes, mewing of Cats, cheeping of Mice, squeaking of Weasles, croaking of Frogs, crowing of Cocks, Kekling of Hens, calling of Partridges, chanting of Swans, chattering of Jays, peeping of Chickens, singing of Larks, creaking of Geese, chirping of Swallows, clucking of Moorfowls, cucking of Cuckows, bumbling of Bees, rammage of Hawks, chirping of Linets, croaking of Ravens, screeching of Owls, wicking of Pigs, gushing of Hogs, curring of Pigeons, grumbling of cushet-doves, howling of Panthers, curkling of Quails, chirping of Sparrows, crackling of Crows, nuzzing of Camels, wheening of Whelps, buzzing of Dromedaries, mumbling of Rabets, cricking of Ferrets, humming of Wasps, mioling of Tygers, bruzzing of Bears, sussing of Kirning, clamring of Scarfes, whimpring of Fullmarts, boing of Buffalos, warbling of Nightingales, quavering of Meavises, drintling of Turkies, coniating of Storks, frantling of Peacocks, clattering of Magpies, murmering of Stock-doves, crouting of Cormorants, cigling of Locusts, charming of Beagles, gnarring of Puppies, sharling of Messues, rantling of Rats, querieting of Apes, snuttering of Monkies, pioling of Pelicans, quecking of Ducks, yelling of Wolves, roaring of Lions, neighing of Horses, crying of Elephants, hissing of serpants, and wailing of Turtles.

FRANÇOIS RABELAIS (1494?–1553),
TRANS. SIR THOMAS URQUHART (1653–94)

1

Wakenings of Life

Everybody said that nothing alive cd. come out of an egg rolled backwards & forwards every day under my fingers – & behold a little Dove!

LETTER TO ELEANOR PAGE BORDMAN,
28 SEPTEMBER 1837, FROM
ELIZABETH BARRETT BROWNING
(1806–61)

Life and Death of Athena – An Owlet from the Parthenon
By Miss F. P. Nightingale (Lady Verney)
Dedicated to
the most constant and true friend
the Protector
and the most ardent admirer
of . . . .
the deceased Athena

This distinguished individual was born (as nearly as can be ascertained) on the fifth of June 1850. Her (future) Mistress . . . when passing under the walls of the Acropolis perceived a little ball of fluff tormented by a group of children. Athena had fallen from her nest. She was rescued for the sum of 6 lepta or one farthing. On what slight accident does fame depend! Athena's brothers and sisters have lived and died unknown! while Athena . . . but her biographer must not anticipate. . . .

One quiet autumn evening, after an absence of ten months, her Mistress mounted on foot the steep hill which led her to her mountain home along the old 'Bracken Lane' among the '77 Ioists' past the 'Milk Dale' up through the garden, softly, softly, and in at the steps of the Drawing room window. After sitting half an hour on the sopha between her Mother and sister she put her hand into her pocket and pulled out a little owlet in a bag! Athena's head alone stuck out, the bag being tied about her neck, she seemed however very happy and very warm. When set at liberty she began while sitting on the table, to curtsey and bow with the greatest urbanity. . . .

She occasionally paid visits among the cottages in her Mistress's pocket, to assist her by all the means in her power, tho' not in general much addicted to 'doing good' (her life being one of luxurious tho' intellectual ease). On one occasion known she assisted greatly in the cure of a little burnt child, who suffered dreadfully when her wounds were dressed, but in the contemplation of Athena's bows and curtseys opposite her bed (brought for that especial purpose) forgot her woes, and lay quite still while she was doctored daily, the charm continuing undiminished till the cure was complete.

She was of much interest also to an old Great Aunt, who with all the pent up energies, the strong unregulated feelings, the large uncultivated powers of the brave days of old, when woman's work was so often to sit and wait, had worn out her life under the fearful discipline of repression, and now in her 80th year blindly and patiently sat in the sunshine, on her terrace seat beside the old yew hedge, under the tangle of vine and jessamine above the 'deep garden' where she knew the roses grew which she could no more see, and talked of the sounds of the river which she could no more hear. '*See* where the moon is shining so beautifully on the water,' she would say, as her warm imagination served

Florence Nightingale and Athena, whose 'love of high places was indeed notable'.

her instead of eyes. She looked upon Athena as a great great niece, and enquired after her visitor's comfort with the same dignified and hospitable care that she showed for her Mistress. And Athena returned the interest by hopping on her knee or her shoulder, as she sat by the fire, and sitting herself comfortably there for a snooze, while the blind old lady never showed the smallest fear or surprise at this strange and unaccustomed invasion of the quiet routine of her life. . . .

Her favourite pastime was to fly at a nosegay, capture a dahlia, and retire to sit upon it and pick it in pieces. Her conscience tho' was a very lively one and she rarely did crime without barking vigorously to inform the company.

To have knocked down a jar of roses and carried off the finest to the top of the room was a feat which greatly delighted her. Her love of high places was indeed notable. She would, occasionally mount to the top of her Mistress's head and crow loud and triumphantly as being the most noble and conspicuous position which she could find. . . .

The biographer cannot recall without tears her little run across the room, like the step of a partridge her elegant manners, coming downstairs sitting on a finger quite free, but without any inclination to fly away, her talents and virtues. Indeed 'il ne lui manquait que la parole'.

It will thus be seen that with the exception of a slight tendency to theft, murder, ill temper and conceit, this remarkable person may justly lay claim to be called, so the Italians have it, 'una persona compita'. . . .

We now come to the last and tragical chapter in the life of one whose fame the world will not willingly let die!

Her Mistress had been home after an exceedingly hard summer's work in London with the Cholera added to her usual labours.

She had been exceedingly unwell during that too short visit, and had been confined to her bed and her room, for the chief part of it. Athena was her constant companion, when she could bear no one else of larger size, Athena was welcome; she sat on the bed and talked to her, she ran races all round the room after imaginary mice, every meal she considered to have been brought for her especial use, and she accordingly appropriated the bread and butter or pounced upon the chicken. 'Mademoiselle le gate' [Mademoiselle is spoiling it] was the warning voice, but her Mistress let it be.

At length after a little fortnight her Mistress returned to town, and the day after she arrived began to consider the possibility of her great expedition. . . . The house . . . was packed and everything was in haste and confusion. Athena was put into a room by herself, she had a stout little heart of her own, but the grief, the cold and the isolation were too much for her, she fell down in a fit, there were none at hand to succour her, and when found she was lying dead on her little side, on the very day that her Mistress was to have left England. The departure was delayed for two days, till the expedition could be got together, Athena's body was sent up to be embalmed. Her Mistress asked to see her again, and the only tear she should shed through that tremendous week was when ... put the little bodie into her hands

    'Poor little beastie, it was odd how much I loved you.'

So let her lie so wept.

LADY VERNEY, SISTER OF FLORENCE
NIGHTINGALE, *c.* 1860

... Have I not before anounced to you the birth of a new little cockney dove.? I am *dreadfully* proud of it, – & in consequence of its appearance I opine that there must be a development upon my cranium, of the organ of self-esteem. Everybody said that nothing alive cd. come out of an egg rolled backwards & forwards every day under my fingers – & behold a little Dove! – It is just like a ball of floss silk (the effect of the rolling) & its present acquirements are confined to eating & drinking (which I am sorry to say it was not precocious in acquiring by its own act) & sleeping. As to sleeping, it combines all the talents of the seven sleepers in one! When it has eaten it sleeps! When it has drunk, it sleeps! When I have kissed it, it sleeps! When I have given it a lesson in flying, it sleeps! When the sun shines, it goes to sleep in it! And when the wind blows, it goes to sleep *from* it. In short, it very seldom has an eye open – & when it has, it is scarcely ever more than one at a time! This is no poetical license or exaggeration.

LETTER TO ELEANOR PAGE BORDMAN, 28
SEPTEMBER 1837, FROM ELIZABETH
BARRETT BROWNING (1806–61)

This distinguished individual was born at Huntingdon, on the 25th of April, in the year 1599, and baptized four days after, in the parish church of St Johns; his uncle, Sir Oliver, for whom he was named, appearing in the capacity of godfather. If we may trust to the gossip of his more ancient biographers, his childhood did not pass without many remarkable occurrences, which seemed to indicate that an uncommon fortune awaited his riper years. For example, they say that his grandfather, Sir Henry Cromwell, having sent for him when an infant in the nurse's arms, to Hinchinbrooke, a monkey took him from the cradle, and bolting from a window, ran with him upon the leads which covered part of the roof. Alarmed at the danger to which the young visitor was exposed, the family brought beds upon which to receive him, supposing that the creature would drop him from its paws; but, it is added, the sagacious animal, appreciating the value of its treasure, brought the 'Fortune of England' down in safety, and replaced him in his bed.

Constable's Miscellany, Vol. XLVII:
Life of Oliver Cromwell, 1829

When first bought, Mr Alligator was in a very seedy condition indeed, terribly thin and wan-looking – in fact, half starved. His skin was all in cracks, and his coat of mail had to be oiled every morning by means of a flannel on the end of a stick. This acted like a Turkish bath to our friend, and did his constitution good. For many days, even weeks, he sulkily refused to eat, and lay quiet and still like a stuffed thing. At last he took – all of a gulp – a live pigeon, and ever afterwards he fed well. The secret of getting him to eat was temperature – temperature the old story, the key to so many fishery problems, whether of salmon, oysters, or alligators. Hot-water pipes were introduced under the floor of his den, and Mr Alligator, feeling the agreeable heat to his gouty toes and elegant figure, fancied, I suppose, he was back again in the tropics, so he woke up and began to eat; and what more tasteful beginning could there be than a nice live pigeon with feather sauce?

Anxious to show me his pet feeding, the curator offered to give the alligator more supper; he had already devoured his proper supper. The curator got on to the top of the cage and touched him gently with an iron rod. I was surprised to see the activity of the rascal; he opened his eyes with a jerk, up went his head like a run-away hansom cab horse, he gave an indignant whisk with his tail like a lady picking up her skirts when a clumsy fellow puts his foot on the pet lace, and, to my surprise, began to puff himself up. Gradually he became larger, larger, and larger, like the blowing up of a football; his armour glittered, and the bony studs stood well out from the soft intermediate skin. I confess, when at his full I longed to run a pin into him to save his life, as I saw he had a chance of meeting with the same fate as the foolish frog in Æsop's Fables, who vainly puffed himself up trying to become as big as the ox, with whom he was having an argument. Just, however, as he came to the bursting point, *Alligatorus Rex alligatorum* suddenly relaxed himself, and his steam escaped, I suppose, through his larynx and nose. Anyhow, he began a most sonorous hiss. 'H-i-s-s, h-i-s-s;' I can hear it now – just the noise a dragon ought to make. It was like no hiss I ever heard before, much deeper and louder sounding than any snake. As he continued his hissing he became thinner and thinner, till he looked quite the skeleton of his former pretty self. Then he began to blow himself up again, for (I could see it) the iron rod was getting up the monkey of Mr Alligator.

A chicken's head and neck were then suddenly thrown into the bath; in an instant Leviathan forgot his rage. (Mem.: when a *homo bipes implumes*, one of our own noble species, loses his temper, give him a dinner, and he will be all right, showing once again that 'the nearest way to the heart is down the mouth.') However, hearing the chicken's head fall splash into the water, the alligator – he should be called Uncle Tom – was after it in an instant, and seized it just as a dog catches up a running rat in his mouth. He first of all bit it spitefully as though to kill it, if it happened to be a live thing; and then – one, two, three, and away –

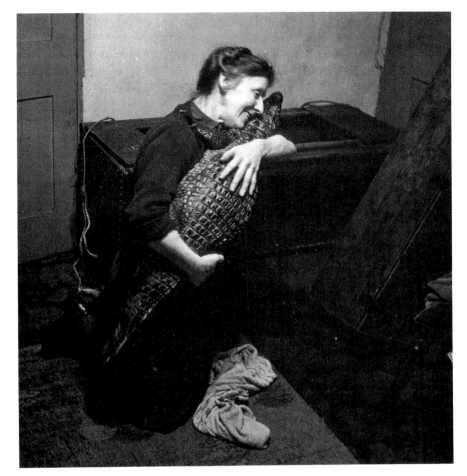

Two maiden ladies share their four room Chertsey cottage with two perfectly mannered alligators and a crocodile. One of the ladies, Miss Roberts, brushes the teeth of her pets, trims their nails and massages them with oil. The other lady, Miss Davis, is seen here kissing one of the alligators goodnight. She thinks alligators are the nicest of pets and prefers them to cats.'

(ALLIGATORS AND
OLD LACE, 1948)

chump; back went his head, down his throat went the chicken's head in a moment. Reader, hold your nose and swallow a pill before the looking-glass, and you will understand how Uncle Tom swallowed the chicken's head. His blackship then gave a gulp, and, like the 'Oh the poor workhouse boy' in the song, asked for more. Three chickens' heads and a bit of beef, extra rations, did Uncle Tom get that evening, and all on my account. . . .

Really, 'Uncle Tom' is a grand beast; he is growing so fast that he is to have a new drawing-room and dining-room, and then he will have space to swish his tail; he has not much room for his tail just now. I wonder how it is that in the 'struggle for existence' his tail has not begun to curl; may be his descendants in one hundred thousand years will have their tails curled up like a pug-dog's. By the way, why do some pigs wear straight tails, some curly tails? There's a problem for you.

Besides big Uncle Tom, there are a number of smaller alligators. Close to the end of the Uncle's cage is a charming family of baby alligators, from ten inches to one foot long. These little boys and girls have a nice hot nursery, heated from underneath, and a flannel blanket over their dear little heads. They are as active as blackbeetles, and when their counterpane is taken off, scuttle away in all directions. If I reckoned right, there were twenty or thirty of these little fellows. Several of the ladies in Southport have purchased pets from among them, and it may be that no Southport lady will consider her establishment perfect without a baby alligator to bask on the hearth-rug, and go out for a walk on the promenade with her. When the pet defuncts, he can be stuffed, gilt, and put in the hat for an ornament, don't you know?

However, those in the Aquarium are growing fast; they gorge like charity children at a 'tea and bun' festival. The keeper cuts up fish into small bits, and throws them into the cage; they scramble for them famously, and apparently love each other in that disinterested, charitable, and unselfish manner which may be seen by a careful observer who throws down handfuls of coppers among the London *gamins* and street Arabs in the crowd when waiting for the Lord Mayor to pass through Fleet Street on 'All Sprats' Day, November 9.

But there are yet more of Uncle Tom's relations at Southport. A huge box, looking like a gigantic coal-scuttle, stands near the boilers in the engine-room. Open *sesame*! and lo and behold a nest of young alligators of all sizes and shapes, like the ladies' bonnets and hats in a Regent Street shop! The curator dives his hand in and picks them out one by one, holding them aloft like an old fishwife in the Edinburgh market selling Scotch haddies. The lot are not yet presentable. They have not yet received the certificates of the School Board, and their tempers and appetites are not sufficiently mollified by the furnace fire to go into the glass apartments which are getting ready for them, so they remain at their ease, toast themselves before the engine-room fire, while the engine-driver consoles their minds by whistling to them 'Tommy, make room for your uncle,' and feeds them with bread and cheese, which they will not eat.

FRANK BUCKLAND (1826–80),
Notes and Jottings from Animal Life, 1882

In those days I was abysmally ignorant about seals. I had no idea what I was letting myself in for when I ordered that the creature be brought aboard my raiding schooner. The last thing that crossed my mind when I determined to adopt Panayoti was that I would pass the closing phases of the war in the Aegean playing nursemaid to an inexperienced seal and have to face something approaching mutiny among my crew. All that I knew was that it was intolerable to watch a helpless little creature left to die of thirst under a pile of stones in a sun-baked courtyard. My blood boiled at the thought that six brawny fishermen were too superstitious to dare to kill the little animal outright, but had to immure it collectively so that the bad luck would be spread among them. . . .

Panayoti simply refused to have anything to do with the fish which I bought for his dinner. Every time I offered him one he would turn his head away and start crying. The noise he made was heartbreaking – something between the wail of a baby and the bleat of a lamb. I tried everything. I rubbed his nose with the fish, but he only cried louder. I pushed the fish into his mouth. He spat it out. Thinking that perhaps the fish was too big for him to swallow, I cut it up into small pieces and tried to force them down his throat. He spat them out too. I decided that the only thing to do was to let Panayoti go on his way without a meal, since probably he preferred to catch his own fish.

I dropped Panayoti overboard, expecting to see him streak off. The next thing I knew he was drowning: his head went down and his tail flippers came up out of the water. They beat wildly for a time, then more and more feebly. A stream of bubbles rose from his mouth. There could be no doubt that Panayoti did not know how to swim. I dove overboard and fished him out, which took some doing, since he was very slippery and I didn't know then that the best way to pick up a seal is from under the flippers, the way one picks up a baby. Besides, I was still not at all sure he wouldn't bite.

I needn't have worried. The poor creature was more dead than alive when I got him aboard again. I was afraid he might die and gave him artificial respiration as best I could. After a while he revived and started crying again weakly, moving his head from side to side in the most dejected manner. In a few minutes he seemed completely recovered and went crawling about the deck in that hopelessly inept, broken-boned way seals have out of the water. But at least he was clean now and his fur was fluffing up in the sunshine, turning from black to a soft dove gray on top and ivory underneath.

By then it had belatedly occurred to me that Panayoti was only a baby. I know it sounds stupid that I had not thought of it earlier, but I had never seen a seal before and I had no basis for comparison. A length of nearly three feet did not seem to me conclusive one way or the other. I began suspecting it when I found out, on trying to feed him more fish, that he had no teeth. So I tried milk. First I gave him some in a saucer, but he only turned away and spilled it with his flippers. Then I made a feeding bottle out of an empty gin bottle with a nipple made from the little finger of a rubber glove which we used for handling the smoke-screen apparatus. It was a complete flop. Panayoti didn't even want to look at it. The mere sight sent him into tantrums. He became hysterical when I tried to force the nipple into his mouth and went dragging himself round and round in circles across the deck screaming his head off.

More in self-defense than from compassion I opened a can of New Zealand butter, of which we had a lot, and stuffed his mouth full of it.

He spat most of it out, but a good deal stuck and for a time he was too busy choking and spitting to make much noise. He was so quiet that I almost forgot about him, and when I looked for him later I discovered to my delight that he was engaged in licking the butter off his nose. I spent the rest of the day carrying the can of butter around with me and smearing Panayoti's nose every time I passed him. . . .

From that day on the feeding problem ceased to exist and he started gaining rapidly in size and strength. Our relationship also became much more intimate, not to say exclusive. Panayoti had at last become convinced that he had a friend in the world, and Panayoti was not a seal to do things by halves. His gratitude and affection were embarrassing. He would not let me out of his sight. When I was standing on deck, day or night, he would come and rest his head on my bare feet. If I moved he gave forth pathetic little moans. To avoid hurting his feelings I often found myself rooted to the same spot for what seemed like hours. To stumble into him, as I often did on dark nights, was a major calamity. It upset him terribly, and he would go on whimpering until he was picked up and comforted.

As his strength and devotion grew, he would not even let me go down into my cabin unescorted. . . . Sliding down into the cabin and setting up a to-do until he was taken out was nuisance enough, particularly if it happened at night when we were sailing and had to keep our ears open for the first, faint sound of enemy engines. But when he took to falling overboard every time I rowed away in the dinghy or went swimming over the side, it became really insufferable; for in spite of several unpleasant experiences, Panayoti had learned neither how to swim nor the obvious truth that he couldn't. He had to be rescued from drowning every time.

In desperation I decided to teach him how to swim. Every morning after the schooner was moored against the rocks and camouflaged under nets, after the machine guns were emplaced and the lookouts posted, I took Panayoti in the dinghy to some secluded cove. First we began in the shallows with Panayoti barely awash. I would walk away on the beach following the edge of the sea, and Panayoti would try to wriggle after me, splashing and thrashing about in the foot-deep water. As often as not he would give up his allegedly native element and crawl out onto the sand. If he didn't, the lesson usually ended up with his swallowing a mouthful of water and choking. . . .

In spite of the disappointing results, I persevered with the lessons, using every method that I had ever heard of. . . . The trouble was his head. Whenever I let go of him it sank like a piece of lead and his tail beat uselessly in the air. I even strapped a Mae West on him, but he was so slippery in the water that after a few minutes it worked loose, and he slid out of it head-first to assume his usual perpendicular position.

*

Then one night we had an unpleasant experience. We were making a long crossing to one of the more distant islands, in company with another schooner, and we were caught in the open by a German Ems Craft. Being shelled at night on the sea by a more powerful enemy vessel against which you have no means of retaliating is very unnerving. We spent a desperate half-hour dodging right and left until the Germans lost us in the darkness. In the morning when we had made port the atmosphere on the schooner was distinctly ominous. Much as I tried to laugh it off it was clear that the crew attributed the previous night's alarms to the presence of a seal aboard. This was a serious matter, for a discontented crew could mean disaster. I was a worried man when my fellow captain hailed me and invited me to go swimming. I put Panayoti in the dinghy and rowed across.

My friend and I discussed our narrow escape and my crew's disaffection as we swam slowly round and round, with Panayoti propped up in the dinghy's stern, watching our every movement. We got so engrossed in our conversation we did not realize the boat was drifting further and further away, until with a splash Panayoti, unable to stand the separation any longer, slipped overboard. We dashed to rescue him, but he had disappeared. We dived and dived. There wasn't a trace of him. Then my friend pointed with his hand and shouted, 'There he is!' A hundred yards away a little black head was bobbing up and down among the waves at the mouth of the cove. I was torn between a terrible sense of loss and the joy of knowing that Panayoti had come into his own at last. After all, I thought, this was the best, the only way, for it to end. I waved to him, and he disappeared.

My friend and I stood treading water, both of us wondering, I suppose, whether Panayoti was well enough equipped to face the challenge of the open sea, when, with a flurry of churning waves, his glossy head bobbed up between us. His face wore the broadest, most triumphant smile I have ever seen. He kept on looking from one of us to the other, his whiskers twitching with excitement, his round eyes opened wide as if to say, 'You see, I've done it.' Then, as though he needed to prove the point, he dived, nipped me playfully in the calf and was off at such speed that I had hardly time to turn around before he had served my friend in the same way.

We tried to catch him but he always slipped between our fingers. He darted through our legs and brushed across our backs; he dived, he leaped out of the water, and every now and then he surfaced to look at us and make sure that we were enjoying it too. Then he'd be off again like a cockeyed torpedo. We spent a wildly exciting quarter of an hour until he got tired and came to me and put his flippers on my shoulders, wanting to be lifted back into the dinghy. From that time on the days became one long delight. We were in the water whenever we had a chance. From early morning Panayoti would start worrying me to take him swimming, and by nightfall we were both exhausted.

Fortunately the war in the Aegean was drawing to its close . . . I tried to coax Panayoti into starting a new life on his own, but he would have none of it. So having no choice I decided to take him to Athens with me. Taxis were scarce and when we finally got one, seven of us piled into it. I sat in front with three others and with Panayoti on my lap. He was very curious and kept sticking his head out of the window the way dogs do.

When we arrived at my aunt's apartment the maid who opened the door nearly fainted at the sight of the seal. My aunt too only managed to overcome her revulsion for a rather grimy Panayoti because of her great pleasure at seeing me return from the wars. But she put her foot down when I suggested that Panayoti should be bathed in the bathtub. We struck a compromise eventually, and a tin hip bath was brought out to the balcony overlooking the street. Panayoti hated getting soap in his eyes and started an awful ruckus. The strange bleatings attracted the attention of the people sitting on their balconies in the flats above my aunt's, and they leaned over to see what it was all about. The sight of so many people staring at something that was going on above the street intrigued the passers-by and soon a large enough crowd had formed in the street below to block the traffic. Drivers, after honking their horns futilely for a while, got out and joined the others. The people in the street started calling up to those in the balconies above, 'What is it?' The people in the balconies answered, 'We're not sure. We think it's a seal,' and presently the whole crowd started shouting in unison, 'Show us the seal!'

When I had rinsed Panayoti, I picked him up from behind and he bowed to the people of Athens right and left as if he were a young prince. He was roundly applauded. Perhaps for many of those people he was as good a symbol as any for the end of the occupation. A seal being bathed on a balcony may not be what one ordinarily conceives of as a return to normalcy, but it was certainly a departure from the grim mood of the previous years.

A few days later I was ordered to go on a patrol to the north where the Germans were still fighting a rear-guard action. I took Panayoti to my mother and left him with her in our home in the country. We put him in a kennel next to the pool so he could take a swim.

When I returned a few weeks later, after the last German had been driven out of Greece, my mother told me that Panayoti was dead. The weather had suddenly turned bitter, and she thought that he had died of a cold. She had had him buried in the garden at the foot of a young cypress tree. I went to say good-by and standing next to the little grave with the icy wind moaning mournfully in the branches, I couldn't help thinking that Panayoti had died of loneliness.

ALEXIS LADAS, *Harper's Magazine*, 1957

This puppy have I called the Lord of Life because I cannot conceive of a more complete embodiment of vitality, curiosity, success, and tyranny. Vitality first and foremost. It is incredible that so much pulsating quicksilver, so much energy and purpose, should be packed into a foot and a half of black hide. He is up earliest in the morning, he retires last at night. He sleeps in the day, it is true, but it is sleep that hangs by a thread. Let there be a footfall out of place, let a strange dog in the street venture but to breathe a little louder than usual, let the least rattle of plates strike upon his ear, and his sleep is shaken from him in an instant. From an older dog one expects some of this watchfulness. But when an absurd creature of four months with one foot still in the cradle is so charged with vigilance, it is ridiculous.

. . . What kind of a view of human life a dog, even a big dog, acquires, I have sometimes tried to imagine by kneeling or lying full length on the ground and looking up. The world then becomes strangely incomplete: one sees little but legs. Of course the human eye is set differently in the head, and a dog can visualize humanity without injuring his neck as I must do in that grovelling posture; but none the less the dog's view of his master standing over him must be very partial, very fragmentary. Yet this little puppy, although his eyes are within eight inches of the ground, gives the impression that he sees all. He goes through the house with a microscope.

E. V. LUCAS (1868–1938),
'The More I See of Men . . .', 1927

Four years ago . . . were born in a traveling menagerie – which happened then to be in New York – two families of baby Lions, five in all.

The collection of animals being broken up while they were still very young, they fell into the hands of Mrs Lincoln's husband; and he – as the most sensible thing he could do – handed the queer family of babies to his wife to care for.

They were pretty little creatures, about the size of three-months' old kittens, and looked very much like young tabby cats. Their parents were entirely wild.

Mrs Lincoln had no children, and she devoted herself to her strange babies; but not knowing exactly how to feed them, three died before she found out what was best for them. The two that were left she fed on goat's milk, using a regular baby's nursing-bottle, and she soon had the satisfaction of seeing them grow strong and well.

She named them Willie and Martha, and they were playful as kittens, rolling and tumbling over each other on the floor, chasing each other around the room, and even coaxing the cat to join in their romps.

They slept on their mistress's bed, ran all over the house, and were fond of a ball as a cat. Even now, though he is four years old and grown

The lions at home in Boston.

up, the play is not all out of Willie. He enjoys a frolic with a cocoa-nut as much as puss enjoys a marble, and in the same way.

Whenever they wished, these two wild babies went into the yard in the heart of Boston, where they live, and played by the hour, the cat being generally of the party. If tired they would lie on the lounge, or before the fire – when they had grown too big for their mistress's lap; and if hungry they would cry for something to eat, not exactly as a human baby – as some travelers have said – nor like a cat, but with a peculiar hoarse sound, like nothing in the world but a Lion. In every way they acted like domestic animals.

OLIVE THORNE MILLER,
Queer Pets and Their Doings, 1880

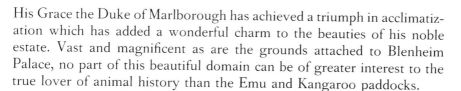

His Grace the Duke of Marlborough has achieved a triumph in acclimatiz-ation which has added a wonderful charm to the beauties of his noble estate. Vast and magnificent as are the grounds attached to Blenheim Palace, no part of this beautiful domain can be of greater interest to the true lover of animal history than the Emu and Kangaroo paddocks.

. . . the Duke handed me over to the care of old Long, the special guardian of the Emus. Passing through the well-kept grounds, we soon came to a rustic temple, commanding a glorious view of the lake and the surrounding woods. . . .

Turning short to the left after leaving this temple, Long conducted me to the Emu paddock. 'I will show you the chicks first,' said he. Throwing open the door of a shed, I saw a great lot of straw on the ground, and, in one corner, what in the dim light appeared to be a feather bed with a long neck. 'That's the old bird,' said he, 'but he is precious artful; he don't want to be disturbed.' The old Emu was evidently following his instincts by keeping as quiet as possible to avoid observation. 'It's always the male bird as sits,' says Long; 'he has sat on them eggs for eleven weeks, and I knows they is good for nothing. There's a dead chick in every one on 'em. I knows by the shake and the weight they ain't no account. I be sure on it, and he'v' a'got three chicks as he's a nussing as well as the eggs, and fine chicks they be.'

'Get up, Tom; get up, you rascal,' said the old man, giving the bird a slight touch with his stick. So Tom, seeing that disguise was of no further use, jumped up like a jack-in-the-box, shook out his beautiful feathers, gave a peck at me as he passed, and stalked solemnly out of the building. To my great delight three little emus appeared squatting on the ground as quiet as granite blocks, as their magnificent nurse rose up from the nest. Seeing their father go out, the dear little things ran after him. They were all small editions of their parents. Imagine an Emu seen through the wrong end of a telescope, and you have a good idea of an emu chick, but their colour does not resemble that of their parents; they are striped with black and white, and look not unlike a bit of animated zebra's skin.

'The old Tom' (said Long) 'will take care on 'em till they be big ones, and the t'others can't run 'em. He picks up the food for 'em, and drops it just like a hen does for her chicks, and he calls 'em in his way. When he takes 'em round the field he goes along with 'em, and he picks up his great long legs so as to be careful not to tread on 'em. They don't want for nothing, does them chicks, and they are healthy and well. I feeds 'em now, their mother don't care nothing about 'em; if she was to come into the paddock she would throw the chicks up into the air with her head as high as them trees. I had a job to find out how to feed 'em. I've reared thousands of pheasants in my time.

'I am seventy-two years old, and I've never been further away from Blenheim than fourteen miles, and then I went a-shooting with the pre-sent Duke's grandfather, and I recollect his great grandfather. I've three sons gamekeepers to the Duke now, and my sons knows how to rear pheasants as well as I does. When the emus first come from Australia the Duke put 'em under my charge, and when the first lot of chicks was hatched I could not tell how to feed 'em. So the Duke gives me a book; but this ere book ain't no use – not a bit on't. If I had agone according to this book I should never have reared "nairn" ' (Oxfordshire for 'never a one'). 'This book said I was to give 'em peaches and apricots, and them sort of things, but where was I to get peaches and apricots in May? so I lays in bed of a night and studies. I knows young pheasants likes anything milky, so all at once I thinks of spinach. The next morning at sunrise I was off with some spinach to the emus, and they pecked at it as though they would eat the ground. So I says to myself, What the old ones likes the young ones likes; and now I feeds 'em principally on spinach. I sits upon the stone water trough, and if I says to chicks, "Come," they comes along directly. I cuts up the spinach fine for 'em with my knife, and I pads off the old Tom with my stick while the young ones feeds; they likes bread, too, and cake, but spinach is what they is most partial to.'

'Make them run, Long,' said I. 'Run, sir! they would run if *I* could run, sir, but I can't; I got the gout in one foot and the rheumatics in the t'other knee. I'm lame of both legs, sir, and I can't make 'em run; and they knows *I* can't run as well as I knows myself; but, bless me, sir, you should have seen 'em when Lord Randolph's harriers got into the paddock. *They run then* – such a job; but the harriers never catched e'er a one, and his lordship he run till his nose bled. It *was* a job, and no mistake; but no harm was done, and the Duke he never said nothing about it.'

FRANK BUCKLAND (1826–80), *Log Book of a Fisherman and Zoologist*, 1875

On King Charles II

He took delight to have a number of little spaniels follow him, & lie in his bed-Chamber, where often times he suffered to bitches to puppy & give suck, which rendered it very offensive, & indeede made the whole court nasty & stinking.

JOHN EVELYN, *Diary*, 6 FEBRUARY 1685

It was a blustering February day of last year ... We were seated before a blazing fire, and we were not expecting visitors, when the bell rang.

The door being opened, an expressman was literally blown into the hall, carrying a cigar-box, with the explanation that it had come from Florida, and had been delayed several days on the road. From Florida! Many were the wonderings as to its contents. Everything, from Indian relics to orange-blossoms, was suggested; but as it was being opened, a label was discovered on the side, bearing the words, 'Live Alligators from Florida.' This sounded quite alarming, and the ladies present moved to a safe distance, as the cover was gingerly raised.

An apprehensive glance within disclosed a pair of widely opened jaws, which, armed with rows of needle-like teeth, belonged to a very angry little alligator. The poor little fellow had had an unpleasant journey, and, resenting his numerous grievances, seemed, in his ludicrously defiant attitude, ready to fight all comers. The ladies summoned courage to take a nearer view, and one of them declared that he 'looked just like the Evil One'; the others agreed, and to this uncomplimentary remark the little alligator owes his name.

He was tempted with all imaginable dainties, but refused everything offered, and seemed determined to starve to death to show his profound disgust with the world in general, very much as certain youngsters, after being punished for some misdeed, take a heroic revenge by refusing to come to the table.

Finally a letter arrived, explaining that Nature has provided that alligators shall lie in a dormant state through the winter months, and not taste food until spring. So the weeks passed, and although he grew thinner and thinner, he kept up his fast, spending Lent in a most exemplary manner.

But a bright April day wrought a marvelous change in 'Beelzebub.' He evidently thought he had been transported to his native land, such was the warmth of the sun. He had heard our approach and stood with head raised, his bright eyes scintillating with an eager light, and his whole appearance much like that of a pet dog when told to 'speak.' . . .

He has passed through many adventures which doubtless were thrilling, from his point of view. One day he slipped from my hand to the floor, landing head-downward with a thump. A few feeble flops of the tail and he lay apparently lifeless. I picked him up, and he was perfectly limp, with every muscle relaxed. Laying him sorrowfully on the floor, I left the room to announce his fate to the family. The mourners arrived just in time to see his Satanic Majesty scuttling away at a remarkable pace, very much resembling a gentleman in tight boots hurriedly crossing over cobblestones. At another time he was left for a while in a pan covered with cotton netting. During the night he became restless, and, breaking through the netting, started on a tour of exploration. The stairs being in his line of march, the dauntless little 'gator started down them to the lower regions. When found by the terrified cook, he had traveled down two flights of stairs, and was standing at the head of the third, when, seeing that he was discovered, down he went tumbling, sliding, and rolling until he reached the bottom. He was captured and returned to his home in the attic.

Who shall say that General Putnam's ride down the long flight of stone steps, in escaping from the British soldiers, showed more daring than our venturesome Beelzebub's tumbles?

He was of a patient disposition, and would bear a certain amount of teasing without objecting; but at times he showed that a Spartan can fight as well as show endurance. 'Tabby,' the cat, manifested great curiosity, not unmixed with jealousy, when Beelzebub was installed as a family pet; and she acquired the unkind habit of walking up to him at every opportunity, and showing her displeasure by deliberately cuffing him with her paw. Then she would retire showing evident satisfaction as if she had performed a duty.

This was done once too often; for Beelzebub had evidently harbored in his memory her former insults, and this last one proved too much for his injured spirit. His eyes flashed with a yellowish light, and, when Tabby was walking away, he scrambled after, seized her tail, and clung to it viciously. This frightened the bully, and she started on a race around the room, taking aërial flights over chairs and tables, with Beelzebub desperately clinging to her tail. When we released the panic-stricken Tabby, we were surprised to find that Beelzebub was none the worst for his wild experience, and with widely distended jaws, he breathed a general defiance; but Tabby had received a lesson, and she never molested Beelzebub again. . . .

I wish I could add, in closing, that Beelzebub 'lived happily for ever after,' as the story-books say, for he certainly deserved a peaceful life after so many trials, even though they were self-inflicted. But one morning we found he had suddenly vanished. His tank seemed perfectly secure in every way, but Beelzebub was not inside, and how he escaped is still a mystery.

We live not far from a river which Longfellow has immortalized in verse; and to me it seems probable that Beelzebub's instinct has led him to it, and that by this time he is swimming contentedly about. He might survive the winters by burying himself in the mud; and who knows but, in years to come, some duck or swan floating on the river will suddenly disappear, to every one's astonishment – and perhaps nobody but the little alligator will ever know what became of it. But then, again, Beelzebub himself may have met his fate, and all that remains of him now may be only his skin made up into articles useful to mankind.

ALBERT CARLTON PEARSON, 'BEELZEBUB',
St Nicholas Magazine, JANUARY 1895

'Articles useful to mankind.'

2

The Nature of the Beast

It has been calculated by an American man of science that if the mule had the same proportionate power in his hind legs as has the flea, he could kick an ordinary-sized man 33 miles 1004 yards and 21 inches.

G. A. HENTY (1832–1902),
Those Other Animals, 1891

One of my dogs has a petition to offer, say a lap to be occupied. I am conscious of a supplicant standing, rigid, on the invisible but established line of prohibition – (which, nevertheless, is the point from which the proposed assault can be projected) – bright eyes of concentrated resolve assert their will to conquer, pricked questioning ears demand a reply. I feel resistance failing. Instantly the weakness is discerned. The ears flicker and flow backwards. The eyes soften and glow, the little expectant face is flooded by a smile. Before I have time actually to acknowledge defeat, there is an effortless spring, a feather-weight pressure on the knees. The position is stormed.

SOMERVILLE AND ROSS (1858–1949 and
1862–1915), 'Happy Days!', 1946

The majestic metal wire armature of a crinoline, which belonged to some lady of the Czarist court, is exhibited at the Kremlin Museum. From the waistband, or rather from the horrifying metal hoop that serves as a waistband, hang two small tubes made of china, with the shape and size of specimen vials used by chemists; one reads on the description that they were traps for fleas. A teaspoon of honey was put at the bottom of the vial; the fleas, in their peregrinations between one fold of cloth and another, were attracted by the smell of the honey, entered the vial, slipped down its smooth sides, fell to the bottom and were stuck.

This is a chapter in a novel which describes the interminable struggle between two forms of cunning: the conscious, short-term cunning of man who must defend himself from parasites, invent his stratagems in the course of a few generations, and the evolutionary cunning of the parasite which required millions of years but attains results that astound us. . . .

Man's fleas . . . are no longer fashionable and nobody regrets their loss, but precisely now we are witnessing a mysterious revival of the louse, and so we must be on guard. It will help to remember that the flea, besides being a vehicle for epidemics, was only a few decades ago part of European civilisation and folklore, patronised all social classes (as shown by the above-described crinoline) and was often described by literary men. Bernardin de Saint-Pierre, who had boundless faith in Providence, declared that fleas are dark and are attracted by white cloth, so that men can catch them: 'but for the instinct for whiteness of these small, light, black, and nocturnal animals it would be impossible for us to see them and catch them'. Giuseppe Gioacchino Belli, in a sonnet of 1835, paints this strangely sensual miniature of the 'flea-catcher', for whom there is no delight equal to that of catching fleas:

> Everyone has his favourite delight.
> I have the fleas. So there, I love
> To crack them and hear those little pops.

In Balzac's *Droll Stories* the nuns of the merry monastery of Poissy explain to a naïve novice how one must go about telling whether the captured flea is male, female, or virgin, but finding a virgin flea is extremely rare 'because these beasts are unmannerly, they are all lascivious sluts, who give themselves to the first comer'.

In the popular mind the flea, as for that matter also the fly, is related to the devil. In *Faust*, in the inn at Auerbach, Mephistopheles is applauded by everyone when he starts to sing the song of the king who had a huge flea, loved it like a son (not a daughter: *Floh*, in German is masculine), and kept him as a son and had a silk and velvet suit cut for him.

PRIMO LEVI (1919–87), *Other People's Trades*,
1985, TRANS. RAYMOND ROSENTHAL

It has been calculated by an American man of science that if the mule had the same proportionate power in his hind legs as has the flea, he could kick an ordinary-sized man 33 miles 1004 yards and 21 inches. Mankind has therefore good reason for congratulating itself upon the fact that the flea has not, in the course of his career, had any ambition in the direction of size, and that the smallest and most active only survived in the struggle for existence. . . .

The flea is capable of being tamed, and of affording amusement to man by various little tricks. The first step in the process is to restrain his natural inclination to jump. This is done by placing him in a low, flat box with a glass lid. The flea, supposing that he has an open space overhead, jumps, strikes the glass with great violence, and falls half-stunned. This discourages him, but, unable to account for the phenomenon, he tries again and again, until at last, after some days, he arrives at the conclusion that there is something altogether wrong with the atmosphere, and that jumping must be abandoned. After this the rest is easy. He can be taught to drag a little carriage, to sit on the box, to fire a tiny cannon, or to perform other feats. He never, however, recovers thoroughly from the effect of his terrible blows against the glass. His heart and his spirit appear to be alike broken. Like a caged eagle he mopes out his life, and seldom lives more than a month or six weeks after his education is completed.

His is, in fact, the true gipsy spirit. Free, he will make himself happy under any circumstances, and although he may have his preferences, can get on anywhere. He loves the young and the tender, but does not despise age. Free, he is joyous, lively, and daring: a captive and chained, he pines and dies. It is a pity that no one will do for him what Sir John Lubbock has done for the ant. Such an investigator would no doubt be able to rehabilitate the flea in public estimation. Although he may be forced to live in dirty places, he is himself perfectly clean, taking great pains to clean himself with his hind legs, as does the fly. He is clad in shining armour, which is wonderfully tough and strong; his eyes are lively and prominent. Even in his most joyous moments, he is never noisy; his attentions to man are unwearied, and the gentle irritation thereby caused affords means of occupation and excitement to the lazy mendicant, the indolent native of the South, and the contemplative Oriental, and rouses them from the lethargy in which they might otherwise sink. Fully and properly understood, the flea might take high rank among the benefactors of man.

G. A. HENTY (1832–1902),
Those Other Animals, 1891

The Comte de Buffon supposed the Weesel to be untameable; but Mademoiselle de Laistre, in a letter on this subject, gives a very pleasing account of the education and manners of a Weesel which she took under her protection. This she fed with fresh meat and milk, the latter of which it was very fond of. It frequently ate from her hand, and seemed to be more delighted with this manner of feeding than any other. 'If I pour (says this lady) some milk into my hand, it will drink a good deal; but if I do not pay it this compliment, it will scarcely take a drop. When it is satisfied, it generally goes to sleep. My chamber is the place of its residence; and I have found a method of dispelling its strong smell by perfumes. By day, it sleeps in a quilt, into which it gets by an unsewn place which it had discovered on the edge: during the night, it is kept in a wired box or cage; which it always enters with reluctance, and leaves with pleasure. If it be set at liberty before my time of rising, after a thousand little playful tricks, it gets into my bed, and goes to sleep in my hand or on my bosom. If I am up first, it spends a full half-hour in caressing me; playing with my fingers like a little dog, jumping on my head and on my neck, and running around on my arms and body, with a lightness and elegance which I never found in any other animal. If I present my hands at the distance of three feet, it jumps into them without ever missing. It shows a great deal of address and cunning in order to compass its ends, and seems to disobey certain prohibitions merely through caprice. During all its actions, it seems solicitous to divert, and to be noticed; looking, at every jump, and at every turn, to see whether it is observed or not. If no notice be taken of its gambols, it ceases them immediately, and betakes itself to sleep; and even when awaked from the soundest sleep, it instantly resumes its gaiety, and frolics about in as sprightly a manner as before. It never shows any ill humour, unless when confined, or teased too much; in which case it expresses its displeasure by a sort of murmur, very different from that which it utters when pleased.

'In the midst of twenty people, this little animal distinguishes my voice, seeks me out, and springs over every body to come at me. His play with me is the most lively and caressing; with his two little paws he pats me on the chin, with an air and manner expressive of delight. This, and a thousand other preferences, show that his attachment to me is real. When he sees me dressed for going out, he will not leave me, and it is not without some trouble that I can disengage myself from him; he then hides himself behind a cabinet near the door, and jumps upon me as I pass, with so much celerity that I often can scarcely perceive him.

'He seems to resemble a squirrel in vivacity, agility, voice, and his manner of murmuring. During the summer, he squeaks and runs about all night long; but since the commencement of the cold weather I have not observed this. Sometimes, when the sun shines while he is playing on the bed, he turns and tumbles about and murmurs for a while.

'From his delight in drinking milk out of my hand, into which I pour a very little at a time, and his custom of sipping the little drops and edges of the fluid, it seems probable that he drinks dew in the same manner. He very seldom drinks water, and then only for want of milk, and with great caution, seeming only to refresh his tongue once or twice, and to be even afraid of that fluid. During the hot weather, it rained a good deal. I presented to him some rain water in a dish, and endeavoured to make him go into it, but could not succeed. I then wetted a piece of linen cloth in it, and put it near him; when he rolled upon it with extreme delight.

'One singularity in this charming animal is his curiosity; it being impossible to open a drawer or a box, or even to look at a paper, but he will examine it also. If he gets into any place where I am afraid of permitting him to stay, I take a paper or a book, and look attentively at it; when he immediately runs upon my hand, and surveys with an inquisitive air whatever I happen to hold. I must further observe, that he plays with a young cat and dog, both of some size; getting about their necks, backs, and paws, without their doing him the least injury.'

The method of taming these creatures is, according to M. de Buffon, to stroke them gently over the back; and to threaten, and even to beat them, when they attempt to bite. Aldrovandus tells us, that their teeth should be rubbed with garlic, which will take away all their inclination to bite!

The last-mentioned author quotes from Strozza the following part of an elegy on the death of a tame Weesel:

> Loving and lov'd! thy master's grief!
> Thou couldst th' uncounted hours beguile;
> And, nibbling at his fingers soft,
> Watch anxious for th' approving smile:
> Or stretching forth the playful foot,
> Around in wanton gambols rove;
> Or gently sip the rosy lip,
> And in light murmurs speak thy love.
>
> REV. W. B. BINGLEY (1774–1823),
> *Animal Biography*, 1813

'A fierce little fellow.'

Uncle Karl gave a new pet to Marcy. It was a Florida chameleon, or rather, a small lizard which is called so, though it is really, in the books, the green Carolina lizard. . . .

Marcy was much interested in the lively little creature, named it Snap, because of the way he seized a fly, and tried to keep him supplied with food. She also found a book in the library that had an account of some of the same family that were kept by an American gentleman, and she made him a home, as near as she could, like the one his pets lived in. . . .

The same gentleman tells of a lady in Florida who had four of these little creatures as pets. She kept them fastened to her head by silk cord, and let them run over her hair and shoulders as they pleased. . . .

The real chameleon, after which this little American is named because of his changing colors, is quite a different animal, and has often been kept as a pet. . . .

I have read of a tame chameleon, belonging to a lady in Egypt, that would drink from a cup, lifting its head like a chicken, and also enjoyed mutton broth. This one lived on its mistress' head and shoulder for months, fastened by a silk cord to a button, and was a fierce little fellow to others of its kind, biting off their legs and tails when shut up with them. He had notions, too; he did not like to have faces made at him. If a person opened the mouth at him, he would puff and turn black, and sometimes hiss.

OLIVE THORNE MILLER,
Queer Pets and Their Doings, 1880

I often used to take the chameleon out in the garden, and there watch its strange habits.

The reptile could get over the ground at a tolerable pace; that is to say, it would win a race against a tortoise. Its mode of progression can scarcely be named; it was not walking, nor running, nor sprawling, nor waddling, but a unique mixture of them all. It bore about the same relation to the walk of ordinary animals as does the hobble of a man with two wooden legs to the stately march of a drum-major in the Guards. . . .

Any one can walk like a chameleon by employing the following rules: – Let him stoop and rest upon his hands and the tips of his toes, taking care to spread the hands and to hold them with the thumbs pointing directly backwards and the fingers forwards, as to get his elbows well out. In lieu of a tail, let him put together the first three joints of a trolling rod, and push the butt up the back of his waistcoat as he kneels. Then let him try to run a stipulated distance within a given time, and he will find himself imitating to a nicety the action of a chameleon while on the ground. If the path could be strewed thickly with brushwood and large flints, it would add to the accuracy of the personation.

REV. J. G. WOOD (1827–89),
Petland Revisited, 1890

The bell of Monastier was just striking nine as I got quit of these preliminary troubles and descended the hill through the common. As long as I was within sight of the windows, a secret shame and the fear of some laughable defeat withheld me from tampering with Modestine. She tripped along upon her four small hoofs with a sober daintiness of gait; from time to time she shook her ears or her tail; and she looked so small under the bundle that my mind misgave me. We got across the ford without difficulty – there was no doubt about the matter, she was docility itself – and once on the other bank, where the road begins to mount through pine-woods, I took in my right hand the unhallowed staff, and with a quaking spirit applied it to the donkey. Modestine brisked up her pace for perhaps three steps, and then relapsed into her former minuet. Another application had the same effect, and so with the third. I am worthy the name of an Englishman, and it goes against my conscience to lay my hand rudely on a female. I desisted, and looked her all over from head to foot; the poor brute's knees were trembling and her breathing was distressed; it was plain that she could go no faster on a hill. God forbid, thought I, that I should brutalize this innocent creature; let her go at her own pace, and let me patiently follow.

What that pace was, there is no word mean enough to describe; it was something as much slower than a walk as a walk is slower than a run; it kept me hanging on each foot for an incredible length of time; in five minutes it exhausted the spirit and set up a fever in all the muscles of the leg. And yet I had to keep close at hand and measure my advance exactly upon hers; for if I dropped a few yards into the rear, or went on a few yards ahead, Modestine came instantly to a halt and began to browse. The thought that this was to last from here to Alais nearly broke my heart. Of all conceivable journeys, this promised to be the most tedious. I tried to tell myself it was a lovely day; I tried to charm my foreboding spirit with tobacco; but I had a vision ever present to me of the long, long roads, up hill and down dale, and a pair of figures ever infinitesimally moving, foot by foot, a yard to the minute, and, like things enchanted in a nightmare, approaching no nearer to the goal.

In the meantime there came up behind us a tall peasant, perhaps forty years of age, of an ironical snuffy countenance, and arrayed in the green tail-coat of the country. He overtook us hand over hand, and stopped to consider our pitiful advance.

'Your donkey,' says he, 'is very old?'

I told him, I believed not.

Then, he supposed, we had come far.

I told him, we had but newly left Monastier.

'*Et vous marchez comme ça!*' cried he; and, throwing back his head, he laughed long and heartily. I watched him, half prepared to feel offended, until he had satisfied his mirth; and then, 'You must have no pity on these animals,' said he; and, plucking a switch out of a thicket, he began to lace Modestine about the sternworks, uttering a cry. The rogue pricked up her ears and broke into a good round pace, which she kept up without flagging, and without exhibiting the least symptom of distress, as long as the peasant kept beside us. Her former panting and shaking had been, I regret to say, a piece of comedy.

My *deus ex machina*, before he left me, supplied some excellent, if inhumane, advice; presented me with the switch, which he declared she would feel more tenderly than my cane; and finally taught me the true cry or masonic word of donkey-drivers, 'Proot!' All the time, he regarded me with a comical, incredulous air, which was embarrassing to confront; and smiled over my donkey-driving, as I might have smiled over his orthography, or his green tail-coat. But it was not my turn for the moment.

I was proud of my new lore, and thought I had learned the art to perfection. And certainly Modestine did wonders for the rest of the forenoon, and I had a breathing space to look about me. . . .

I hurried over my mid-day meal, and was early forth again. But, alas, as we climbed the interminable hill upon the other side, 'Proot!' seemed to have lost its virtue. I prooted like a lion, I prooted mellifluously like a sucking-dove; but Modestine would be neither softened nor intimidated. She held doggedly to her pace; nothing but a blow would move her, and that only for a second. I must follow at her heels, incessantly belabouring. A moment's pause in this ignoble toil, and she relapsed into her own private gait. I think I never heard of anyone in as mean a situation. I must reach the lake of Bouchet, where I meant to camp, before sundown, and, to have even a hope of this, I must instantly maltreat this uncomplaining animal. The sound of my own blows sickened me. Once, when I looked at her, she had a faint resemblance to a lady of my acquaintance who formerly loaded me with kindness; and this increased my horror of my cruelty.

ROBERT LOUIS STEVENSON (1850–94),
Travels with a Donkey in the Cévennes, 1879

Toby, when full grown, was a strong coarse dog: coarse in shape, in countenance, in hair, and in manner. I used to think that, according to the Pythagorean doctrine, he must have been, or been going to be, a Gilmerton carter. He was of the bull-terrier variety, coarsened through much mongrelism and a dubious and varied ancestry. His teeth were good, and he had a large skull, and a rich bark as of a dog three times his size, and a tail which I never saw equalled – indeed it was a tail *per se*; it was of immense girth and not short, equal throughout like a policeman's baton; the machinery for working it was of great power, and acted in a way, as far as I have been able to discover, quite original. We called it his ruler.

When he wished to get into the house, he first whined gently, then growled, then gave a sharp bark, and then came a resounding, mighty stroke which shook the house; this, after much study and watching, we found was done by his bringing the entire length of his solid tail flat upon the door, with a sudden and vigorous stroke; it was quite a *tour de force* or a *coup de queue*, and he was perfect in it at once, his first *bang* authoritative, having been as masterly and telling as his last.

With all this inbred vulgar air, he was a dog of great moral excellence – affectionate, faithful, honest up to his light, with an odd humour as peculiar and as strong as his tail. My father, in his reserved way, was very fond of him, and there must have been very funny scenes with them, for we heard bursts of laughter issuing from his study when they two were by themselves: there was something in him that took that grave, beautiful, melancholy face. One can fancy him in the midst of his books, and sacred work and thoughts, pausing and looking at the secular Toby, who was looking out for a smile to begin his rough fun, and about to end by coursing and *gurrin'* round the room, upsetting my father's books, laid out on the floor for consultation, and himself nearly at times, as he stood watching him – and off his guard and shaking with laughter. Toby had always a great desire to accompany my father up to town; this my father's good taste and sense of dignity, besides his fear of losing his friend (a vain fear!), forbade, and as the decision of character of each was great and nearly equal, it was often a drawn game. Toby, ultimately, by making it his entire object, triumphed. He usually was nowhere to be

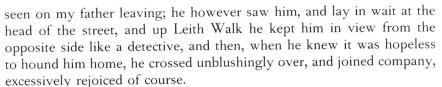
seen on my father leaving; he however saw him, and lay in wait at the head of the street, and up Leith Walk he kept him in view from the opposite side like a detective, and then, when he knew it was hopeless to hound him home, he crossed unblushingly over, and joined company, excessively rejoiced of course.

One Sunday he had gone with him to church, and left him at the vestry door. The second psalm was given out, and my father was sitting back in the pulpit, when the door at its back, up which he came from the vestry, was seen to move, and gently open, then, after a long pause, a black shining snout pushed its way steadily into the congregation, and was followed by Toby's entire body. He looked somewhat abashed, but snuffing his friend, he advanced as if on thin ice, and not seeing him, put his fore-legs on the pulpit, and behold there he was, his own familiar chum. I watched all this, and anything more beautiful than his look of happiness, of comfort, of entire ease when he beheld his friend – the smoothing down of the anxious ears, the swing of gladness of that mighty tail, – I don't expect soon to see. My father quietly opened the door, and Toby was at his feet and invisible to all but himself; had he sent old George Peaston, the 'minister's man,' to put him out, Toby would probably have shown his teeth, and astonished George. He slunk home as soon as he could, and never repeated that exploit.

JOHN BROWN (1810–82), *Horae Subsecivae*, 1897

I doubt very much whether any Brighton lassie could find a prettier object to put in her hat than a live herring. The herrings at the Brighton Aquarium are perfectly beautiful. As they swim about, their lovely scales glisten and glitter with gold, silver, and ruby colours all intermixed. I think it would puzzle even Rolfe – the Landseer of fishes – with all his talent, to paint a live herring. These herrings, too, seem to me to have artful-looking faces. They always swim together, and it appears to me they have a leader, who shows them the road. They swim up to the glass, halt like cavalry, then 'threes about,' and away they go again. They are rather restless; they want to go somewhere. Where *do* they want to go, my dear fish? Far, far away, no doubt, into the deep ocean's sea-weed forests. But we have got you now, and we intend to make you tell us some of your family secrets.

FRANK BUCKLAND (1826–80), *Log Book of a Fisherman and Zoologist*, 1875

D is for Darwin, who, recognising the mental capacity of worms, devoted some thirty-five pages to the subject.

NEVILLE BRAYBROOKE, 'An Animal Alphabet', *The Tablet*, 21/28 DECEMBER 1991

JEUNE ÉLÉPHANT DE X. ANS.

According to Pliny, it was not uncommon to see elephants dance upon a rope, their steps being so practised and certain, that four of them traversed a rope, bearing a litter which contained one of their companions, who feigned to be sick.

Divers Exercices des Éléphants, et détails sur l'usage de leur trompets.

The Parlour Menagerie, DEDICATED TO THE
BARONESS BURDETT-COUTTS (1814–1906)

The cat is domestic only as far as suits its own ends; it will not be kennelled or harnessed nor suffer any dictation as to its goings out or comings in. Long contact with the human race has developed in it the art of diplomacy, and no Roman Cardinal of mediæval days knew better how to ingratiate himself with his surroundings than a cat with a saucer of cream on its mental horizon. But the social smoothness, the purring innocence, the softness of the velvet paw may be laid aside at a moment's notice, and the sinuous feline may disappear, in deliberate aloofness, to a world of roofs and chimney-stacks, where the human element is distanced and disregarded. Or the innate savage spirit that helped its survival in the bygone days of tooth and claw may be summoned forth from beneath the sleek exterior, and the torture-instinct (common alone to human and feline) may find free play in the death-throes of some luckless bird or rodent. It is, indeed, no small triumph to have combined the untrammelled liberty of primeval savagery with the luxury which only a highly developed civilization can command; to be lapped in the soft stuffs that commerce has gathered from the far ends of the world; to bask in the warmth that labour and industry have dragged from the bowels of the earth; to banquet on the dainties that wealth has bespoken for its table, and withal to be a free son of nature, a mighty hunter, a spiller of life-blood. This is the victory of the cat.

SAKI (1870–1916),
The Achievement of the Cat

I resolved to have a horse to ride. I had never seen such wild, free, magnificent horsemanship outside of a circus as these picturesquely-clad Mexicans, Californians and Mexicanized Americans displayed in Carson streets every day. How they rode! Leaning just gently forward out of the perpendicular, easy and nonchalant, with broad slouch-hat brim blown square up in front, and long *riata* swinging above the head, they swept through the town like the wind! The next minute they were only a sailing puff of dust on the far desert. If they trotted, they sat up gallantly and gracefully, and seemed part of the horse; did not go jiggering up and down after the silly Miss-Nancy fashion of the riding-schools. I had quickly learned to tell a horse from a cow, and was full of anxiety to learn more. I was resolved to buy a horse.

While the thought was rankling in my mind, the auctioneer came skurrying through the plaza on a black beast that had as many humps and corners on him as a dromedary, and was necessarily uncomely; but he was 'going, going, at twenty-two! – horse, saddle and bridle at twenty-two dollars, gentlemen!' and I could hardly resist.

A man whom I did not know (he turned out to be the auctioneer's brother) noticed the wistful look in my eye, and observed that that was a very remarkable horse to be going at such a price; and added that the saddle alone was worth the money. It was a Spanish saddle, with ponderous *tapidaros*, and furnished with the ungainly sole-leather covering with the unspellable name. I said I had half a notion to bid. Then this keen-eyed person appeared to me to be 'taking my measure'; but I dismissed the suspicion when he spoke, for his manner was full of guileless candor and truthfulness. Said he:

'I know that horse – know him well. You are a stranger, I take it, and so you might think he was an American horse, maybe, but I assure you he is not. He is nothing of the kind; but – excuse my speaking in a low voice, other people being near – he is, without the shadow of a doubt, a Genuine Mexican Plug!'

I did not know what a Genuine Mexican Plug was, but there was something about this man's way of saying it, that made me swear

inwardly that I would own a Genuine Mexican Plug, or die.

'Has he any other – er – advantages?' I inquired, suppressing what eagerness I could.

He hooked his forefinger in the pocket of my army-shirt, led me to one side, and breathed in my ear impressively these words:

'He can out-buck anything in America!'

'Going, going, going – at *twent–ty*-four dollars and a half, gen –'

'Twenty-seven!' I shouted, in a frenzy.

'And sold!' said the auctioneer, and passed over the Genuine Mexican Plug to me.

I could scarcely contain my exultation. I paid the money, and put the animal in a neighboring livery-stable to dine and rest himself.

In the afternoon I brought the creature into the plaza, and certain citizens held him by the head, and others by the tail, while I mounted him. As soon as they let go, he placed all his feet in a bunch together, lowered his back, and then suddenly arched it upward, and shot me straight into the air a matter of three or four feet! I came as straight down again, lit in the saddle, went instantly up again, came down almost on the high pommel, shot up again, and came down on the horse's neck – all in the space of three or four seconds. Then he rose and stood almost straight up on his hind feet, and I, clasping his lean neck desperately, slid back into the saddle, and held on. He came down, and immediately hoisted his heels into the air, delivering a vicious kick at the sky, and stood on his forefeet. And then down he came once more, and began the original exercise of shooting me straight up again. The third time I went up I heard a stranger say:

'Oh, *don't* he buck, though!'

While I was up, somebody struck the horse a sounding thwack with a leathern strap, and when I arrived again the Genuine Mexican Plug was not there. A Californian youth chased him up and caught him, and asked if he might have a ride. I granted him that luxury. He mounted the Genuine, got lifted into the air once, but sent his spurs home as he descended, and the horse darted away like a telegram. He soared over three fences like a bird, and disappeared down the road toward the Washoe Valley.

I sat down on a stone, with a sigh, and by a natural impulse one of my hands sought my forehead, and the other the base of my stomach. I believe I never appreciated, till then, the poverty of the human machinery – for I still needed a hand or two to place elsewhere. Pen cannot describe how I was jolted up. Imagination cannot conceive how disjointed I was – how internally, externally and universally I was unsettled, mixed up and ruptured. There was a sympathetic crowd around me, though.

One elderly-looking comforter said:

'Stranger, you've been taken in. Everybody in this camp knows that horse. Any child, any Injun, could have told you that he'd buck; he is the very worst devil to buck on the continent of America. You hear *me*. I'm Curry. *Old* Curry. Old *Abe* Curry. And moreover, he is a simon-pure, out-and-out, genuine d——d Mexican plug, and an uncommon mean one at that, too. Why, you turnip, if you had laid low and kept dark, there's chances to buy an *American* horse for mighty little more than you paid for that bloody old foreign relic.'

I gave no sign; but I made up my mind that if the auctioneer's brother's funeral took place while I was in the Territory I would postpone all other recreations and attend it.

After a gallop of sixteen miles the Californian youth and the Genuine Mexican Plug came tearing into town again, shedding foam-flakes like the spume-spray that drives before a typhoon, and, with one final skip over a wheelbarrow and a Chinaman, cast anchor in front of the 'ranch.'

Such panting and blowing! Such spreading and contracting of the red equine nostrils, and glaring of the wild equine eye! But was the imperial beast subjugated? Indeed he was not. His lordship the Speaker of the House thought he was, and mounted him to go down to the Capitol; but the first dash the creature made was over a pile of telegraph poles half as high as a church; and his time to the Capitol – one mile and three quarters – remains unbeaten to this day. But then he took an advantage – he left out the mile, and only did the three quarters. That is to say, he made a straight cut across lots, preferring fences and ditches to a crooked road; and when the Speaker got to the Capitol he said he had been in the air so much he felt as if he had made the trip on a comet.

MARK TWAIN (1835–1910),
Roughing It, 1875

'Poor dogs! I've a strange feeling about the dumb things, as if they wanted to speak, and it was a trouble to 'em because they couldn't. I can't help being sorry for the dogs a lump, though perhaps there's no need. But they may well have more in them than they know to make us understand, for we can't say half what we feel with all our words.'

GEORGE ELIOT (1819–80),
Adam Bede, 1859

So accustomed are we to the pig in his sty that we are apt to forget that he is naturally one of the most valiant of animals, a sturdy and desperate fighter, able to hold his own against most wild beasts, and ready to face man and to die, fighting valiantly to the last, in defence of his wife and offspring. Whether the pig has improved or deteriorated under the hand of man depends upon the point of view from which he is regarded. Those engaged in consuming the succulent ham, or the crisp rasher, would, doubtless, reply in the affirmative; while the Indian officer, on his return from a morning spent in the fierce and hazardous sport of pig-sticking, would utter as decided a negative. Between the wild boar and the domestic pig the difference is as wide as between the aboriginal Briton and the sleek alderman. . . .

The pig is an eminently vocal animal, and even in the bosom of his family he maintains a steady, if to man monotonous, conversation. He possesses a large variety of notes, in this respect far surpassing any other animal. The cat has an extensive register, but principally among the high notes; while the pig's tones embrace the whole gamut, from the deep grunt of discontent to the wild shriek of despair. Properly educated, the pig should be capable of vocal triumphs of a very high kind, its upper notes being as clear and no more unpleasant than the corresponding ones of an operatic soprano, while the lower ones would be the envy of a basso profundo. It is a little singular that no persistent effort should have been made to utilise the pig's vocal powers in this direction, although he has at times been taught to spell and to perform other feats requiring as high an intelligence as that of singing.

. . . No one has yet, so far as we are aware, adopted the pig as a drawing-room pet; and yet, if tended, with the same care bestowed upon the lap-dog, there is no reason why he should not shine in that capacity. His tail is fully as curly as that of the pug, his skin may compare not unfavourably with that of the shaved poodle, while in point of sprightliness he is, at any rate in his younger days, superior to the bulldog. He would not run up curtains like a kitten, nor knock down valuable ornaments from the chimney-piece; while he might, doubtless, be trained with very little trouble into becoming an efficient guard in the house. He is certainly capable of affection, and, as all acquainted with his habits are aware, has pronounced likes and dislikes.

G. A. HENTY (1832–1902),
Those Other Animals, 1891

It is said that the famous British biologist, J. B. Haldane . . . asked by a churchman what his concept of God was, answered: 'He is inordinately fond of beetles.' . . . Since many environments and many geographic areas have not yet been explored by the specialists, it is calculated that at present there exist one and a half million species of coleoptera. Now, we mammals, with our pride as the crown of creation, do not number more than five thousand species; at the very most a few dozen new ones might be discovered, while many existing species are rapidly becoming extinct. . . .

But the coleoptera's armour is an admirable structure: to be admired, unfortunately, only in the glass cases of natural history museums. It is a masterpiece of natural engineering and reminds one of the all-iron armours of medieval warriors. It has no gaps: although not welded, head,

Steeds of metal and muscle for a steeple-chase in earnest.

Phantoms foot it to the Death-watch drum.

neck, thorax and abdomen form a squat, practically invulnerable block, the tenuous antennae can be retracted into grooves and the legs' articulations are protected by flanges that recall the shinguards in the *Iliad*. The resemblance between a beetle that advances pushing aside the grass, slow and powerful, and a tank is so striking that it immediately calls to mind a metaphor in both senses: the insect is a small Panzer, the Panzer is an enormous insect. And the back of the beetle is heraldic: convex or flat, opaque or shiny, it is a noble escutcheon, even if its aspect has no symbolic relation to the 'trade' of its bearer, that is, its manner of escaping aggressors, reproducing and nourishing itself.

Here the Eternal's fondness for beetles has truly unleashed all his imaginative powers. There is no organic material, living, dead, or decomposed, that hasn't an enthusiast among the coleoptera. Many of them are omnivorous, others feed at the expense of a single animal or vegetable species. There are those who eat snails exclusively and have turned themselves into a tool suited to this purpose: they are living syringes, their abdomen is voluminous, but head and chest have an elongated and penetrating shape. They plant themselves in the victim's soft body, inject it with digestive juices, wait for the tissues to disintegrate, and then suck them up.

The very beautiful cetonias or rosechafers (dear to the poet Gozzano: 'desperate cetonias overturned', one of the most beautiful verses ever composed in our [Italian] language), feed only on roses, and the no less sacred scarabs only on bovine excrement: the male makes a small ball of it, clasps it between his hind tarsi as if between two pivots and takes off in reverse gear, pushing and rolling it until he finds a terrain suitable for burying it; at that point the female makes her entrance and deposits on it a single egg. The larva will feed on the matter (by now no longer ignoble) to which the farsighted couple has devoted so much effort, and after the moulting a new scarab will emerge from the tomb: indeed, according to some ancient observers, the same scarab as before, risen from death like the phoenix.

Other beetles can be found in slow or stagnant waters. They are splendid swimmers: some, inexplicably (who knows why?), swim in narrow circles or complicated spirals, others point in a straight line at an invisible prey. None of them, however, has lost the faculty of flying, for often necessity forces them to abandon a pond that has dried up to find another body of water, perhaps even far away. Once, travelling at night along a highway lit by the moon, I heard the windows and the roof of my car bombarded as if by hail: it was a swarm of diving beetles, shiny, brown, and edged with orange, as big as half a nut, who had mistaken the asphalt on the road for a river, and vainly tried to land on it. These beetles, for hydrodynamic reasons, have achieved a compactness and simplicity of shape which I believe to be unique in the animal kingdom: looked at from the back, they are perfect ellipses from which protrude only the legs transformed into oars.

In eluding dangers and aggressions also these insects 'think of everything'. Some exotic species, as large as a fava bean, are endowed with incredible muscular strength. If enclosed in a hand they force their way out between the fingers; if swallowed by a toad (by mistake, but toads will swallow any small object they see moving on a horizontal line), they do not follow the strategy of Jonah gulped down by the whale, nor that of Pinocchio and Geppetto in the shark's entrails, but simply, with the providential assistance of their front legs built to act as earth-movers, dig their way out through the body of the aggressor. . . .

All these modes of behaviour evoke a complex range of impressions: amazement, curiosity, admiration, horror, and laughter, but it seems to me that over them all predominates the sensation of extraneousness or alienation. These small flying fortresses, these portentous little machines, whose instincts were programmed one hundred million years ago, have nothing at all to do with us, they represent a totally different solution to the survival problem. To some extent, or even only symbolically, we humans recognise ourselves in the social structures of ants and bees; in the industriousness of the spider, in the dance of butterflies: but nothing really ties us to the beetle, not even parental concerns, because among coleoptera it is very rare for a mother (and much less a father) to see its offspring before dying. They are the different ones, the aliens, the monsters. Kafka's atrocious hallucination is not chosen by chance – the story in which the travelling salesman Gregor 'awakening one morning from agitated dreams' finds himself changed into an enormous beetle, so inhuman that no one in his family can bear his presence.

So then, these different ones have shown how marvellous capacities of adjustment to all climates, have colonised all ecological niches and eat everything: some even perforate lead and tinfoil. They have elaborated an armour with extraordinary resistance to impact, compression, chemical agents and radiation. Some of them have dug shelters in the ground that are a metre deep. In the case of a nuclear catastrophe they would be the best candidates to be our successors (not the tumble bugs, who eat excrement, and these because of the lack of raw materials).

On top of everything else, their technology is ingenious but rudimentary and instinctive; after the planet becomes theirs, many millions of years will have to pass before a beetle particularly loved by God, at the end of its calculations will find written on a sheet of paper in letters of fire that energy is equal to the mass multiplied by the square of the velocity of light. The new kings of the world will live tranquilly for a long time, confining themselves to devouring each other and being parasites among each other on a cottage industry scale.

PRIMO LEVI (1919–87), *Other People's Trades*,
1985, TRANS. RAYMOND ROSENTHAL

Two Ceylonese Elephants, a male and a female, each about two years and a half old, were, in 1786, brought into Holland, a present to the Stadtholder from the Dutch East India Company. They had been separated, in order to be conveyed from the Hague to Paris; where, in the Museum of Natural History, a spacious hall was prepared for their reception. This was divided into two apartments, which had a communication by means of a large door resembling a portcullis. The enclosure round these apartments consisted of very strong wooden rails. The morning after their arrival, they were conveyed to this habitation. The male was first brought. He entered the apartment with suspicion, reconnoitred the place, and then examined each bar separately with his trunk, and tried their solidity by shaking them. He attempted to turn the large screws on the outside, which held them together, but was not able. When he arrived at the portcullis, which separated the apartments, he observed that it was fastened only by a perpendicular iron bar. This he raised with his trunk, then pushed up the door, and entered the second apartment, where he received his breakfast. – These two animals had been parted (but with the utmost difficulty) for the convenience of carriage, and had not seen each other for some months; and the joy they experienced, on meeting again, after so long a separation, is scarcely to be expressed. They immediately rushed towards each other, and sent forth cries of joy so animated and loud as to shake the whole hall. They breathed also through their trunks with such violence, that the blast resembled an impetuous gust of wind. The joy of the female was the most lively. She expressed it by quickly flapping her ears, which she made to move with astonishing velocity, and drew her trunk over the body of the male with the utmost tenderness. She particularly applied it to his ear, where she kept it a long time; and, after having drawn it over his whole body, often moved it affectionately towards her own mouth. The male did the same over the body of the female, but his joy was more steady. He seemed, however, to express it by his tears, which fell from his eyes in abundance. Since this time they have occupied the same apartment; and their mutual tenderness, and natural affection, have excited the admiration, and even the esteem, of all who have visited them.

REV. W. B. BINGLEY (1774–1823),
Animal Biography, 1813

3

Admirable Antics

My little sparrow rose to fame during the dark days of the blitz, for he became an actor and, though his career was a short one, he gave genuine pleasure to many of London's weary citizens while it lasted.... I can say with complete truth that no sparrow ever served his country so faithfully and so efficiently as he did in those terrible months.

CLARE KIPPS, *Sold for a Farthing*, 1953

LE CERF COCO et LE CERF AZOR

OF THE DEXTERITY OF SOME MEN IN THE INSTRUCTION OF SEVERAL CREATURES

Man is seldom so fortunate a teacher as when he hath himself got his scholar; but he should employ at home that ingenuity and industry which he sometimes makes use of abroad, what a wonderful proficient would he be in all kind of virtue! for there is scarce any-thing that may seem so difficult, but his care and constancy has overcome; as the following examples will be sufficient to account for.

1. The Count of Stolberg in Germany, had a deer, which he bestowed on the Emperor Maximillian the Second, that would receive a rider on his back, and a bridle in his mouth, and would run a race with the fastest horse that came to the field, and outstrip him too. . . .

7. Cardinal Ascanio had a parrot that was taught to repeat the Apostles' creed verbatim, in Latin and in the Court of Spain there was one that could sing the gamut perfectly. If at any time he was out, he would say 'nove bueno' that is not well; but when he was right, he would say 'bueno va' now it is well. As John Barnes, an English friar, relates in his book De Aequivocatione.

8. The Elephant is a creature of a very docile and capable nature to

learn almost any thing: they have been taught by their keepers 'to adore the King' says Aristotle, to dance, to throw stones at a mark; to cast up arms in the air, and catch them again in their fall; to walk up on ropes, which Galba was the first that exhibited at Rome, says Suetonius. And these things they learn with that care, that they have often been found practicing in the night what had been taught them in the day. 'They write too,' says Pliny, 'speaking of one who wrote in the Green tongue *ipsi ego haec scripsi, et spolia Celtica dicavi.*' 'I myself saw' says Aelian, 'one of them writing Roman letters upon a tablet with his trunk, and the letters he made were not ragged, but straight and even; and his eyes were fixed upon the tablet, as one that was serious and intent upon his work.' In the plays that Germanicus Caesar showed in Rome, in the reign of Tiberius, there were twelve elephants, six males and six females; these were clothed as men and women. At the command of their keeper they danced and performed all the gestures of a mimic. At last they were brought where they were to feast; a table was covered with all kinds of dainties, and goblets of gold, with other little cups of wine, placed and beds covered with purple carpets after the manner of Roman eating, for them to lie upon: upon these they laid down, and at the signal given, they reached out their trunks to the table, and with great modesty, fell to eating, and eat and drank as a sort of civil men would.

NATHANIEL WANLEY (1634–80),
The Wonders of the Little World, 1678

In an American scientific journal, there is a well-authenticated account of a strange and overpowering sensibility to music, as evinced by a mouse. It says, 'that one evening, as a few officers on board a British man-of-war, in the harbour of Portsmouth, were seated round the fire, one of them began to play a plaintive air upon the violin. He had scarcely performed ten minutes, when a mouse, apparently frantic, made its appearance in the centre of the floor. The strange gestures of the little animal strongly excited the attention of the officers, who, with one consent, resolved to suffer it to continue its singular actions unmolested. Its exertions appeared to be greater every moment; it shook its head, leapt about, and exhibited signs of the most ecstatic delight. It was observed that in proportion to the gradations of the tones to the soft point, the feelings of the animal appeared to be increased. After performing actions, which an animal so diminutive would, at first sight, seem incapable of, the little creature, to the astonishment of the delighted spectators, suddenly ceased to move, fell down, and expired without evincing any symptoms of pain.'

MRS R. LEE, *Anecdotes of the Habits and
Instincts of Animals,* 1852

The wisest dog I ever had was what is called the Bull-dog Terrier. I taught him to understand a great many words, insomuch that I am positive that the communication betwixt the canine species and ourselves might be greatly enlarged. Camp once bit the baker, who was bringing bread to the family. I beat him, and explained the enormity of his offence; after which, to the last moment of his life, he never heard the least allusion to the story, in whatever voice or tone it was mentioned, without getting up and retiring into the darkest corner of the room with great appearance of distress. Then if you said, 'The baker was well paid,' or 'The baker was not hurt after all,' Camp came forth from his hiding-place, capered, and barked, and rejoiced. When he was unable, towards the end of his life, to attend me when on horseback, he used to watch for my return, and the servant used to tell him 'his master was coming down the hill, or through the moor,' and although he did not use any gesture to explain his meaning, Camp was never known to mistake him, but either went out at the front to go up the hill, or at the back to get down to the moor-side. He certainly had a singular knowledge of spoken language.

SIR WALTER SCOTT (1771–1832)

The rat is even more unconventional as a pet than is the mouse, and yet I believe that there are few animals which will better repay petting.

It is intellectual to a degree, as any professional rat-catcher can witness, for to catch an old rat is a feat which taxes human intellect to the utmost. . . .

I have seen and handled a pair of tame rats belonging to some young friends, and prettier, more playful, and more intelligent pets could not be imagined. They were accustomed to run about on the table at meal times. They never stole food; but when anything was offered them, they sat up on their hind-legs, held the morsel between their fore-paws, and ate it daintily.

They were very fond of a game which I saw them play.

The rats were put into the boy's cap which was hung on the hat-stand in the hall. The boy and his sister then went to the top of the house and whistled. At the sound of the whistle, the rats jumped out of the cap, scrambled to the floor, and then ran up the stairs and perched on their owner's shoulders.

REV. J. G. WOOD (1827–89),
Petland Revisited, 1890

It is nearly four years since the interest in the condition and fate of Emin Pasha reached its height in this country. The pages of NATURE and the columns of the daily press of the time will afford evidence of the universality and intensity of that interest, and of the reality of the belief that

Emin and his people were in imminent danger of being exterminated by the Mahdists. . . . To rescue and relieve the pioneer of science and of civilization was the one object of the Expedition with the leadership of which Mr Stanley was entrusted. . . .

Mr Stanley gives some natural history notes which he obtained from Emin. Here, for example, is a statement which he gives in Emin's own words, but which notwithstanding is somewhat astounding:–

'The forest of Msongwa is infested with a large tribe of chimpanzees. In summer-time, at night, they frequently visit the plantations of Mswa Station to steal the fruit. But what is remarkable about this is the fact that they use torches to light the way! Had I not witnessed this extraordinary spectacle personally, I should never have credited that any of the Simians understood the art of making fire. . . .'

The importance of this fact with regard to fire-using (it is not stated that they are fire-making) chimpanzees need not be pointed out. We cannot doubt the accuracy of Mr Stanley's report, nor the trustworthiness of Emin's observation; but we should like to have more details.

Nature, 3 JULY 1890

*

Perhaps Mr Stanley or Surgeon Parke, if applied to, could throw some more light on the extraordinary statement made by Emin Pasha, recently referred to, which, if it be true, is the most important statement in the whole book.

It is probable that when Emin Pasha witnessed the torch-bearers, whether chimpanzees or young negroes or dwarfs, he was not alone, and, even though very short-sighted, he would have been able to verify his observation of the torch-bearing animals by reference to those near him. An experienced naturalist like Emin Pasha is not likely to have made the mistake Professor Romanes thinks he did make – but it is possible.

Bearing in mind that a large ape is now undoubtedly acting as a signalman (under direction) on a railway at Natal, who can say what the limits of intelligence are in the tribes of Simians?

Nature, 21 JULY 1890

*

Sir – May I correct an error made in *Nature* 100 years ago – quoted from *Nature* 24 July 1890 in Then and Now of 26 July 1990? Your correspondent, J.F., mentioned a large ape said to be acting as a signalman on a railway in Natal. In fact, the ape – a baboon – acted as a railwayman at Uitenhage in the Cape, *not* in Natal. The story is of some interest and is well-documented.

Mr Wide, a railwayman, had lost both lower legs. He trained a baboon, Jack, to work the lever of the signals and to carry out various other duties, including pushing Wide to work and back on a trolley which the baboon had to put on the rails, pumping water, gardening and locking the door. On one occasion, when Wide had injured his arm, Jack took

Railway Jack.

over all his signal-changing duties. According to a reliable witness, 'Jack knew every one of the various signals and which lever to pull as well as his master himself'. Not unnaturally, the railway passengers objected initially, but 'the baboon never failed during his many years of work; and on several occasions he acted in a manner simply astounding to those who have not had personal experience of the high degree of intelligence possessed by these animals'. Jack also defended Wide in a quarrel, and on another occasion drove off a foreman by beating him with a dirty coal sack.

Mr Wide and Jack were extremely close, a relationship expressed by mutual grooming. The story was documented originally by the Rev. George Howe, in 1890, and then by F. W. Fitzsimons, director of the Port Elizabeth Museum. Fitzsimons' account (of which I have a copy) is published in the *Cape Mercury* of 29 May 1923. It relies on an interview with Wide, the written statements of 25 witnesses, filed in the museum collection, and on Howe's evidence.

LETTER TO *Nature*, 25 OCTOBER 1990, FROM
E. G. NISBET, UNIVERSITY OF SASKATCHEWAN

I ran across a dim photograph of him the other day, going through some old things. He's been dead twenty-five years. His name was Rex and he was a bull terrier. He was big and beautifully made and never lost his dignity even when trying to accomplish the extravagant tasks my brothers and myself used to set for him.

One of these was the bringing of a ten-foot wooden rail into the yard through the back gate. We would throw it out into the alley and tell him to go get it. Rex was as powerful as a wrestler, and there were not many things that he couldn't manage somehow to get hold of with his great jaws and lift or drag to wherever he wanted to put them, or wherever we wanted them put. He could catch a rail at the balance and lift it clear of the ground and trot with great confidence towards the gate. Of course, since the gate was only four feet wide or so, he couldn't bring the rail in broadside. He found that out when he got a few terrific jolts, but he wouldn't give up. He finally figured out how to do it, by dragging the rail holding on to one end, growling. He got a great, wagging satisfaction out of his work.

JAMES THURBER (1894–1961)

Remembrances of Youthful Days. – A parrot had been caught young, and trained by a Spanish lady, who sold it to an English sea-captain. For a time the bird seemed sad among the fogs of England, where birds and men all spoke to her in a foreign tongue. By degrees, however, she learned the language, forgot her Spanish phrases, and seemed to feel at home. Years passed on, and found Pretty Poll the pet of the Captain's family. At last her brilliant feathers began to turn grey with age, she could take no food but soft pulp, and had not strength enough to mount her perch. But no one had the heart to kill the old favourite, she was entwined with so many pleasant household recollections. She had been for some time in this feeble condition, when a Spanish gentleman called one day to see her master. It was the first time she had heard the language for many years. It probably brought back to memory the scenes of her youth in that beautiful region of vines and sunshine. She spread forth her wings with a wild scream of joy, rapidly ran over the Spanish phrases, which she had not uttered for years, and fell dead.

The Parlour Menagerie, DEDICATED TO THE
BARONESS BURDETT-COUTTS (1814–1906)

Not only is the pig naturally clever, but it is capable of instruction, and has been taught to perform duties that belong to other animals. It will be unnecessary to relate here the well-known history of 'Slut,' the famous pointer-pig, and I will merely mention that the animal was regularly broken in by a gamekeeper, just as he would have broken in a pointer.

The pig always accompanied him to the field, and learned to point as well as any dog. Indeed, in several cases, the pig discovered birds which the dog had missed. Immediately after discovering a covey of birds, she was rewarded by a little piece of pudding which her master kept ready in his pocket. She seemed to enjoy the sport quite as much as any dog. More than that, the keeper could put her to work which dogs could not undertake. It is considered a great crime for a pointer to stand to any other game than that to which it has been especially trained. But Slut has been known to point out partridges, pheasants, black-game, snipes, and rabbits, all in the same day; but never could be taught to notice hares. The animal was not very often taken to the ground, because the dogs were too aristocratic in their habits to approve of such a companion; but she often went out voluntarily, and joined the sportsmen, remaining with them for several hours. So fond was she of the sport, that she used to go backwards and forwards, from one of her keepers to the other, a distance of seven miles, in order to find some one who was going out shooting.

REV. J. G. WOOD (1827–89),
Petland Revisited, 1890

THE FOUNDLING

July the first, nineteen hundred and forty. . . . I was returning from a long day's duty as an Air-Raid Warden at a neighbouring Post when I saw, on the doorstep of my little bungalow in one of London's suburbs, a tiny bird that had fallen or been thrown from its nest. . . .

Feeling that if a new-born infant is left outside one's doorstep something should be done about it, I picked it up, wrapped it in warm flannel and, sitting over the kitchen fire, endeavoured for several hours to revive it. After I had succeeded in opening its soft beak – an operation that required a delicate touch and immense patience to avoid injury – I propped it open with a spent match and dripped one drop of warm milk every few minutes down the little throat. At the end of half-an-hour, though the bird was still quite cold, I noticed a slight movement of one skinny wing, so, after adding a little soaked bread to the last feed, I put it gently into a small pudding-basin lined and covered with wool, which I deposited in the airing-cupboard. Then, fully expecting it to die in the night, I went to bed.

To my astonishment, early next morning I heard a faint, continuous sound coming from that airing-cupboard – an incredibly thin yet happy sound, the kind of noise a pin would make if it could sing; and there was the little creature, still in his porcelain cradle, but warm and alert and crying for his breakfast.

After that his mouth was rarely shut; and, as he required constant feeding, I took him with me in his basin to the Warden's Post, where he began to serve his country by providing us with endless amusement during the long hours of waiting. I fed him on soaked bread mixed with Bemax, hard-boiled yolk of egg, and one drop of halibut-liver oil, given frequently in small quantities and pushed gently down his throat with the carefully-pointed end of a match. Though the children of the neighbourhood constantly brought along caterpillars and worms in matchboxes tied with blue ribbon, I kept him strictly to this vegetarian diet; and he thrived and grew into a lusty and importunate fledgling. . . .

My intention of course had been to set him free as soon as he could fly and feed himself, but as the wing-feathers developed, a tragedy was revealed, and it looked as though he would never be able to fly freely or to any height that would be reasonably safe. The left wing appeared to be normal, but the right was obviously deformed, the primary feathers standing upright above his back like a little fan. It had a curious effect, this fan-wing, particularly when he fluttered it at my approach in the most engaging manner. However, he learned to jump and to use his wings, in the manner of fledglings, to help his feet as he scrambled after me from room to room. The left foot was also faulty and had a deformed and curled hind toe. . . .

As soon as he was able to feed himself I left him at home, shut safely into a small room with food and milk in each corner of the floor. He soon began to recognise my voice and step, and even the sound of my key in the door, and gave me a vociferous welcome on my return. The moment I opened the door of his boudoir there would be a rush of flying feet, and he would scramble up my leg, over my knee and on to my shoulder, chattering excitedly, before tucking himself under my chin or just inside my collar. . . .

After breakfast, if the siren allowed, came the 'Morning Scrap.' The bed would be cleared for action and I would sit at one end and the sparrow, looking like a miniature eagle, at the other. Then he would rush at me, tail spread and wings outstretched, and hold down my hand with one tiny claw while he hammered it with his beak like a miner with a pickaxe. He would then retreat only to return in fury to the attack – pecking, pinching, tumbling and scolding as the wild sparrows do in the hedgerows. . . .

When left alone in the house he seemed quite content. I often watched through the window to satisfy my mind that he was not fretting in my absence, but apparently, as soon as he realised that I had gone, he settled down and amused himself with his food and toys. I had provided him with a great variety of playthings, but the only ones that ever appealed to him were hairpins, patience-cards and matches which he would carry about in his cage by the hour. Once he knew I was in the house, however, toys were discarded and I occupied his whole attention. He was a most persistent little person. He couldn't bear me out of his sight for one moment, and the sound of his voice and the patter of his little feet seemed to fill every room until I found it difficult to believe that I had not adopted a whole nestful of birds.

Yet, in spite of all his youthful energy and high spirits, he was ready at any hour of the day to share my slumbers. When I was confined to bed for a fortnight with a severe attack of the measles he lived in a state of bliss. Every day was field-day and life became 'one constant round of pleasure.' He shared my food, and cuddled under the bedclothes more or less all day, though occasionally he climbed back into his cage to attend to his toilet and to take a nibble between meals as children love to do, and then rushed back to me with a chirrup of joy. He fought, scolded and bullied the District Nurse and amused her so intensely that she brought her patients to see for themselves what they had refused to believe, with the result that he took on the whole crowd and fought them *en masse* until he had vanquished them all. . . .

HIS LIFE AS AN ACTOR

. . . my little sparrow rose to fame during the dark days of the blitz, for he became an actor and, though his career was a short one, he gave genuine pleasure to many of London's weary citizens while it lasted. . . . He never showed any fear at any time during the raids, although they annoyed him at night and he would ring his bell and rattle his bars if the noise was louder than usual. . . .

I can say with complete truth that no sparrow ever served his country so faithfully and so efficiently as he did in those terrible months. People who had lost their homes and all their possessions forgot their troubles, at least for the time being; terrified children became merry and carefree, and those who had obstinately refused to allow their gas-masks to be fitted held up their heads at once if promised a game with the sparrow. Indeed he became quite an important member of the Civil Defence, doing useful and valiant work as an entertainer when the blitz was at its height, and he very seldom disappointed an audience. Even at houses where people were hostile and where I had been sullenly refused admittance, the doors were often opened with a smile and a cheery word if the young actor accompanied me. Little stories about him, and greeting-cards with a sketch on the cover, were bought eagerly in aid of the Red Cross and found their way into homes and hospitals, not only in England but overseas in the more remote corners of the earth.

He began his performance by sitting sedately in his historic pudding-basin, where he was fed with hemp-seeds by favoured ticket-holders from the front row of the stalls. Then, as gay and light of foot as a ballet-dancer, he would leap out and, suddenly transforming himself into an Infant Hercules with set brows and straining muscles, he would

*'I can turn round ten
times without dropping
it. It's not easy. Try it!'*

engage me in a tug-of-war with a hairpin, holding it tenaciously in his
beak, pulling with all his might until I allowed him to win and carry his
trophy in triumph to his cage. After a curtain and a brief interval he
would re-appear in the role of conjuror and pick a card from a hand
presented to him, usually the one chosen by the audience if I pointed to
it or pushed it very slightly forward. When he tired of this he took a
patience-card in his beak, and turned it round ten or twelve times without
dropping it as he rounded the corners. This, I believe, was his favourite
trick, for it was self-taught and he amused himself with it for years after
he had left the footlights and had forgotten all the others. . . .

His most popular number, however, was his famous 'Air-Raid-Shelter
Trick' which never failed to bring down the house and gave him many
curtains. I had taught him to sit down in my left hand – at first by putting
hemp-seeds there – while I cupped it with my right. Thereafter it was
comparatively easy to associate this action with the repetition of certain
words and before long I had only to say: 'Siren's gone!' and he would
run into this improvised shelter, sit quite motionless for several minutes
and then poke out his head as if enquiring if the All-Clear had yet
sounded. . . .

The question of a name for the sparrow arose about this time, so the
matter was discussed at the next Warden's Meeting. An actor of his
ability must have a name if only to blazon on his posters! My original
intention had been to christen him 'Clarissa,' for, on account of his pale
grey chest and the absence of male markings, I had mistaken him for a
hen; but after his first moult he suddenly appeared with an Old School
Tie, that patch of black or very dark brown under the chin which is the
Sparrow's Badge of Manhood. So feminine an appellation, therefore,

had become an insult. 'Let's call him Clarence!' said the children, and, preposterous as it may seem, Clarence he became, though he never owned it, answering only to the name of 'Boy'. . . .

After his first moult he was really a beautiful little bird. Much daintier, more graceful in his movements, sleeker and more streamlined than his male relatives in garden and gutter, he also displayed a good deal more colour. In addition to a decidedly yellowish collar, he sported a saffron waistcoat and primrose pants. This unusual brilliance of plumage was probably due to colour-feeding with yolk of egg since, after this useful article of food became almost obsolete and the mere mention of it an anachronism, it faded very noticeably. His beak and toe-nails were like polished ebony, and even the 'fan-wing' was decorative and certainly distinctive. . . .

HIS LIFE OF MUSIC

. . . I cannot remember precisely when he began to sing, as I was not the first person privileged to hear him; but it must have been sometime in January 1941, when he was only six months old. I was already aware of the fact that, although many of the sounds he uttered were the common language of sparrows, many of them were not. The range and variety of his notes and calls were remarkable, and he was continually adding to them until – as a mere stripling – he must have wielded (like Mr Winston Churchill on a higher intellectual level) the most formidable vocabulary of his generation. His famous 'Hitler Speeches,' as the children called them, which he had delivered for our edification – his fan-wing lifted high as if in Nazi salute – ever since he was out of his cradle, had increased in length until they lasted, with only a few short breaks, for nearly three minutes and a half. It is true they bore some resemblance to the chattering of sparrows in the hedgerows; yet they were distinctly different and rose, like oratory, from a solemn and impressive statement, by gradual crescendos, to a fiery and impassioned climax. But these perorations gave me no reason to suspect that the little master of rhetoric possessed any latent musical talent.

Ever since he was out of, or rather into, his petticoats, and whenever I was off duty in the early morning, I had put him on my shoulder and, carrying him to the grand piano, had played to him for over an hour. He showed, almost from the first day, that he was affected and excited by the music. Not only his wing but his whole body would quiver, and he would pinch my neck, twisting little bits of the flesh until my scales became suddenly staccato, as if his emotions were too strong for him. Whether this was an expression on his part of pleasure or pain I cannot say: it may have been both, but I never dreamed for one moment that he would learn to sing.

. . . Suddenly one morning as I was running water from the bathroom tap, I heard it myself – a strange little song – clearly and unmistakably coming from the locked room. It began with twitterings; then there was a little turn, an attempt at melodic outline, a high note (far above the vocal register of a sparrow) and then – wonder of wonders! – a little trill. . . .

I began . . . playing to him whenever there was an opportunity, and my joy was great when he began to follow me to the piano of his own accord, climb on to my shoulder and sing to my accompaniment.

One day in early Spring, as by this time the young musician could usually be relied upon to honour his engagements, I arranged a little informal concert so that he could make his debut as an operatic tenor. Six or seven people accepted the invitation and, after tea, the stage was set for the First Appearance of the Infant Prodigy. The audience was seated, breathless and expectant, at some distance from the piano, and the doors, both of the music-room and of an apartment opposite which served as the Artist's Room, were opened wide. I took my seat at the keyboard, and all eyes were focused on the floor at the threshold.

For several minutes after I began to play nothing happened. There was neither sight nor sound of the artist, and my heart sank. Then someone said in a stage-whisper, 'Hush! He's coming!' and a moment later a minute figure appeared in the doorway. I cannot truthfully say that his entrance was a success. It was not impressive and it lacked style. Perhaps he felt some tenseness in the atmosphere that he did not understand. However, after stopping on the hearthrug to arrange his coat-tails or, to be more exact, to draw the feathers of his tail one by one through his beak, he half ran and half flew across the room – as if the cat were after him – and scrambled up my leg and on to my shoulder. The silence could be felt. Once again there was an anti-climax and it looked as if the money for the tickets would have to be returned, for he sat there for several minutes, quietly preening his feathers.

At long last, after more rapid trilling and rippling from me in the upper register of the piano, he began to tune up and suddenly burst into full song to the accompaniment of the Black Study. This, alas! was his swan-song as a Concert Artist, for the applause so terrified him that he disappeared down my neck and never sang in public again.

Nevertheless, our own private and intimate recitals continued to be a joy to me for several years. He loved music that trilled, and scales played rapidly in the treble; and, though I do not for a moment suggest that he knew one piece from another, some undoubtedly had a special appeal and inspired him to more spontaneous outbursts. I always think that he learned to trill from Chopin's Berceuse, but that of course is impossible to prove. He never sang so well as in the early morning and, as I played faster and faster, and higher and higher in the treble, he would pour out his soul in an ecstacy as great, if not as melodious, as any skylark. . . .

THE LAST PHASE

. . . My Sparrow died on August 23rd, 1952, four months after this little book was written. He was nearly blind, though his hearing was still acute. Too weak to stand, though he made two gallant attempts to do so, he settled himself quietly in my warm hand and lay motionless for several hours. Suddenly he lifted his head, called to me in his old intimate way and was gone. He had lived twelve years, seven weeks and four days, and was courageous, intelligent and apparently conscious to the end. The cause of death was extreme old age.

His remains – and what a tiny morsel of tattered feathers was all that was left of him – repose in a small Hoptonwood tomb sacred to the memory of

CLARENCE
THE FAMOUS AND BELOVED SPARROW

CLARE KIPPS, *Sold for a Farthing*, 1953

4

Our Dumb Companions

His mistress alone he [the duck] distinguished by holding her skirt in his beak as he walked beside her, and on her return from a holiday he would greet her with extravagant manifestations of delight, jumping around her, rubbing his head against her, and making little snatches at her dress.

MURIEL KENNY, QUOTED IN *Birds, Dogs and Others*, HUGH MASSINGHAM, 1921

Attachment of the Lion to his Keeper. – In Martin's 'History of the British Colonies,' it is related that Prince, a tame lion on board HMS *Ariadne*, was much attached to his keeper, a seaman, who, having got drunk one day, was ordered to be flogged. The grating was rigged on the main deck opposite Prince's den, a large barred-up place, the pillars very strong, and cased with iron. When the keeper began to strip, to undergo this disgusting and savage punishment, the lion rose gloomily from his couch, and got as near to his friend as possible. On beholding his bare back, he walked hastily round the den; and when he saw the boatswain inflict the first lash, his eyes sparkled with fire, and his sides resounded with the strong and quick beating of his tail. At last, when the blood began to flow down the poor fellow's back, and the clotted cat-o'-nine-tails jerked its gory knots close to the lion's cage, the animal's fury became tremendous; he roared with a voice of thunder, shook the strong bars as if they were osiers, but finding his efforts to break them unavailing, he rolled and roared in a manner so terrific, that the captain, fearing the consequences, ordered the unfortunate keeper to be cast off and to enter the den. It is impossible to describe the joy evinced by the lion. He licked with care the mangled and bleeding back of his keeper, caressed him with his paws, and even folded them around him as if to protect him from similar treatment; and it was only after several hours that the lion would allow the keeper to return among those wretches who had so ill-used him.

The Parlour Menagerie, DEDICATED TO THE
BARONESS BURDETT-COUTTS (1814–1906)

In your review of a book on runner ducks you mention that they have been found 'as intelligent as a parrot,' and regret that the author has given no illustration of their quality. I can fully endorse his opinion. Humphry, an Indian runner drake, was given to us when six weeks old, and proved through three years as intelligent, affectionate, and companionable a pet as we have ever known. He had no bird-companions and quickly developed a passion for human society. He would escort one round the garden in embarrassingly close attendance, maintaining a low conversational twitter of unutterable content. His mistress alone he distinguished by holding her skirt in his beak as he walked beside her, and on her return from a holiday he would greet her with extravagant manifestations of delight, jumping around her, rubbing his head against her, and making little snatches at her dress. He had a special salutation for well-known friends, a sudden profound curtsey accompanied by loud quacking and raising of the crest feathers. He and the gardener were devoted comrades, and spent many an hour pacing solemnly side by side behind the roller or mowing machine. But his most constant playmate was the kitten with whom he grew up, and into mature life cat and duck continued their favourite pastime of mock mortal combat, rolling over and over, locked in apparently deadly embrace, a huge delight to both. Humphry's primary mission in life was to clear the garden of snails and slugs, and I have known him eat as many as thirty large snails at a meal, shells and all. The vegetable foods he selected were invariably the treasures of the garden. Much had to be forgiven him, and no form of punishment was ever discovered that made the slightest impression on the culprit; he plainly took a droll delight in doing anything he had once clearly understood he was not desired to do. He loved to investigate the house, and would find his way in at any open door and straight to any room where he heard human voices. Visitors grew accustomed to being welcomed in the drawing-room by a vociferous duck, but neighbours were apt to return him with contumely when he was discovered ascending their front stairs. Alone or accompanied, he was bent on exploring the world; and he habitually walked to bathe in a pool at half a mile's distance, following or preceding his escort along the public road with the greatest composure and enjoyment, the cynosure of every eye. He came to spend a great part of his time outside the gate, fearlessly making advances to every passer-by and numbering a wide circle of admiring friends. I fear his roving propensities and a too confiding disposition led to his end, for there came a day when Humphry disappeared and was never heard of more. Doubtless he became a dinner, and we can only console ourselves with the conviction that he was tough.

Muriel Kenny
July 13, 1918

NOTE – And with the hope that he gave the diners dyspepsia ever afterwards.

QUOTED IN *Birds, Dogs and Others*, HUGH
MASSINGHAM, 1921

The Chimpanzee . . . belonged to a Zoological Garden in Berlin, and his name was Apollo.

He was named – I believe – after his keeper, of whom he was extremely fond; and in the garden was a statue of another Apollo, the famous Apollo Belvedere, near which was placed a favorite seat, as you see.

The Chimpanzee was a great pet among visitors, showing a partiality to ladies. He always slept in a bed, which he made up nicely himself, spreading the blanket and arranging the pillow. He drank like a person, ate at a table, and used a napkin at meals. He had his own tastes in food, and his pet dainty was a honey-cake, which would even cheer him when he was sad. . . .

His constant playmate was a Baboon. . . . So warmly were they

The three Apollos.

attached to each other, that when Apollo died, after some years, his companion could not live without him. He died very soon, it is supposed of grief.

OLIVE THORNE MILLER,
Queer Pets and Their Doings, 1880

I am eighty-nine and have time on my hands, so for the last two winters I have kept a single house fly. His name is Freddy the Fearless. . . .

LETTER TO THE *Daily Telegraph,*
NOVEMBER 1985

Pomero belonged to the poet Walter Savage Landor (1775–1864)

Pomero, I should say, was a Pomeranian; but let me quote Sir Sidney Colvin's charming sentences upon both man and dog: 'With Pomero, Landor would prattle in English and Italian as affectionately as a mother with a child. Pomero was his darling, the wisest and most beautiful of his race; Pomero had the brightest eyes and the most wonderful yaller tail ever seen. Sometimes it was Landor's humour to quote Pomero in speech and writing as a kind of sagacious elder brother whose opinion had to be consulted on all subjects before he would deliver his own. This creature accompanied his master wherever he went, barking "not fiercely but familiarly" at friend and stranger, and when they came in would either station himself upon his master's head to watch the people passing in the street, or else lie curled up in his basket until Landor, in talk with some visitor, began to laugh, and his laugh to grow and grow, when Pomero would spring up and leap upon and fume about him, barking and screaming for sympathy until the whole street resounded. The two together, master and dog, were for years to be encountered daily on their walks about Bath and its vicinity, and there are many who perfectly well remember them; the majestic old man, looking not a whit the less impressive for his rusty and dusty brown suit, his bulging boots, his rumpled linen, or his battered hat; and his noisy, soft-haired, quick-glancing, inseparable companion.'

Landseer, one feels, should have painted them: Dignity and Fidelity, Unreason and Understanding, Lion and Pomeranian. Since he did not, we must go to Forster's extracts from the letters to fill in the picture. Here is Landor . . . in 1844:

'. . . Pomero is sitting in a state of contemplation, with his nose before the fire. He twinkles his ears and his feathery tail at your salutation. He now licks his lips and turns round, which means "Return mine." The easterly wind has an evident effect upon his nerves. Last evening I took him to hear Luisina de Sodre play and sing. She is my friend the Countess de Molande's granddaughter. . . . Pomero was deeply affected, and lay close to the pedal on her gown, singing in a great variety of tones, not always in time. It is unfortunate that he always will take a part where there is music, for he sings even worse than I do.'

E. V. LUCAS (1868–1938),
The More I See of Men . . .', 1927

To my thinking the greatest blessing life can bestow . . . is the gift of winning and keeping affection. Yet, in this connection, wise words of a very dear friend of mine, one old enough to be my mother, often come back to me now: 'If the old feel like the young', she wrote, *'tant pis*; let them be silent!' And what she meant, I think, is, that though when old age is your portion you may, and let us hope you will, love certain people dearly, a sense of the fitness of things warns you to be careful about demonstration. You must find some other way of bringing home to them that your affection is not only deep but warm; founded firmly on the rock of personality but also in its essence a fiery particle.

Now as regards your dog, although to make a fool of oneself is never permissible, not even in private, no such reticencies need cramp your style.

For instance, one of Pan's peculiarities was to march downstairs in front of you and suddenly sit down on one of the steps. Whereupon, if the mood happened to be on me, I would drop down on the step just below him, throw an arm round his firm, woolly body, bury my face in his fur, and ejaculate: 'Pan, I *do* love you so . . . *but don't lick my face.*'

Now suppose, though I allow it is improbable, one of my friends acquired this stair-sitting habit. I can think of one who might – a comely matron between forty-five and fifty (which is about the equivalent of eight years in a dog); or, to keep to Pan's sex, let it be a certain relation of mine, a portly fellow of about fifty, who is on the Stock Exchange, and as golfer affects curiously shaggy tweeds. We are attached to each other, and when one's handicap (in years I mean) is over seventy, it need not be added that the attachment is strictly platonic. Suppose that as you go down to breakfast an impersonally expansive mood of sheer gratitude for the boon of being alive comes over you, the sort of mood that made Schiller and Beethoven cry out in the Ninth Symphony: 'Be embraced, O ye millions! I kiss the whole world!'; a mood that can be induced, if on that day you are sensitively attuned, by almost anything: a ray of sunshine shooting at you through the branch of the birch tree, a charming letter from someone you are fond of, or even satisfactory news from the Bank concerning your balance will do it.

Now in such a mood of incandescent benevolence, could you, as elderly lady, encircle the Norfolk jacket of the figure on the step above you, hide your face on his shaggy shoulder, and ejaculate 'James! I *do* love you so' (omitting of course the latter half of the phrase as it would be addressed to Pan)?

I fear not. Even with my distant relation, James, it would be difficult to pull off this scene with ease and elegance. True, there is an old friend of mine, alas well over fifty now, who I know loves me, my devotion to whom would permit me to take almost any risk with him. But then he never comes down here, nor does he wear shaggy tweeds. Also it would make him feel rather uncomfortable. But how natural, how soul-satisfying to both parties is such an action in the case of one's dog! Doubtless grandmothers indulge in such effusions with their baby-grandchildren, but the lonely hard-working spinster I have in mind has presumably no such outlet for her emotions.

In all seriousness, it seems to me a pity to block up any sources of tenderness that advancing years have left us. Why become prematurely desiccated, why turn into a fossil if the presence of a dog may hinder the process? Luckily as long as life lasts nothing can debar us from the bliss of expending affection, and I think the love you bestow on your dog not only blesses him, the receiver, but yourself, the giver.

ETHEL SMYTH, DBE (1858–1944),
Inordinate(?) Affection, A Story for Dog Lovers, 1936

As a boy, living in Corfu with his family, Gerald Durrell was taught at home surrounded by a rich and varied menagerie

Roger, of course, thought that I was simply wasting my mornings. However, he did not desert me, but lay under the table asleep while I wrestled with my work. Occasionally, if I had to fetch a book, he would wake, get up, shake himself, yawn loudly, and wag his tail. Then, when he saw me returning to the table, his ears would droop, and he would walk heavily back to his private corner and flop down with a sigh of resignation. George [the tutor] did not mind Roger being in the room, for he behaved himself well, and did not distract my attention. Occasionally, if he was sleeping very heavily and heard a peasant dog barking, Roger would wake up with a start and utter a raucous roar of rage before realizing where he was. Then he would give an embarrassed look at our disapproving faces, his tail would twitch, and he would glance round the room sheepishly.

For a short time Quasimodo [a pigeon] also joined us for lessons, and behaved very well as long as he was allowed to sit in my lap. He would drowse there, cooing to himself, the entire morning. It was I who banished him, in fact, for one day he upset a bottle of green ink in the exact centre of a large and very beautiful map that we had just completed. I realized, of course, that this vandalism was not intentional, but even so I was annoyed. Quasimodo tried for a week to get back into favour by sitting outside the door and cooing seductively through the crack, but each time I weakened I would catch a glimpse of his tail-feathers, a bright and horrible green, and harden my heart again.

Achilles [a tortoise] also attended one lesson, but did not approve of being inside the house. He spent the morning wandering about the room and scratching at the skirting-boards and door. Then he kept getting wedged under bits of furniture and scrabbling frantically until we lifted the object and rescued him. The room being small, it meant that in order to move one bit of furniture we had to move practically everything else. After a third upheaval George said that . . . he thought Achilles would be happier in the garden.

So there was only Roger left to keep me company. It was comforting, it's true, to be able to rest my feet on his woolly bulk while I grappled with a problem, but even then it was hard to concentrate, for the sun would pour through the shutters, tiger-striping the table and floor, reminding me of all the things I might be doing.

GERALD DURRELL (b. 1925), *My Family and Other Animals*, 1956

In the year 1774, being much indisposed both in mind and body, incapable of diverting myself either with company or books, and yet in a condition that made some diversion necessary, I was glad of anything that would engage my attention, without fatiguing it. The children of a neighbour of mine had a leveret given them for a plaything; it was at that time about three months old. Understanding better how to tease the poor creature than to feed it, and soon becoming weary of their charge, they readily consented that their father, who saw it pining and growing leaner every day, should offer it to my acceptance. I was willing enough to take the prisoner under my protection, perceiving that, in the management of such an animal, and in the attempt to tame it, I should find just that sort of employment which my case required. It was soon known among the neighbours that I was pleased with the present; and the consequence was, that in a short time I had as many leverets offered to me as would have stocked a paddock. I undertook the care of three, which it is necessary that I should here distinguish by the names I gave them – Puss, Tiney, and Bess. Notwithstanding the two feminine appellatives, I must inform you that they were all males. Immediately commencing carpenter, I built them houses to sleep in; each had a separate apartment so contrived that their ordure would pass through the bottom of it; an earthen pan placed under each received whatsoever fell, which being duly emptied and washed, they were thus kept perfectly sweet and clean. In the day-time they had the range of a hall, and at night retired each to his own bed, never intruding into that of another.

Puss grew presently familiar, would leap into my lap, raise himself upon his hinder feet, and bite the hair from my temples. He would suffer me to take him up, and carry him about in my arms, and has more than once fallen fast asleep upon my knee. He was ill three days, during which time I nursed him, and kept him apart from his fellows, that they might not molest him (for, like many other wild animals, they persecute one of their own species that is sick), and by constant care, and trying him with a variety of herbs, restored him to perfect health. No creature could be more grateful than my patient after his recovery; a sentiment which he most significantly expressed by licking my hand, first the back of it, then the palm, then every finger separately, then between all the fingers, as if anxious to leave no part of it unsaluted; a ceremony which he never performed but once again upon a similar occasion. Finding him extremely tractable, I made it my custom to carry him always after breakfast into the garden, where he hid himself generally under the leaves of a cucumber vine, sleeping or chewing the cud till evening; in the leaves also of that vine he found a favourite repast. I had not long habituated him to this state of liberty, before he began to be impatient for the return of the time when he might enjoy it. He would invite me to the garden by drumming upon my knee, and by a look of such expression, as it was not possible to misinterpret. If this rhetoric did not immediately succeed, he would take the shirt of my coat between his teeth, and pull at it with all his force. Thus Puss might be said to be perfectly tamed, the shyness of his nature was done away, and on the whole it was visible by many symptoms, which I have not room to enumerate, that he was happier in human society, than when shut up with his natural companions.

Not so Tiney: upon him the kindest treatment had not the least effect. He too was sick, and in his sickness had an equal share of my attention; but if, after his recovery, I took the liberty to stroke him, he would grunt, strike with his fore-feet, spring forward, and bite. He was, however, very entertaining in his way; even his surliness was matter of mirth; and in his play he preserved such an air of gravity, and performed his feats with such a solemnity of manner, that in him too I had an agreeable companion.

Bess, who died soon after he was full grown, and whose death was occasioned by his being turned into his box, which had been washed, while it was yet damp, was a hare of great humour and drollery. Puss was tamed by gentle usage; Tiney was not to be tamed at all; and Bess had a courage and confidence that made him tame from the beginning. I always admitted them into the parlour after supper, when, the carpet affording their feet a firm hold, they would frisk, and bound, and play a thousand gambols, in which Bess, being remarkably strong and fearless, was always superior to the rest, and proved himself the Vestris of the party. One evening the cat being in the room, had the hardiness to pat Bess upon the cheek, an indignity which he resented by drumming upon her back with such violence, that the cat was happy to escape from under his paws, and hide herself.

I describe these animals as having each a character of his own. Such they were in fact, and their countenances were so expressive of that character, that, when I looked only on the face of either, I immediately knew which it was. It is said, that a shepherd, however numerous his flock, soon becomes so familiar with their features, that he can, by that indication only, distinguish each from all the rest; and yet to a common observer, the difference is hardly perceptible. I doubt not that the same discrimination in the cast of countenances would be discoverable in hares, and am persuaded that among a thousand of them, no two can be found exactly similar: a circumstance little suspected by those who have not had opportunity to observe it. These creatures have a singular sagacity in discovering the minutest alteration that is made in the place to which they are accustomed, and instantly apply their nose to the examination of a new object. A small hole being burnt in the carpet, it was mended with a patch, and that patch in a moment underwent the strictest scrutiny. They seem too to be very much directed by the smell in the choice of their favourites: to some persons, though they saw them daily, they could never be reconciled, and would even scream when they attempted to touch them; but a miller coming in engaged their affections at once; his powdered coat had charms that were irresistible. It is no wonder that my intimate acquaintance with these specimens of the kind has taught me to hold the sportsman's amusement in abhorrence; he little knows what amiable creatures he persecutes, of what gratitude they are capable, how cheerful they are in their spirits, what enjoyment they have of life, and that, impressed as they seem with a peculiar dread of man, it is only because man gives them peculiar cause for it. . . .

Bess, I have said, died young; Tiney lived to be nine years old, and died at last, I have reason to think, of some hurt in his loins by a fall; Puss is still living, and has just completed his tenth year, discovering no signs of decay, nor even of age, except that he is grown more discreet, and less frolicsome than he was. I cannot conclude without observing, that I have lately introduced a dog to his acquaintance, a spaniel that had never seen a hare, to a hare that had never seen a spaniel. I did it with great caution, but there was no real need of it. Puss discovered no token of fear, nor Marquis the least symptom of hostility. There is, therefore, it should seem, no natural antipathy between dog and hare, but the pursuit of the one occasions the flight of the other, and the dog pursues because he is trained to it; they eat bread at the same time out of the same hand, and are in all respects sociable and friendly.

I should not do complete justice to my subject, did I not add, that they have no ill scent belonging to them, that they are indefatigably nice in keeping themselves clean, for which purpose nature has furnished them with a brush under each foot; and that they are never infested by any vermin.

WILLIAM COWPER (1731–1800), *Poems by William Cowper Esq.*, 1840

'A Turkey carpet was their lawn whereon they loved to bound.'

EPITAPH ON A HARE

Here lies, whom hound did ne'er pursue,
Nor swifter greyhound follow,
Whose foot ne'er tainted morning dew,
Nor ear heard huntsman's halloo.

Old Tiney, surliest of his kind,
Who nursed with tender care,
And to domestic bounds confined,
Was still a wild jack-hare.

Though duly from my hand he took
His pittance every night
He did it with a jealous look,
And, when he could, would bite.

His diet was of wheaten bread,
And milk, and oats, and straw;
Thistles, or lettuces instead,
With sand to scour his maw.

On twigs of hawthorn he regaled;
On pippins' russet peel,
And, when his juicy salads fail'd,
Sliced carrot pleased him well.

A Turkey carpet was his lawn,
Whereon he loved to bound,
To skip and gambol like a fawn,
And swing his rump around.

His frisking was at evening hours,
For then he lost his fear,
But most before approaching showers,
Or when a storm drew near.

Eight years and five round rolling moons
He thus saw steal away,
Dozing out all his idle noons,
And every night at play.

I kept him for his humour's sake,
For he would oft beguile
My heart of thoughts, that made it ache,
And force me to a smile.

But now beneath his walnut shade
He finds his long last home,
And waits, in snug concealment laid,
Till gentler Puss shall come.

He, still more aged, feels the shocks,
From which no care can save,
And, partner once of Tiney's box,
Must soon partake his grave.

WILLIAM COWPER (1731–1800)

In the narrative of the execution of Mary Queen of Scots, indorsed in Lord Burghley's hand, and which was forwarded from Fotheringay to the Court, it is recorded:–

'Then one of the executioners, pulling off her garters, espied her little dogg which was crept under her clothes, which could not be gotten forth but by force, yet afterwards would not departe from the dead corpse, but came and lay betweene her head and her shoulders, which being imbrued with her bloode, was caryed away and washed, as all things ells were that had any bloode was either burned or clean washed.'

LORD BURGHLEY (1520–98), *Ellis's Letters,*
Vol. III, WR. 1586

A Watchful Gander. – Mr Staveley, of Clifton, Yorkshire, was almost invariably accompanied by a gander belonging to a farmer residing in that township. The bird, every morning about five o'clock, came from his own domicile to Mr Staveley's residence, and, by its cackling noise, called the old gentleman up. It then accompanied him in his rambles during the day, and was frequently to be seen in the busy streets, close at his heels, utterly heedless of the throng around, and the crowds of children by which the pair were often accompanied. When Mr Staveley sat down to rest himself, the gander immediately sat down at his feet. There were several places at which the old gentleman had been in the habit of resting; and just before he arrived at them, his feathered companion starting off, arrived at the spot a little before him. If any one molested the old gentleman, the gander chattered, and tried to bite the intruder. If Mr S. went into a public-house, it entered also, if permitted; and stood behind him while he drank his glass of ale, sometimes partaking of the refreshing beverage.

The Parlour Menagerie, DEDICATED TO THE
BARONESS BURDETT–COUTTS (1816–1906)

The otter and I enjoyed the Consul-General's long-suffering hospitality [at Basra in Iraq] for a fortnight. The second night Mijbil came on to my bed in the small hours and remained asleep in the crook of my knees until the servant brought tea in the morning, and during that day he began to lose his apathy and take a keen, much too keen, interest in his surroundings. I fashioned a collar, or rather a body-belt, for him, and took him on a lead to the bathroom, where for half an hour he went wild with joy in the water, plunging and rolling in it, shooting up and down the length of the bath underwater, and making enough slosh and splash for a hippo. This, I was to learn, is a characteristic of otters; every drop of water must be, so to speak, extended and spread about the place; a bowl must at once be overturned, or, if it will not overturn, be sat in and

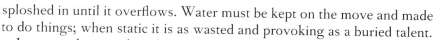
splashed in until it overflows. Water must be kept on the move and made to do things; when static it is as wasted and provoking as a buried talent.

It was only two days later that he escaped from my bedroom as I entered it, and I turned to see his tail disappearing round the bend of the corridor that led to the bathroom. By the time I had caught up with him he was up on the end of the bath and fumbling at the chromium taps with his paws. I watched, amazed by this early exhibition of an intelligence I had not yet guessed; in less than a minute he had turned the tap far enough to produce a dribble of water, and, after a moment or two of distraction at his success, achieved the full flow. (He had in fact, been fortunate to turn the tap the right way; on subsequent occasions he would as often as not try with great violence to screw it up still tighter, chittering with irritation and disappointment at its failure to co-operate.) . . .

The days passed peacefully at Basra but I dreaded dismally the unpostponable prospect of transporting Mij to England, and to his ultimate destination, Camusfeàrna. BOAC would not fly livestock at all, and there was then no other line to London. Finally I booked a Trans-World flight to Paris, with a doubtful Air France booking on the same evening to London. Trans-World insisted that Mij should be packed into a box of not more than eighteen inches square, and that this box must be personal luggage, to be carried on the floor at my feet. . . .

Dinner was at eight, and I thought that it would be as well to put Mij into the box an hour before we left, so that he would become accustomed to it before the jolting of the journey began to upset him. I manoeuvred him into it, not without difficulty, and he seemed peaceful when I left him in the dark for a hurried meal.

But when I returned, with only barely time for the Consulate car to reach the airport for the flight, I was confronted with an appalling spectacle. There was complete silence from inside the box, but from its airholes and the chinks around the hinged lid, blood had trickled and dried on the white wood. I whipped off the padlock and tore open the lid, and Mij, exhausted and blood-spattered, whimpered and tried to climb up my leg. He had torn the zinc lining to shreds, scratching his mouth, his nose and his paws, and had left it jutting in spiky ribbons all around the walls and the floor of the box. When I had removed the last of it, so that there were no cutting edges left, it was just ten minutes until the time of the flight, and the airport was five miles distant. It was hard to bring myself to put the miserable Mij back into that box, that now represented to him a torture chamber, but I forced myself to do it, slamming the lid down on my fingers as I closed it before he could make his escape. Then began a journey the like of which I hope I shall never know again.

I sat in the back of the car with the box beside me as the Arab driver tore through the streets of Basra like a ricochetting bullet. Donkeys reared, bicycles swerved wildly, out in the suburbs goats stampeded and poultry found unguessed powers of flight. Mij cried unceasingly in the

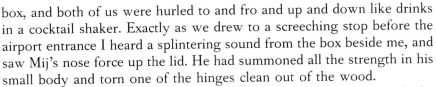

box, and both of us were hurled to and fro and up and down like drinks in a cocktail shaker. Exactly as we drew to a screeching stop before the airport entrance I heard a splintering sound from the box beside me, and saw Mij's nose force up the lid. He had summoned all the strength in his small body and torn one of the hinges clean out of the wood.

The aircraft was waiting to take off; as I was rushed through the customs by infuriated officials I was trying all the time to hold down the lid of the box with one hand, and with the other, using a screwdriver purloined from the driver, to force back the screws into the splintered wood. But I knew that it could be no more than a temporary measure at best, and my imagination boggled at the thought of the next twenty-four hours.

It was perhaps my only stroke of fortune that the seat booked for me was at the extreme front of the aircraft, so that I had a bulkhead before me instead of another seat. The other passengers, a remarkable cross-section of the orient and occident, stared curiously as the dishevelled late arrival struggled up the gangway with a horrifyingly vocal Charles Addams-like box, and knowing for just what a short time it could remain closed I was on tenterhooks to see what manner of passenger would be my immediate neighbour. I had a moment of real dismay when I saw her to be an elegantly dressed and *soignée* American woman in early middle age. Such a one, I thought, would have little sympathy or tolerance for the draggled and dirty otter cub that would so soon and so inevitably be in her midst. For the moment the lid held, and as I sat down and fastened my safety belt there seemed to be a temporary silence from within.

The port engines roared, and then the starboard and the aircraft trembled and teetered against the tug of her propellers, and then we were taxiing out to take off, and I reflected that whatever was to happen now there could be no escape from it, for the next stop was Cairo. Ten minutes later we were flying westwards over the great marshes that had been Mij's home, and peering downward into the dark I could see the glint of their waters beneath the moon.

I had brought a brief-case full of old newspapers and a parcel of fish, and with these scant resources I prepared myself to withstand a siege. I arranged newspapers to cover all the floor around my feet, rang for the air hostess, and asked her to keep the fish in a cool place. I have retained the most profound admiration for that air hostess, and in subsequent sieges and skirmishes with otters in public places I have found my thoughts turning towards her as a man's mind turns to water in desert wastes. She was the very queen of her kind. I took her into my confidence; the events of the last half hour together with the prospect of the next twenty-four had shaken my equilibrium a little, and I daresay I was not too coherent, but she took it all in her graceful sheer nylon stride, and she received the ill-wrapped fish into her shapely hands as though I were travelling royalty depositing a jewel case with her for safe keeping. Then she turned and spoke with her countrywoman on my left. Would I not prefer, she then enquired, to have my pet on my knee? The animal would surely feel happier there, and my neighbour had no objection. I could have kissed her hand in the depth of my gratitude. But, not knowing otters, I was quite unprepared for what followed.

I unlocked the padlock and opened the lid, and Mij was out like a flash. He dodged my fumbling hands with an eel-like wriggle and disappeared at high speed down the fuselage of the aircraft. As I tried to get into the gangway I could follow his progress among the passengers by a wave of disturbance amongst them not unlike that caused by the passage of a stoat through a hen run. There were squawks and shrieks and a flapping of travelling-coats, and half-way down the fuselage a woman stood up on her seat screaming out, 'A rat! A rat!' Then the air hostess reached her, and within a matter of seconds she was seated again and smiling benignly. That goddess, I believe, could have controlled a panic-stricken crowd single-handed. . . .

I heard the ripple of flight and pursuit passing up and down the body of the aircraft behind me, but I could see little. I was craning my neck back over the seat trying to follow the hunt when suddenly I heard from my feet a distressed chitter of recognition and welcome, and Mij bounded on to my knee and began to nuzzle my face and neck. In all the strange world of the aircraft I was the only familiar thing to be found, and in that first spontaneous return was sown the seed of the absolute trust that he accorded me for the rest of his life.

GAVIN MAXWELL (1914–69),
Ring of Bright Water, 1960

'Comfortably tucked up
in bed between us.'

The cowman, who was also the pig man, warned us to keep away from the palings round the sty, for the old sow was not too good-tempered and had been known to bite. There was a black hen in the yard with one foot missing. It had been bitten off by the sow, so the pig man said. The hen got about quite nimbly on one leg, but we were much concerned for her and made a special bed of straw in a corner by the barn, but she scorned this, preferring to roost with her fellows. Later on she became much attached to me, and would follow us indoors and even try to mount the stairs. Unknown to Father and Mother, I carried her upstairs and she spent several nights in our bedroom; until one evening, going up after we were asleep, Mother discovered her comfortably tucked up in bed between us. And that was the end of that.

ERNEST SHEPARD (1879–1976),
Drawn from Memory, 1957

At his home in Chelsea the Pre-Raphaelite painter and poet
Dante Gabriel Rossetti (1828–82) kept a curious
menagerie

... there is a rather charming story concerning a Mrs Tebbs (she was the wife of a solicitor, Mr Virtue Tebbs) who sat for him. When the sitting was over, she was unable to find her new hat. There was consternation, and it was only too obvious that the wombat had eaten it, and with relish, since only a few pieces of straw were left. The only consolation Gabriel had to give his indignant visitor was: 'Oh, poor wombat! It is so indigestible!'

GALE PEDRICK, *Life with Rossetti*, 1964

M. Bergeret set to work again and plunged head foremost into his *Virgilius nauticus*. He loved the work; it rested his thoughts, and became a kind of game that suited him, for he played it all by himself. On the table beside him were several boxes filled with pegs, which he fixed into little squares of cardboard to represent the fleet of Æneas. Now while he was thus occupied he felt something like tiny fists tapping at his legs. Riquet, whom he had quite forgotten, was standing on his hind legs patting his master's knees, and wagging his little stump of a tail. When he tired of this, he let his paws slide down the trouser leg, then got up and began his coaxing over again. And M. Bergeret, turning away from the printed lore before him, saw two brown eyes gazing up at him lovingly.

'What gives a human beauty to the gaze of this dog,' he thought, 'is probably that it varies unceasingly, being by turns bright and vivacious or serious and sorrowful; because through these eyes his little dumb soul finds expression for thought that lacks nothing in depth nor sequence. My father was very fond of cats, and, consequently, I liked them too. He used to declare that cats are the wise man's best companions, for they respect his studious hours. Bajazet, his Persian cat, would sit at night for hours at a stretch, motionless and majestic, perched on a corner of his table. I still remember the agate eyes of Bajazet, but those jewel-like orbs concealed all thought, that owl-like stare was cold, and hard, and wicked. How much do I prefer the melting gaze of the dog!'...

He had a sudden way of appearing to find something of interest beneath the chairs and tables, and would sniff long and noisily; then he would walk aimlessly about or sit down in a corner with an air of great humility, like the beggars who are to be seen in church porches. Finally he began to bark at a cast of Hermes which stood upon the mantel-shelf, whereupon M. Bergeret addressed him in words full of just reproach.

'Riquet! such vain agitation, such sniffing and barking were better suited to a stable than to the study of a professor, and they lead one to suppose that your ancestors lived with horses whose straw litters they shared. I do not reproach you with that. It is only natural you should have inherited their habits, manners, and tendencies as well as their close-cropped coat, their sausage-like body, and their long, thin nose. I do not speak of your beautiful eyes, for there are few men, few dogs even, who can open such beauties to the light of day. But, leaving all that aside, you are a mongrel, my friend, a mongrel from your short, bandy legs to your head. Again I am far from despising you for that. What I want you to understand is that if you desire to live with me, you will have to drop your mongrel manners and behave like a *scolar*, in other words, to remain silent and quiet, to respect work, after the manner of Bajazet, who of a night would sit for four hours without stirring, and watch my father's pen skimming over the paper. He was a silent and tactful creature. How different is your own character, my friend! Since you came into this chamber of study your hoarse voice, your unseemly snufflings and your whines, that sound like steam whistles, have constantly confused my thoughts and interrupted my reflections. And now you have made me lose the drift of an important passage in Servius, referring to the construction of one of the ships of Æneas. Know then, Riquet, my friend, that this is the house of silence and the abode of meditation, and that if you are anxious to stay here you must become literary. Be quiet!'

Thus spoke M. Bergeret. Riquet, who had listened to him with mute astonishment, approached his master, and with suppliant gesture placed a timid paw upon the knee, which he seemed to revere in a fashion that savoured of long ago. Then a kind thought struck M. Bergeret. He picked him up by the scruff of his neck, and put him upon the cushions of the ample easy chair in which he was sitting. Turning himself round three times, Riquet lay down, and then remained perfectly still and silent. He was quite happy. M. Bergeret was grateful to him, and as he ran through Servius he occasionally stroked the close-cropped coat, which, without being soft, was smooth and very pleasant to the touch. Riquet fell into a gentle doze, and communicated to his master the generous warmth of his body, the subtle, gentle heat of a living, breathing thing. And from that moment M. Bergeret found more pleasure in his *Virgilius nauticus*.

ANATOLE FRANCE (1844–1924),
The Amethyst Ring, 1899

Newton's temper was so mild and equal, that scarce any accident disturbed him. One instance in particular, which is authenticated by a person now living (1780), brings this assertion to a proof. Sir Isaac being called out of his study to a contiguous room, a little dog called Diamond, the constant but incurious attendant of his master's researches, happened to be left among the papers, and by a fatality not to be retrieved, as it was in the latter part of Sir Isaac's days, threw down a lighted candle, which consumed the almost finished labours of some years. Sir Isaac, returning too late but to behold the dreadful wreck, rebuked the author of it with an exclamation 'O Diamond! Diamond! thou little knowest the mischief done!' without adding a single stripe.

JEAN-BAPTISTE BIOT, *Life of Newton*, 1822

ON THE COLLAR OF MRS DINGLEY'S LAP DOG

Pray steal me not, I'm Mrs Dingley's,
Whose heart in this four-footed thing lies.

JONATHAN SWIFT (1667–1745)

Should you, while wandering in the wild sheepland about the twin Pikes, happen on moor or in market upon a very perfect gentle knight clothed in dark grey habit, splashed here and there with rays of moon; free by right divine of the guild of gentlemen, strenuous as a prince, lithe as a rowan, graceful as a girl, with high king-carriage, motions and manners of a fairy queen; should he have a noble breadth of brow, an air of still strength born of right confidence, all unassuming; last, and most unfailing test of all, should you look into two snowcloud eyes, calm, wistful, inscrutable, their soft depths clothed on with eternal sadness – yearning, as is said, for the soul that is not theirs – know then, you look upon one of the line of most illustrious sheepdogs of the North.

Such is one; such are all. And such was Owd Bob o' Kenmuir – owd, young though he was, by reason of that sprinkling shower of snow upon the dome of his head.

. . . there came a day when from grey dawn to greyer eve neither James Moore nor Owd Bob stirred out into the wintry white. And the Master's face was hard and set, as it always was in time of trouble.

Outside, the wind screamed down the Dale; while the snow fell relentlessly, softly fingering the windows, blocking the doors, and piling deep against the walls. Inside the house there was a strange quiet; no sound save for hushed voices, and upstairs the shuffling of muffled feet.

Below, all day long, Owd Bob patrolled the passage like some silent grey spectre.

Once there came a low knocking at the door; and David, his face, hair and cap smothered in the all-pervading white, entered in with an eddy of snow. He patted Owd Bob, and moved on tiptoe into the kitchen. To him came Maggie, shoes in hand, big-eyed, white-faced. The two whispered anxiously awhile like brother and sister as they were; then the boy crept softly away, only a little pool of water on the floor, and wet treacherous foot-dabs towards the door testifying to his visit.

Towards evening the wind died down, but the mourning flakes still fell.

With the darkening of night Owd Bob retreated to the porch and lay down on his blanket. The light from the lamp at the head of the stairs shone through the crack of open door on his dark head and the eyes that never slept.

The hours passed; and still the grey knight kept his vigil. Alone in the darkness, alone, it almost seemed, in the house, he watched. His head lay motionless along his paws; but the steady, grey eyes never flinched or drooped.

Time tramped on on leaden foot, and still he waited; and ever the pain of hovering anxiety was stamped deeper in the grey eyes.

At length it grew past bearing. The hollow stillness of the house overcame him. He rose, pushed open the door, and softly pattered across the passage.

At the foot of the stairs he halted, his forepaws on the first step, his grave face and pleading eyes uplifted, as though he was praying. The dim light fell on the raised head; and the white escutcheon on his breast shone out like the snow on Salmon.

At length, with a sound like a sob, he dropped to the ground, and stood listening, his tail drooping and head raised. Then he turned, and began softly pacing up and down like some velvet-footed sentinel at the gate of Death.

Up and down, up and down, softly as the falling snow, for a weary, weary while.

Again he stopped and stood listening intently at the foot of the stairs; and his grey coat quivered as though there were a draught.

Of a sudden, the deathly stiffness of the house was broken. Upstairs, feet were running hurriedly. There was a cry and again silence.

A life was coming in; a life was going out.

The minutes passed; hours passed; and, at the sunless dawn, a life passed.

And all through that night of age-long agony the grey figure stood, still as a statue, at the foot of the stairs. Only when, with the first chill breath of the morning, a dry quick-quenched sob of a strong man sorrowing for the help-meet of a score of years, and a tiny cry of a new-born child wailing because its mother was not, came down to his ears, the Grey Watchman dropped his head upon his bosom, and with a little whimpering note crept back to his blanket.

A little later, the door above opened and James Moore tramped down the stairs. He looked taller and gaunter than his wont, but there was no trace of emotion on his face.

At the foot of the stairs Owd Bob stole out to meet him. He came crouching up, as though guilty of the deadly sin, head and tail down, in a manner no man ever saw before or since. At his master's feet he stopped and whined pitifully.

Then, for one short moment, James Moore's whole face quivered.

'Weel, lad,' he said, quite low, and his voice broke – 'she's awa'.'

That was all: for they were an undemonstrative couple.

Then they turned and went out together into the bleak morning.

ALFRED OLLIVANT (1874–1927),
Owd Bob: The Grey Dog of Kenmuir, 1934

The noise of my chaise had disturbed the quiet of the establishment. Out sallied the warder of the establishment, a black greyhound, and leaping on one of the blocks of stone, began a furious barking. This alarm brought out the whole garrison of dogs, all open-mouthed and vociferous. In a little while, the lord of the castle himself made his appearance. I knew him at once, by the likenesses that had been published of him. He came limping up the gravel walk, aiding himself by a stout walking staff, but moving rapidly and with vigour. By his side jogged along a large iron-grey staghound, of most grave demeanour, who took no part in the clamour of the canine rabble, but seemed to consider himself bound, for the dignity of the house, to give me a courteous reception. . . .

As we sallied forth, every dog in the establishment turned out to attend us. . . . In our walks, he would frequently pause in conversation to notice his dogs, and speak to them as if rational companions; and indeed, there appears to be a vast deal of rationality in these faithful attendants on man, derived from their close intimacy with him. Maida [the old staghound] deported himself with a gravity becoming his age and size, and seemed to consider himself called upon to preserve a great degree of dignity and decorum in our society. As he jogged along a little distance ahead of us, the young dogs would gambol about him, leap on his neck, worry at his ears, and endeavour to tease him into a gambol. The old dog would keep on a long time with imperturbable solemnity, now and then seeming to rebuke the wantonness of his young companions. At

length he would make a sudden turn, seize one of them, and tumble him in the dust, then giving a glance at us, as much as to say, 'You see, gentlemen, I can't help giving way to this nonsense', would resume his gravity, and jog on as before. Scott amused himself with these peculiarities. 'I make no doubt,' said he, 'when Maida is alone with these young dogs, he throws gravity aside, and plays the boy as much as any of them; but he is ashamed to do so in our company . . .'

While we were discussing the humours and peculiarities of our canine companions, some object provoked their spleen, and produced a sharp and petulant barking from the smaller fry; but it was some time before Maida was sufficiently roused to ramp forward two or three bounds, and join the chorus with a deep-mouthed *bow-wow*. It was but a transient outbreak, and he returned instantly, wagging his tail, and looking up dubiously in his master's face, uncertain whether he would receive censure or applause. 'Ay, ay, old boy!' cried Scott, 'you have done wonders; you have shaken the Eildon hills with your roaring; you may now lay by your artillery for the rest of the day. Maida,' continued he, 'is like the great gun at Constantinople; it takes so long to get it ready, that the smaller guns can fire off a dozen times first; but when it does go off, it plays the very devil.'

WASHINGTON IRVING (1783–1859),
DESCRIBING A VISIT TO SIR WALTER SCOTT
AT ABBOTSFORD IN 1817

Love me, love my dog.

AESCHYLUS (525–456 BC)

5

Outrage, Revenge and Betrayal

Others have taken a Hedgehogge, and tying it straite by the foot under the horses taile, the Hidiousness of the crie of that little beast, will make the Horse not only go forward, but also run away violentlie.

GERVASE MARKHAM (1568–1637), *Cavelarice, or The English Horseman*, 1607

A CHILD'S PET

When I sailed out of Baltimore
With twice a thousand head of sheep,
They would not eat, they would not drink,
But bleated o'er the deep.

Inside the pens we crawled each day,
To sort the living from the dead;
And when we reached the Mersey's mouth,
Had lost five hundred head.

Yet every night and day one sheep,
That had no fear of man or sea,
Stuck through the bars its pleading face,
And it was stroked by me.

And to the sheep-men standing near,
'You see', I said, 'this one tame sheep:
It seems a child has lost her pet
And cried herself to sleep'.

So every time we passed it by,
Sailing to England's slaughter house,
Eight ragged sheep-men – tramps and thieves –
Would stroke that sheep's black nose.

W. H. DAVIES (1871–1940)

One day I was standing a watch in the nursery. That is to say, I was
asleep on the bed. The baby was asleep in the crib, which was alongside
the bed, on the side next the fireplace. It was the kind of crib that has a
lofty tent over it made of a gauzy stuff that you can see through. The
nurse was out, and we two sleepers were alone. A spark from the wood-
fire was shot out, and it lit on the slope of the tent. I suppose a quiet
interval followed, then a scream from the baby woke me, and there was
that tent flaming up toward the ceiling! Before I could think, I sprang to
the floor in my fright, and in a second was halfway to the door; but in
the next half-second . . . I was back on the bed again. I reached my head
through the flames and dragged the baby out by the waist-band, and
tugged it along, and we fell to the floor together in a cloud of smoke; I
snatched a new hold, and dragged the screaming little creature along and
out at the door and around the bend of the hall, and was still tugging
away, all excited and happy and proud, when the master's voice shouted –

'Begone, you cursed beast!' and I jumped to save myself; but he was wonderfully quick, and chased me up, striking furiously at me with his cane, I dodging this way and that, in terror, and at last a strong blow fell upon my left foreleg, which made me shriek and fall, for the moment, helpless; the cane went up for another blow, but it never descended, for the nurse's voice rang wildly out, 'The nursery's on fire!' and the master rushed away in that direction, and my other bones were saved.

The pain was cruel, but, no matter, I must not lose any time; he might come back at any moment; so I limped on three legs to the other end of the hall, where there was a dark little stairway leading up into a garret where old boxes and such things were kept, as I had heard say, and where people seldom went. I managed to climb up there, when I searched my way through the dark amongst the piles of things, and hid in the secretest place I could find. It was foolish to be afraid there, yet still I was; so afraid that I held in and hardly even whimpered, though it would have been such a comfort to whimper, because that eases the pain, you know. But I could lick my leg, and that did me some good.

For half an hour there was a commotion downstairs, and shoutings, and rushing footsteps and then there was quiet again. Quiet for some minutes and that was grateful to my spirit, for then my fears began to go down; and fears are worse than pain – oh, much worse. Then came a sound that froze me! They were calling me – calling me by name – hunting for me!

It was muffled by distance, but that could not take the terror out of it, and it was the most dreadful sound to me that I had ever heard. It went all about, everywhere, down there; along the halls, through all the rooms, in both storeys, and in the basement and the cellar; then outside, and farther and farther away – then back, and all about the house again, and I thought it would never, never stop. But at last it did, hours and hours after the vague twilight of the garret had long ago been blotted out by black darkness.

Then in that blessed stillness my terrors fell little by little away, and I was at peace and slept. It was a good rest I had, but I woke before the twilight had come again. I was feeling fairly comfortable, and I could think out a plan now. I made a very good one, which was to creep down, all the way down the back stairs, and hide behind the cellar door, and slip out and escape when the iceman came at dawn, whilst he was inside filling the refrigerator; then I would hide all day, and start on my journey when night came; my journey to – well, anywhere where they would not know me and betray me to the master. I was feeling almost cheerful now; then suddenly I thought, why, what would life be like without my Puppy!

That was despair. There was no plan for me. I saw that; I must stay where I was; stay, and wait, and take what might come – it was not my affair; that was what life is – my mother had said it. Then – well, then the calling began again! All my sorrows came back. I said to myself, the master will never forgive. I did not know what I had done to make him so bitter and so unforgiving, yet I judged it was something a dog could not understand, but which was clear to a man and dreadful.

They called and called – days and nights, it seemed to me. So long that the hunger and thirst near drove me mad, and I recognized that I was getting very weak. When you are this way you sleep a great deal, and I did. Once I woke in an awful fright – it seemed that the calling was right there in the garret! And so it was; it was Sadie's voice, and she was crying; my name was falling from her lips all broken, poor thing, and I could not believe my ears for the joy of it when I heard her say,

'Come back to us, – oh, come back to us, and forgive – it is all so sad without our . . .'

I broke in with *such* a grateful little yelp, and the next moment Sadie was plunging and stumbling through the darkness and the lumber, and shouting for the family to hear, 'She's found! she's found!'

The days that followed – well, they were wonderful. The mother and Sadie and the servants – why, they just seemed to worship me. They couldn't seem to make me a bed that was fine enough; and as for the food, they couldn't be satisfied with anything but game and delicacies that were out of season; and every day the friends and neighbours flocked in to hear about my heroism – that was the name they called it by, and it means agriculture. I remember my mother pulling it on a kennel once, and explaining it that way, but didn't say what agriculture was, except that it was synonymous with intramural incandescence; and a dozen times a day Mrs Gray and Sadie would tell the tale to newcomers, and say I risked my life to save the baby's, and both of us had burns to prove it, and then the company would pass me around and pet me and exclaim about me, and you could see the pride in the eyes of Sadie and her mother; and when the people wanted to know what made me limp, they looked ashamed and changed the subject, and sometimes when people hunted them this way and that with questions about it, it looked to me as if they were going to cry.

And this was not all the glory; no, the master's friends came, a whole twenty of the most distinguished people, and had me in the laboratory, and discussed me as if I was a kind of discovery; and some of them said it was wonderful in a dumb beast, the finest exhibition of instinct they could call to mind; but the master said, with vehemence, 'It's far above instinct; it's *reason*, and many a man, privileged to be saved and go with you and me to a better world by right of its possession, has less of it than this poor silly quadruped that's fore-ordained to perish,' and then he laughed, and said, 'Why, look at me – I'm a sarcasm! Bless you, with all my grand intelligence, the only thing I inferred was that the Dog had gone mad and was destroying the child, whereas but for the beast's intelligence – it's *reason*, I tell you – the child would have perished!'

They disputed and disputed, and *I* was the very centre and subject of it all, and I wished my mother could know that this grand honor had come to me; it would have made her proud.

Then they discussed optics, as they called it, and whether a certain injury to the brain would produce blindness or not, but they could not agree about it, and said they must test it by experiment by and by; and next they discussed plants, and that interested me, because in the summer Sadie and I had planted seeds – I helped her dig the holes, you know – and after days and days a little shrub or a flower came up there, and it was a wonder how that could happen; but it did, and I wished I could talk – I would have told those people about it and shown them how much I knew, and been all alive with the subject; but I didn't care for the optics; it was dull, and when they came back to it again it bored me, and I went to sleep.

Pretty soon it was spring, and sunny and pleasant and lovely, and the sweet mother and the children patted me and the Puppy good-bye, and went away on a journey and a visit to their kin, and the master wasn't any company for us, but we played together and had good times, and the servants were kind and friendly, and we got along quite happily and counted the days and waited for the family.

And one day those men came again, and said now for the test, and they took the Puppy to the laboratory, and I walked three-leggedly along, too, feeling proud, for any attention shown the Puppy was a pleasure to me, of course. They discussed and experimented, and then suddenly the Puppy shrieked, and they set him on the floor, and he went staggering around with his head all bloody, and the master clapped his hands and shouted:

'There, I've won – confess it! He's as blind as a Bat!'

And they all said,

'It's so – you've proved your theory, and suffering humanity owes you a great debt from henceforth,' and they crowded around him, and wrung his hand cordially and thankfully, and praised him.

But I hardly saw or heard these things, for I ran at once to my little darling, and snuggled close to it where it lay, and licked the blood, and it put its head against mine, whimpering softly, and I knew in my heart it was a comfort to it in its pain and trouble to feel its mother's touch, though it could not see me. Then it drooped down, presently, and its little velvet nose rested upon the floor, and it was still, and did not move any more.

Soon the master stopped discussing a moment, and rang in the footman, and said, 'Bury it in the far corner of the garden,' and then went on with the discussion; and I trotted after the footman, very happy and grateful, for I knew the Puppy was out of its pain now, because it was asleep. We went far down the garden to the farthest end, where the children and nurse and the Puppy and I used to play in the summer in the shade of a great elm, and there the footman dug a hole, and I was sure he was going to plant the Puppy, and I was glad, because it would grow and come up a fine handsome Dog, like Robin Adair, and be a beautiful surprise for the family when they came home, so I tried to help him dig, but my lame leg was no good, being stiff, you know, and you have to have two, or it is no use. When the footman had finished and covered little Robin up, he patted my head, and there were tears in his eyes, and he said, 'Poor little Doggie, you SAVED *his* child.'

I have watched two whole weeks, and he doesn't come up! This last week, a fright has been stealing upon me. I think there is something terrible about this. I do not know what it is, but the fear makes me sick, and I cannot eat, though the servants bring me the best of food; and they pet me so, and even come in the night, and cry, and say, 'Poor Doggie – do give it up and come home; *don't* break our hearts!' and all this terrifies me the more, and makes me sure something has happened. And I am so weak; since yesterday I cannot stand on my feet any more. And within this hour, the servants looking towards the sun where it was sinking out of sight and the night chill coming on, said things I could not understand, but they carried something cold to my heart.

'Those poor creatures! They do not suspect. They will come home in the morning, and eagerly ask for the little Doggie that did the brave deed, and who of us will be strong enough to say the truth to them: The humble little friend is gone where go the beasts that perish.'

MARK TWAIN (1835–1910), *Dog's Tale*, 1903

Let us also repay due gratitude to the ravens the gratitude that is their due, evidenced also by the indignation and not only by the knowledge of the Roman nation. When Tiberius was emperor, a young raven from a brood hatched on the top of the Temple of Castor and Pollux flew down to a cobbler's shop in the vicinity, being also commended to the master of the establishment by religion. It soon picked up the habit of talking, and every morning used to fly off to the Platform that faces the forum and salute Tiberius and then Germanicus and Drusus Caesar by name, and next the Roman public passing by, afterwards returning to the shop; and it became remarkable by several years' constant performance of this function. This bird the tenant of the next cobbler's shop killed, whether because of his neighbour's competition or in a sudden outburst of anger, as he tried to make out, because some dirt had fallen on his stock of shoes from its droppings; this caused such a disturbance among the public that the man was first driven out of the district and later actually made away with, and the bird's funeral was celebrated with a vast crowd of followers, the draped bier being carried on the shoulders of two Ethiopians and in front of it going in procession a flute-player and all kinds of wreaths right to the pyre, which had been erected on the right hand side of the Appian Road at the second milestone on the ground called Rediculus's Plain. So adequate a justification did the Roman nation consider a bird's cleverness to be for a funeral procession and for the punishment of a Roman citizen, in the city in which many leading men had had no obsequies at all, while the death of Scipio Aemilianus after he had destroyed Carthage and Numantia had not been avenged by a single person. The date of this was 28 March, AD 36.

PLINY THE ELDER (AD 23–79),
Natural History, TRANS. H. RACKHAM

THE PUZZLED GAME-BIRDS
(Triolet)

They are not those who used to feed us
When we were young – they cannot be –
These shapes that now bereave and bleed us?
They are not those who used to feed us,
For did we then cry, they would heed us.
– If hearts can house such treachery
They are not those who used to feed us
When we were young – they cannot be!

THOMAS HARDY (1840–1928),
Poems of the Past and Present, 1901

Man is the only animal that blushes, or has reason to.

MARK TWAIN (1835–1910),
Following the Equator, 1897

It is not good that animals are so cheap.

ELIAS CANETTI (b. 1905),
Journals of Canetti, 1942

'One evening,' writes another [Oxford] college friend, 'when I was devoting an hour to coaching him up for his little go, I took care to tuck up my legs, in Turkish fashion, on the sofa for fear of a casual bite from the jackal which was wandering about the room. After a time I heard the animal munching up something under the sofa, and was relieved that he should have found something to occupy him. When our work was finished, I told Buckland that the jackal had found something to eat under the sofa. "My poor guinea-pigs!" he exclaimed; and, sure enough, four or five of them had fallen victims.'

GEORGE C. BOMPAS,
Life of Frank Buckland, 1885

All the entrances to Le Voreux had been sealed, and the only door left open was guarded by sixty soldiers standing with ordered arms. . . .

Maheu was backing his wife up when old Mouque came along from Réquillart. They tried to prevent his going through, but he insisted, saying that his horses went on eating oats just the same and didn't care two hoots about a revolution. Besides, a horse had died, and he had to see about bringing it up. Etienne managed to free the old ponyman, whom the soldiers allowed into the pit. A quarter of an hour later, as the band of strikers was growing larger and more threatening, a wide door opened on the ground floor and some men appeared, hauling out the dead animal, a pitiful carcase still tied up in the rope net, and they left it in the middle of the puddles of melted snow. The sensation was so great that nobody thought of preventing their going in again and barricading the door behind them. For they had all recognized the horse, his head stiff and bent back against his side, and whispers ran round:

'It's Trompette, isn't it? Yes, it's Trompette.'

It was. He had never been able to accustom himself to life underground and had remained dismal and unwilling to work, tortured by longing for the daylight he had lost. In vain had Bataille, the father of the pit, given him friendly rubs with his side and nibbled his neck so as to give him a little of his own resignation after ten years underground. Caresses only made him more doleful and his skin quivered when his friend who had grown old in the darkness whispered secrets in his ear. And whenever they met and snorted together they both seemed to be lamenting – the old one because he could not now remember, and the young one because he could not forget. They lived side by side in the stable, lowering their heads into the same manger and blowing into each other's nostrils, comparing their unending dreams of daylight, their visions of green pastures, white roads, and golden sunlight for ever and ever. Then, as Trompette, bathed in sweat, lay dying on his straw, Bataille had begun to sniff at him with heartbroken little sniffs, like sobs. He felt his friend grow cold, the pit was taking his last joy away, this friend who had come from up there, all fresh with lovely smells that brought back his own young days in the open air. And when he saw that the other was lying still he broke his tether, whinneying with fear.

Mouque had been warning the overman for a week. But who bothered about a sick horse at such a time? . . . The day before, Mouque and two men had spent an hour roping Trompette up. Bataille was harnessed to haul him to the shaft. Slowly the old horse dragged his dead friend along through such a narrow gallery that he had to proceed by little jerks, at the risk of grazing the side off the corpse, and he shook his head in distress as he heard the long brushing sound of this mass of flesh bound for the knacker's yard. When they unharnessed him at pit bottom he watched with melancholy gaze the preparations for the ascent, as the body was pushed on to cross-beams over the sump and the net was tied to the bottom of a cage. Then at last the onsetters rang the meat-call, and he raised his head to see his friend go, gently at first and then with a rush up into darkness – lost for ever up the black hole. There he stood craning his neck, his shaky old memory recalling, maybe, some of the things of the earth. It was all over now, and his friend would never see anything again, and he would be done up himself into a dreadful bundle when the day came for him to go up there. His legs began to shake and the open air from that far country seemed to catch his throat, and he plodded back to the stable unsteadily as though he were drunk.

Up in the yard the miners stood round sadly looking at Trompette's body. One woman said softly:

'At any rate a man only goes down there if he wants to!' . . .

ÉMILE ZOLA (1840–1902), *Germinal*, 1885,
TRANS. L. W. TANCOCK

The rats, often fifty or sixty in number, black, brown, white, and piebald, were usually kept in the cellar. Selected white rats were brought up at evening parties for the amusement or torment of the visitors. Snakes were often brought out on these occasions. Frank would produce them from his pocket, or gliding out of his sleeve. 'Don't be afraid,' said he one evening to a young lady who sat down to play quadrilles; 'they won't hurt you: I've taken out their fangs. Now do be a good girl, and don't make a fuss;' and, after a little more persuasion, proceeded to wreathe one snake round her neck, and one round each arm, with which unwonted ornaments she continued to play the dances. His sisters were so often bedecked with similar reptilian necklaces and armlets, that they became used to the somewhat clammy, crawling sensation, which is a drawback to such ornaments.

Chloroform was at this time a recent invention, and the Dean gave several luncheon parties, at which, with Frank's assistance, the effects of the new anæsthetic were tried on various animals.

The eagle was sent to sleep, and could be lifted up by his feet, like a dead bird; or, when half asleep, was walked round the room by two persons, holding him by his wings. One day the eagle was slowly recovering from his stupor, and walking unsteadily upon the floor, when Jacko was brought in to take his turn. He came in with a suspicious and melancholy expression, expecting that something was going to take place; but when he saw the intoxicated condition of his old enemy the eagle, he jumped out of his master's arms with a scream of delight, and seizing the eagle by the tail, paid off old scores by dragging him about the room backwards in a most ludicrous and undignified manner; nor was Jacko secured again till he had espied the bowl of gold fish and thrown them all about the room. Jacko chattered pitifully, however, when his turn came, and then he succumbed. The gold fish, when blotting paper soaked in chloroform was suspended over their bowl, turned on their backs, appearing lifeless until revived with fresh water; the snakes also seemed to die, and come again to life.

GEORGE C. BOMPAS,
Life of Frank Buckland, 1885

Now there was one inconvenience about our villa ... which it is necessary to mention here. By common consent all the cats of the neighbourhood had selected our garden for their evening reunions. I fancy that a tortoiseshell kitchen cat of ours must have been a sort of leader of local feline society – I know she was 'at home,' with music and recitations, on most evenings. ...

... for some time we were at a loss for a remedy. At last, one day, walking down the Strand, I chanced to see (in an evil hour) what struck me as the very thing – it was an air-gun of superior construction displayed in a gunsmith's window. I went in at once, purchased it, and took it home in triumph; it would be noiseless, and would reduce the local average of cats without scandal – one or two examples, and feline fashion would soon migrate to a more secluded spot.

I lost no time in putting this to the proof. That same evening I lay in wait after dusk at the study window, protecting my mother's repose. As soon as I heard the long-drawn wail, the preliminary sputter, and the wild stampede that followed, I let fly in the direction of the sound. I suppose I must have something of the national sporting instinct in me, for my blood was tingling with excitement; but the feline constitution assimilates lead without serious inconvenience, and I began to fear that no trophy would remain to bear witness to my marksmanship.

But all at once I made out a dark indistinct form slinking in from behind the bushes. I waited till it crossed a belt of light which streamed from the back kitchen below me, and then I took careful aim and pulled the trigger.

This time at least I had not failed – there was a smothered yell, a rustle – and then silence again. I ran out with the calm pride of a successful revenge to bring in the body of my victim, and I found underneath a laurel, no predatory tom-cat, but ... the quivering carcase of the Colonel's black poodle! ...

Then, to my horror, I heard a well-known ringing tramp on the road outside, and smelt the peculiar fragrance of a Burmese cheroot. It was the Colonel himself, who had been taking out the doomed Bingo for his usual evening run.

I don't know how it was exactly, but a sudden panic came over me. I held my breath, and tried to crouch down unseen behind the laurels; but he had seen me, and came over at once to speak to me across the hedge.

He stood there, not two yards from his favourite's body! Fortunately it was unusually dark that evening.

'Ha, there you are, eh?' he began heartily; 'don't rise, my boy, don't rise.' I was trying to put myself in front of the poodle, and did not rise – at least, only my hair did.

'You're out late, ain't you?' he went on; 'laying out your garden, hey?'

I could not tell him that I was laying out his poodle! My voice shook as, with a guilty confusion that was veiled by the dusk, I said it was a fine evening – which it was not.

'Cloudy, sir,' said the Colonel, 'cloudy – rain before morning, I think. By the way, have you seen anything of my Bingo in here?'

This was the turning point. What I *ought* to have done was to say mournfully, 'Yes, I'm sorry to say I've had a most unfortunate accident with him – here he is – the fact is, I'm afraid I've *shot* him!'

But I couldn't. I could have told him at my own time, in a prepared form of words – but not then....

I decided that my only course was to bury the poor animal where he fell and say nothing about it. With some vague idea of precaution I first took off the silver collar he wore, and then hastily interred him with a garden-trowel and succeeded in removing all traces of the disaster....

By-and-by, I thought, I would plant a rose-tree over his remains, and some day, as Lilian and I, in the noontide of our domestic bliss, stood before it admiring its creamy luxuriance, I might (perhaps) find courage to confess that the tree owed some of that luxuriance to the long-lost Bingo....

But that night my sleep was broken by frightful dreams. I was perpetually trying to bury a great gaunt poodle, which would persist in rising up through the damp mould as fast as I covered him up.... Lilian and I were engaged, and we were in church together on Sunday, and the poodle, resisting all attempts to eject him, forbade our banns with sepulchral barks.... It was our wedding-day, and at the critical moment the poodle leaped between us and swallowed the ring.... Or we were at the wedding-breakfast, and Bingo, a grizzly black skeleton with flaming eyes, sat on the cake and would not allow Lilian to cut it. Even the rose-tree fancy was reproduced in a distorted form – the tree grew, and every blossom contained a miniature Bingo, which barked; and as I woke I was desperately trying to persuade the Colonel that they were ordinary dog-roses.

I went up to the office next day with my gloomy secret gnawing my bosom, and, whatever I did, the spectre of the murdered poodle rose before me. For two days after that I dared not go near the Curries, until at last one evening after dinner I forced myself to call, feeling that it was really not safe to keep away any longer.

I never before saw a family so stricken down by a domestic misfortune as the group I found in the drawing-room, making a dejected pretence of reading or working. We talked at first – and hollow talk it was – on indifferent subjects, till I could bear it no longer, and plunged boldly into danger.

'I don't see the dog,' I began. 'I suppose you-you found him all right the other evening, Colonel?' I wondered as I spoke whither they would not notice the break in my voice, but they did not.

'Why, the fact is,' said the Colonel, heavily, gnawing his grey moustache, 'we've not heard anything of him since – he's run off!'

'Gone, Mr Weatherhead; gone without a word!' said Mrs Currie, plaintively, as if she thought the dog might at least have left an address.

'I wouldn't have believed it of him,' said the Colonel; 'it has completely knocked me over. Haven't been so cut up for years – the ungrateful rascal!'

'Oh, Uncle!' pleaded Lilian, 'don't talk like that; perhaps Bingo couldn't help it – perhaps some one has s-s-shot him!'

'Shot!' cried the Colonel, angrily. 'By heaven! if I thought there was a villain on earth capable of shooting that poor inoffensive dog, I'd——— Why *should* they shoot him, Lilian? Tell me that! I – I hope you won't let me hear you talk like that again. *You* don't think he's shot, eh, Weatherhead?'

I said – Heaven forgive me! – that I thought it highly improbable.

'He's not dead!' cried Mrs Currie. 'If he were dead I should know it somehow – I'm sure I should! But I'm certain he's alive. Only last night I had such a beautiful dream about him. I thought he came back to us, Mr Weatherhead, driving up in a hansom cab, and he was just the same as ever – only he wore blue spectacles, and the shaved part of him was painted a bright red. And I woke up with the joy – so, you know, it's sure to come true!'

F. ANSTEY, 'THE BLACK POODLE', FROM
The Black Poodle and Other Stories

The Abbot of Baigne, a man of great wit, and who had the art of inventing new musical instruments, being in the service of Lewis XI of France, was ordered by that prince to get him a concert of swines voices, thinking

it impossible. The abbot was not surprised, but asked money for the performance, which was immediately delivered him and he wrought a thing as singular as ever was seen. For out of a number of hogs, of several ages, which he got together, and placed under a tent or pavillion, covered with velvet, before which he had a table of wood, painted, with a certain number of Kings, he made an organical instrument, and as he played upon the said keys, with little spikes which pricked the hogs he made them cry in such order and consonance as highly delighted the king and all his company.

A new musical instrument, 1883.

NATHANIEL WANLEY (1634–80),
The Wonders of the Little World, 1678

And little Harry, meanwhile, where was he? That sunny afternoon in June, he had wandered away from the house, and losing sight of the familiar building behind the long fringe of trees by the river, he had lost his bearings. Then came the thunder shower which made him seek for shelter. There was nothing about him but level prairie, and the only shelter he could find was a Badger hole, none too wide even for his small form. Into this he had backed and stayed with some comfort during the thunderstorm, which continued till night. Then in the evening the child heard a sniffing sound, and a great, grey animal loomed up against the sky, sniffed at the tracks and at the open door of the den. Next it puts its head in, and Harry saw by the black marks on its face that it was a

Badger. He had seen one just three days before. A neighbour had brought it to his father's house to skin it. There it stood sniffing, and Harry, gazing with less fear than most children, noticed that the visitor had five claws on one foot and four on the other, with recent wounds, proof of some sad experience in a trap. Doubtless this was the Badger's den, for she – it proved a mother – came in, but Harry had no mind to surrender. The Badger snarled and came on, and Harry shrieked, 'Get out!' and struck with his tiny fists, and then, to use his own words, 'I scratched the Badger's face and she scratched mine.' Surely this Badger was in a generous mood, for she did him no serious harm, and though the rightful owner of the den, she went away and doubtless slept elsewhere.

Night came down. Harry was very thirsty. Close by the door was a pool of rainwater. He crawled out, slaked his thirst, and backed into the warm den as far as he could. Then remembering his prayers, he begged God to 'send mamma,' and cried himself to sleep. During the night, he was awakened by the Badger coming again, but it went away when the child scolded it.

Next morning Harry went to the pool again and drank. Now he was so hungry; a few old rose hips hung on the bushes near the den. He gathered and ate these, but was even hungrier. Then he saw something moving out on the plain. It might be the Badger, so he backed into the den, but he watched the moving thing. It was a horseman galloping. As it came nearer, Harry saw that it was Grogan, the neighbour for whom he had such a dislike, so he got down out of sight. Twice that morning men came riding by, but having once yielded to his shy impulse, he hid again each time. The Badger came back at noon. In her mouth she held the body of a Prairie Chicken, pretty well plucked and partly devoured. She came into the den sniffing as before. Harry shouted, 'Get out! Go away.' The Badger dropped the meat and raised her head. Harry reached and grasped the food and devoured it with the appetite of one starving. There must have been another doorway, for later the Badger was behind the child in the den, and still later when he had fallen asleep she came and slept beside him. He awoke to find the warm furry body filling the space between him and the wall, and knew now why it was he had slept so comfortably.

That evening the Badger brought the egg of a Prairie Chicken, and set it down unbroken before the child. He devoured it eagerly, and again drank from the drying mud puddle to quench his thirst. During the night it rained again, and he would have been cold, but the Badger came and cuddled around him. Once or twice it licked his face. The child could not know, but the parents discovered later that this was a mother Badger which had lost her brood and her heart was yearning for something to love.

Now there were two habits that grew on the boy. One was to shun the men that daily passed by in their search, the other was to look to the

Badger for food and protection, and live the Badger's life. She brought him food often not at all to his taste – dead Mice or Ground-squirrels – but several times she brought in the comb of a bee's nest or eggs of game birds, and once a piece of bread almost certainly dropped on the trail from some traveller's lunch bag. His chief trouble was water. The prairie pool was down to mere ooze, and with this he moistened his lips and tongue. Possibly the mother Badger wondered why he did not accept her motherly offerings. But rain came often enough to keep him from serious suffering.

Their daily life was together now, and with the imitative power strong in all children and dominant in him, he copied the Badger's growls, snarls, and purrs. Sometimes they played tag on the prairie, but both were ready to rush below at the slightest sign of a stranger. . . .

One morning he wandered a little farther in search of water, and was alarmed by a horseman appearing. He made for home on all fours – he ran much on all fours now – and backed into the den. In the prairie grass he was concealed, but the den was on a bare mound, and the horseman caught a glimpse of a whitish thing disappearing down the hole. Badgers were familiar to him, but the peculiar yellow of this and the absence of black marks gave it a strange appearance. He rode up quietly within twenty yards and waited.

After a few minutes the grey-yellow ball slowly reappeared and resolved itself into the head of a tow-topped child. The young man leaped to the ground, and rushed forward, but the child retreated far back into the den, beyond reach of the man, and refused to come out. Nevertheless, there was no doubt that this was the missing Harry Service. 'Harry! Harry! don't you know me? I'm your Cousin Jack,' the young man said in soothing, coaxing tones. 'Harry, won't you come and let me take you back to mamma? Come, Harry! Look! here are some cookies!' but all in vain. The child hissed and snarled at him like a wild thing, and retreated as far as he could till checked by a turn in the burrow.

Now Jack got out his knife and began to dig until the burrow was large enough for him to crawl in a little way. At once he succeeded in getting hold of the little one's arm and drew him out struggling and crying. But now there rushed also from the hole a Badger, snarling and angry; it charged at the man, uttering its fighting snort. He fought it off with his whip, then swung to the saddle with his precious burden, and rode away as for his very life, while the Badger pursued for a time. But it was easily left behind, and its snorts were lost and forgotten.

The father was coming in from another direction as he saw this strange sight; a Horse galloping madly over the prairie, on its back a young man shouting loudly, and in his arms a small dirty child, alternately snarling at his captor, trying to scratch his face, or struggling to be free.

The father was used to changing intensity of feeling at these times, but he turned pale and held his breath till the words reached him: 'I have got him, thank God! He's all right,' and he rushed forward shouting, 'My boy! my boy!'

But he got a rude rebuff. The child glared like a hunted Cat, hissed at him, and menaced with hands held claw fashion. Fear and hate were all he seemed to express. The door of the house flung open and the distracted mother, now suddenly overjoyed, rushed to join the group. 'My darling! my darling!' she sobbed, but little Harry was not as when he left them. He hung back, he hid his face in the coat of his captor, he scratched and snarled like a beast, he displayed his claws and threatened fight, till strong arms gathered him up and placed him on his mother's knees in the old, familiar room with the pictures, and the clock ticking as of old, the smell of frying bacon, and the nearness of his father's form; and, above all, his mother's arms about him, her magic touch on his brow, and her voice, 'My darling! my darling! Oh! Harry, don't you know your mother? My boy! my boy!' And the struggling little wild thing in her arms grew quiet, his animal anger died away, his raucous hissing gave place to a short panting, and that to a low sobbing that ended in a flood of tears and a passionate 'Mamma, mamma, mamma!' as the veil of a different life was rolled away, and he clung to his mother's bosom.

But even as she cooed to him, and stroked his brow and won him back again, there was a strange sound, a snarling hiss at the open door. All turned to see a great Badger standing there with its front feet on the threshold. Father and cousin exclaimed, 'Look at that Badger!' and reached for the ready gun, but the boy screamed again. He wriggled from his mother's arms and rushing to the door, cried, 'My Badgie! my Badgie!' He flung his arms around the savage thing's neck, and it answered with a low purring sound as it licked its lost companion's face. The men were for killing the Badger, but it was the mother's keener insight that saved it, as one might save a noble Dog that had rescued a child from the water. . . .

It was long and only by slow degrees that the mother got the story that is written here, and parts of it were far from clear. It might all have been dismissed as a dream or a delirium but for the fact that the boy had been absent two weeks; he was well and strong now, excepting that his lips were blackened and cracked with the muddy water, the Badger had followed him home, and was now his constant friend.

It was strange to see how the child oscillated between the two lives, sometimes talking to his people exactly as he used to talk, and sometimes running on all fours, growling, hissing, and tussling with the Badger. Many a game of 'King of the Castle' they had together on the low pile of sand left after the digging of a new well. Each would climb to the top and defy the other to pull him down, till a hold was secured and they rolled together to the level, clutching and tugging, Harry giggling, the

Badger uttering a peculiar high-pitched sound that might have been called snarling had it not been an expression of good nature. Surely it was a Badger laugh. There was little that Harry could ask without receiving in those days, but his mother was shocked when he persisted that the Badger must sleep in his bed; yet she so arranged it. The mother would go in the late hours and look at them with a little pang of jealousy as she saw her baby curled up, sleeping soundly with that strange beast.

It was Harry's turn to feed his friend now, and side by side they sat to eat. The Badger had become an established member of the family. But after a month had gone by an incident took place that I would gladly leave untold.

Grogan, the unpleasant neighbour, who had first frightened Harry into the den, came riding up to the Service homestead. Harry was in the house for the moment. The Badger was on the sand pile. Instantly on catching sight of it, Grogan unslung his gun and exclaimed, 'A Badger!' To him a Badger was merely something to be killed. 'Bang!' and the kindly animal rolled over, stung and bleeding, but recovered and dragged herself toward the house. 'Bang!' and the murderer fired again, just as the inmate rushed to the door – too late. Harry ran toward the Badger shouting, 'Badgie! my Badgie!' He flung his baby arms around the bleeding neck. It fawned on him feebly, purring a low, hissing purr, then mixing the purrs with moans, grew silent, and slowly sank down, and died in his arms. 'My Badgie! my Badgie!' the boy wailed, and all the ferocity of his animal nature was directed against Grogan.

'You better get out of this before I kill you!' thundered the father, and the hulking halfbreed sullenly mounted his horse and rode away.

A great part of his life had been cut away, and it seemed as though a deathblow had been dealt the boy. The shock was more than he could stand. He moaned and wept all day, he screamed himself into convulsions, he was worn out at sundown and slept little that night. Next morning he was in a raging fever and ever he called for 'My Badgie!' He seemed at death's door the next day, but a week later he began to mend, and in three weeks was strong as ever and childishly gay, with occasional spells of sad remembering that gradually ceased.

He grew up to early manhood in a land of hunters, but he took no pleasure in the killing that was such sport to his neighbours' sons, and to his dying day he could not look on the skin of a Badger without feelings of love, tenderness, and regret.

ERNEST THOMPSON SETON
(1860–1946), *The Boy and The Badger*

ON THE DEATH OF LADY THROCKMORTON'S BULFINCH

Ye nymphs! if e'er your eyes were red
With tears o'er hapless favourites shed
O share Maria's grief!
Her favourite, even in his cage,
(What will not hunger's cruel rage!)
Assassin'd by a thief.

Where Rhenus strays his vines among,
The egg was laid from which he sprung;
And, though by nature mute,
Or only with a whistle bless'd,
Well taught, he all the sounds express'd
Of flageolet or flute.

The honours of his ebon poll
Were brighter than the sleekest mole;
His bosom of the hue
With which Aurora decks the skies,
Whom piping winds shall soon arise,
To sweep away the dew.

Above, below, in all the house,
Dire foe alike of bird and mouse,
No cat had leave to dwell;
And Bully's cage supported stood
On props of smoothest shaven wood,
Large built, and latticed well.

Well-latticed – but the grate, alas!
Not rough with wire of steel or brass,
For Bully's plumage sake.
But smooth with wands from Ouse's side,
With which, when neatly peel'd and dried,
The swains their baskets make.

Night veil'd the pole, all seem'd secure;
When led by instinct sharp and sure,
Subsistence to provide,
A beast forth sallied on the scout,
Long back'd, long tail'd, with whisker'd snout,
And badger-colour'd hide.

'Assassin'd by a thief.'

He, entering at the study door,
Its ample area 'gan t'explore;
And something in the wind
Conjectured, sniffing round and round,
Better than all the books he found,
Food chiefly for the mind.

Just then, by adverse fate impress'd,
A dream disturb'd poor Bully's rest;
In sleep he seem'd to view
A rat fast clinging to the cage,
And, screaming at the sad presage,
Awoke, and found it true.

For, aided both by ear and scent,
Right to his mark the monster went –
Ah, Muse! forbear to speak
Minute the horrors that ensued;
His teeth were strong, the cage was wood –
He left poor Bully's beak.

O had he made that too his prey;
That beak, whence issued many a lay
Of such mellifluous tone,
Might have repaid him well, I wot,
For silencing so sweet a throat,
Fast stuck within his own.

Maria weeps – the Muses mourn –
So when, by Bacchanalians torn,
On Thracian Hebrus' side
The tree enchanter, Orpheus, fell,
His head alone remain'd to tell
The cruel death he died.

WILLIAM COWPER (1731–1800)

CORRECTIONS AGAINST RESTIFNESS AND THE SEVERALL KINDES THEREOF

It hath been the practise of some horsemen, when they could not make their horse go forwarde, to tie a shrewd Cat to a Poale, with her head and feete at libertie, and so thrusting it under the horses bellye, or betwene his legges, to make her scratch, byte, and clawe him by the Coddes, and other tender partes of the bodye: the strange torment and violence whereof, will make any horse starte, and runne away. Others have taken a Hedgehogge, and tying it straite by the foot under the horses taile, the Hidiousness of the crie of that little beast, will make the Horse not only go forward, but also run away violentlie.... This practise I finde allowed both for this and other purposes by *La Broue*, & some other horsemen but as before I said, so I say again, that under the reformation of their better knowledges I neither like nor would have anye man either practise this or any of the other experiments; my reason being that they are al of that crueltie, eyther in outwardly tormenting the bodye with extraordinarie paine, or inwardly appalling the minde with afright and amazement, that they do not so much good in redressing that one faulte; as hurt in breeding manye faultes of much more worse nature.

GERVASE MARKHAM (1568–1637), *Cavelarice,
or The English Horseman*, 1607

To the memory of one, who was remarkably steady
These stones are erected
What he undertook with spirit he accomplished
His deportment was graceful, nay noble:
The Ladies admired and followed him:
By Application he gained Applause
His abilities were so powerful as to draw easily
The Divine the Lawyer and the Statesman
Into his own smooth Tract
Had he lived in the days of Charles the First, the Cavaliers
Would not have refused his assistance; for the Reins
of due Government he was always obedient.
He was a favourite, yet at times, felt
The wanton Lash of Lawless Power.
After a Life of Laborious Services performed, like
Clarendons,
With unimpeached Fidelity;
He, like that Great Man, was turned out of Employment;
Strip't of all his trappings, without place or pension:
Yet (being endowed with a generous forgiving Temper)
Saint-like,
Not dreading Futurity, placidly met the Hand
Appointed to be his Assassin
Thus he died an Example to all Mortals, under
The widespread Canopy of Heaven.

MEMORIAL TO A HORSE, 1765, IN THE
GROUNDS OF HASWELL HOUSE,
GOATHURST, SOMERSET

I am very anxious that kangaroos should be cultivated in English parks. I am sure they would do well, and be very ornamental, as well as forming a new dish for the table. I wish some gentlemen would oblige me by trying the experiment. . . .

By the way, I wonder if the reader knows the origin of the name kangaroo. The story, as told me by my friend the late Mr E. Blyth, runs, that when Captain Cook first discovered Australia, he saw some natives on the shore, one of them holding a dead animal in his hand. The Captain sent a boat's crew ashore to purchase the animal, and, finding, on receiving it, that it was a beast quite new to him, he sent the boatswain back to ask the natives its name. 'What do you call this 'ere animal,' said the sailor to the naked native. The native shook his head and answered, 'Kan-ga-roo,' which means in Australian lingo, 'I don't understand.' When the sailor returned to the ship the Captain said, 'Well, and what's the name of the animal?' The sailor replied, 'Please, sir, the black party says it's a "Kanga-roo."' The beast has kept this name ever since.

FRANK BUCKLAND (1826–80), *Log Book of a Fisherman and Zoologist,* 1875

2858.—ROAST WALLABY.

Ingredients.—Wallaby, forcemeat, milk, butter.

Mode.—In winter the animal may hang for some days, as a hare, which it resembles, but in summer it must, like all other flesh, be cooked very soon after it is killed. Cut off the hind-legs at the first joints, and, after skinning and paunching, let it lie in water for a little while to draw out the blood. Make a good veal force-meat, and after well washing the inside of the wallaby, stuff it and sew it up. Truss as a hare and roast before a bright clear fire from 1¼ to 1¾ hour, according to size. It

ROAST WALLABY.

must be kept some distance from the fire when first put down, or the outside will be too dry before the inside is done. Baste well, first with milk and then with butter, and when nearly done dredge with flour and baste again with butter till nicely frothed.

Time.—1¼ to 1¾ hour.

Sufficient for 6 persons.

Seasonable.—Best in cold weather.

6

Vim and Verve

Hast thou given the horse strength? Hast thou clothed his neck with thunder?

Canst thou make him afraid as a grasshopper? The glory of his nostrils is terrible.

JOB 39: 19–25

'Mustapha seizes the still lit fuse . . . seventy men fall on the spot.'

One reads in an English journal of a trait from which one might infer a degree of reasoning in animals themselves. Mustapha, an alert and sturdy hound, belonged to an artilleryman from Dublin. Brought up from birth in camps, he always accompanied his master and showed no fear in the midst of battle. In the hottest of actions he remained by the cannon and held the fuse in his jaws.

At the memorable battle of Fontenoy at which we broke the Hano-verian square, Mustapha's master was struck by a mortal blow. He, like his fellows, was felled down by artillery at the very moment he was about to fire on the enemy.

Seeing his master lying dead and covered in blood, the hound despairs and sets up an awful howling. At that moment a body of Frenchmen

advances swiftly to take the cannon pointing at them from the top of a small rise.

Who would have believed it had the fact not been attested to by reliable witnesses? Doubtless in order to avenge his master, Mustapha seizes the still lit fuse from his hand and fires the cannon which is loaded with grapeshot: seventy men fall on the spot that very instant and the remainder flee.

After this bold stroke, the dog lies down sadly by his master's corpse; he licks his wounds and so remains for twenty-two hours without drinking or eating. Some comrades of the artilleryman separate them in the end, but with the greatest difficulty imaginable. This courageous hound was brought back to London and presented to George III who gave him a pension, just like to a brave servant.

A. ANTOINE, *Chiens Célèbres*, 1796
TRANS. NICHOLAS DAVID PHILLIPS, 1992

Rambures: That island of England breeds very valiant creatures: their mastiffs are of unmatchable courage.
Duke of Orleans: Foolish curs! that run winking into the mouth of a Russian bear and have their heads crushed like rotten apples. You may as well say that's a valiant flea that dare eat his breakfast on the lip of a lion.
Constable: Just, just; and the men do sympathize with the mastiffs in robustious and rough coming on, leaving their wits with their wives; and then give them great meals of beef and iron and steel, they will eat like wolves and fight like devils.

WILLIAM SHAKESPEARE (1564–1616),
Henry V, ACT III, SCENE VI

The Maltese Cat was a polo pony in British India; the opposing team in this extraordinary match were called the Archangels

Lutyens felt the little chap take a deep breath, and, as it were, crouch under his rider. The ball was hopping towards the right-hand boundary, an Archangel riding for it with both spurs and a whip; but neither spur nor whip would make his pony stretch himself as he neared the crowd. The Maltese Cat glided under his very nose, picking up his hind legs sharp, for there was not a foot to spare between his quarters and the other pony's bit. It was as neat an exhibition as fancy figure-skating. Lutyens hit with all the strength he had left, but the stick slipped a little in his hand, and the ball flew off to the left instead of keeping close to the boundary. Who's Who was far across the ground, thinking hard as he galloped. He repeated, stride for stride, the Cat's manœuvres with another Archangel pony, nipping the ball away from under his bridle, and clearing his opponent by half a fraction of an inch, for Who's Who was clumsy behind. Then he drove away towards the right as the Maltese Cat came up from the left; and Bamboo held a middle course exactly between them. . . . Bamboo and Who's Who shortened stride to give the Maltese Cat room, and Lutyens got the goal with a clean, smooth, smacking stroke that was heard all over the field. But there was no stopping the ponies. They poured through the goal-posts in one mixed mob, winners and losers together, for the pace had been terrific. The Maltese Cat knew by experience what would happen, and, to save Lutyens, turned to the right with one last effort that strained a back-sinew beyond hope of repair. As he did so he heard the right-hand goal-post crack as a pony cannoned into it – crack, splinter, and fall like a mast. It had been sawed three parts through in case of accidents, but it upset the pony nevertheless, and he blundered into another, who blundered into the left-hand post, and then there was confusion and dust and wood. Bamboo was lying on the ground, seeing stars; an Archangel pony rolled beside him, breathless and angry; Shikast had sat down dog-fashion to avoid falling over the others, and was sliding along on his little bobtail in a cloud of dust; and Powell was sitting on the ground, hammering with his stick and trying to cheer. All the others were shouting at the top of what was left of their voices, and the men who had been spilt were shouting too. . . .

But the Maltese Cat stood with his head down, wondering how many legs were left to him; and Lutyens watched the men and ponies pick themselves out of the wreck of the two goal-posts, and he patted the Cat very tenderly.

'I say,' said the captain of the Archangels, spitting a pebble out of his mouth, 'will you take three thousand for that pony – as he stands?'

'No, thank you. I've an idea he's saved my life,' said Lutyens, getting off and lying down at full length. Both teams were on the ground too, waving their boots in the air, and coughing and drawing deep breaths, as the *saises* ran up to take away the ponies, and an officious water-carrier sprinkled the players with dirty water till they sat up.

'My Aunt!' said Powell, rubbing his back and looking at the stumps of the goal-posts, 'That was a game!'

They played it over again, every stroke of it, that night at the big dinner, when the Free-for-All Cup was filled and passed down the table, and emptied and filled again, and everybody made most eloquent speeches. About two in the morning, when there might have been some singing, a wise little, plain little, gray little head looked in through the open door.

'Hurrah! Bring him in,' said the Archangels; and his *sais*, who was very happy indeed, patted the Maltese Cat on the flank, and he limped in to the blaze of light and the glittering uniforms, looking for Lutyens. He was used to messes, and men's bedrooms, and places where ponies are not usually encouraged, and in his youth had jumped on and off a mess-table for a bet. So he behaved himself very politely, and ate bread dipped in salt, and was petted all round the table, moving gingerly; and they drank his health, because he had done more to win the Cup than any man or horse on the ground.

That was glory and honour enough for the rest of his days, and the Maltese Cat did not complain much when the veterinary surgeon said that he would be no good for polo any more. When Lutyens married, his wife did not allow him to play, so he was forced to be an umpire; and his pony on these occasions was a flea-bitten gray with a neat polo-tail, lame all round, but desperately quick on his feet, and, as everybody knew, Past Pluperfect Prestissimo Player of the Game.

RUDYARD KIPLING (1865–1936),
'THE MALTESE CAT', FROM
The Day's Work, 1898

The almighty who gave the dog to be companion of our pleasure and our toils hath invested him with a nature both noble and incapable of deceit. He forgets neither friend nor foe; he remembers with accuracy, both benefit and injury. He hath a share of man's intelligence but no share of man's falsehood. You may bribe an assassin to slay a man, or a witness to take his life by false accusation, but you cannot make a dog tear his benefactor. He is the friend of man, save when man justly incurs his enmity.

SIR WALTER SCOTT (1771–1832),
ON HIS BULL TERRIER CAMP

Meg was against my putting an L on my car and teaching the ape to drive, but to my mind driving was a crucial test of brain, and I persevered. . . .   It had a tendency to speed, pressing on the accelerator with all its might, and also continuously on the horn, which it enjoyed greatly. I forbade it to do this except when something was in the road in front of it, but I was never sure that it understood about this, though it listened attentively to me and I could see it was trying to follow, by the way it frowned and ground its teeth. I showed it how, when I was at the wheel, I hooted, but only slightly, when anything was in the way, but not at other times, and I hoped it took this in, but it had a great tendency to exaggeration and overdoing, and I saw that it would be a hooting driver. The postman and the newspaper boy on their bicycles, who used the field path, were rather alarmed by its style of driving, and the people in the village began to talk, but I assured them that everything was under control. One day it got alone into the car, which was aunt Dot's small Morris and was standing in the drive with the engine key in, and it started it up and drove off alone through the gate on to the field path, where it put on a tremendous speed and rushed along with the horn blaring after the gardener, who was pushing a wheelbarrow full of leaves for burning. The gardener barely had time to leap aside on to the grass edge, leaving the barrow, and the Morris crashed right into it and broke it up and bent a wing. The gardener was very angry about this, and so was I. Meg and I came back from a walk to find a scene of rage going on, the gardener shaking a stick at the ape, the ape blaring the horn and gibbering at the gardener, who did not dare to get very near it.

The gardener told me and Meg that the ape had no right to be on the road; Meg rather agreed, but I explained that the road was a private one and I said I would pay for the barrow. The gardener said, 'Mrs ffoulkes-Corbett won't be best pleased about her car, nohow,' and I was not best pleased either, as the insurance did not cover accidents caused by L drivers driving alone, and I should have to pay.

ROSE MACAULAY (1881–1958),
The Towers of Trebizond, 1956

A noble sacrifice.

'Tich' (Desert Rat)
Served 1940 to 1945. 1st Battn KRRC
Mongrel. Awarded VC for Gallantry.
Born 1940 – Died 1959
Sleep well little girl.
MEMORIAL IN ILFORD PET CEMETERY

Mrs Balfour tells a beautiful story of a soldier and his horse, in the 'British Workman.' She states:– 'I was walking in Crawford Street, Mary-le-bone, and an empty cab was standing by the pavement; a Life-guardsman was walking along the other side of the way, who suddenly stopped, looked across with a startled gaze, presently came over, and rushed up to the horse in the cab. 'I know him! I know him!' he shouted, in a voice of emotion, regardless of the passers-by; 'It's my own old Valiant, my dear old fellow.' The poor horse seemed to know the caressing hand and voice, for he visibly trembled, laid back his ears, and pushed his nose against the soldier. There certainly was not only a flush on the man's face, but a mist of tears, I think, was gathering in his eyes. The cabman, who was on the pavement, looked up to him evidently touched with the meeting. After a few moments of caresses, the Life-guardsman foraged in his pocket, muttering as he did so, 'He shall have it, if it was my last, the dear old boy. Yes; it isn't much, but it's enough for a feed of corn, and "by George" I'll treat him.' There was a corn-chandler's near, and in a few minutes out came the soldier with a good feed of corn in the nose-bag; and to see him undo the head-gear, and put on the bag, the horse meanwhile seeming to whinny and paw with pleasure, made so hearty a picture, that it was no wonder some little boys cried 'hurrah!' and the astonished cabman called out in his Cockney lingo, 'Give us your flipper, mate; I ain't met with a cove like you, I dunno the day ven.' 'Be good to him, use him well, he's as good a bit of stuff as ever was in harness,' said the poor soldier, in reply; and, while the horse enjoyed his meal, the men were exchanging inquiries, and the soldier learned where the present stable of his four-footed friend was situated, and then he reluctantly went on his way, his pocket lightened, it may be of his last coin, but his heart warmed; and I have no doubt, from the look of the man's kind face, that as often from that time as he could, he looked in to see after old Valiant. And a better lesson on kindness to animals than he had given to the cabman, it would be difficult to present. He had fed both horse and man – the one with corn, the other with a good example; wholesome grain, both.'

REV. T. JACKSON, *Our Dumb Companions*, c. 1863

RIGHT ROYAL

In a race-course box behind the Stand
Right Royal shone from a strapper's hand.
A big dark bay with a restless tread,
Fetlock deep in a wheat-straw bed;
A noble horse of a nervy blood,
By O Mon Roi out of Rectitude.

Something quick in his eye and ear
Gave a hint that he might be queer.
In front, he was all to a horseman's mind;
Some thought him a trifle light behind.
By two good points might his rank be known,
A beautiful head and a Jumping Bone . . .

And now, as proud as a King of Spain,
He moved in his box with a restless tread,
His eyes like sparks in his lovely head,
Ready to run between the roar
Of the stands that face the Straight once more;
Ready to race, though blown, though beat,
As long as his will could lift his feet;
Ready to burst his heart to pass
Each gasping horse in that street of grass . . .

And smiting the turf to clods that scattered
Was the rush of the race, the thing that mattered,
A tide of horses in fury flowing,
Beauty of speed in glory going,
Kubhadar pulling, romping first,
Like a big black fox that had made his burst.

And away and away and away they went,
A visible song of what life meant.
Living in houses, sleeping in bed,
Going to business, all seemed dead,
Dead as death to that rush in strife,
Pulse for pulse with the heart of life . . .

With a rush and a crashing Right Royal went over
With the stride of a stalwart and the blood of a lover,
He landed on stubble now pushing with clover,

And just as he landed, the March sun shone bright
And the blue sky showed flamelike and the dun clouds turned white;
The little larks panted aloft their delight,
Trembling and singing as though one with the light.

And Charles, as he rode, felt the joy of their singing,
While over the clover the horses went stringing,
And up from Right Royal the message came winging,
'It is my day to-day, though the pace may be stinging,
Though the jumps be all danger and the going all clinging.'

The white, square church-tower with its weather-cock swinging
Rose up on the right above grass and dark plough,
Where the elm trees' black branches had bud on the bough . . .

Then they strode the green grass for the Lost Lady's grave,
And Charles felt Right Royal rise up like a wave,
Like a wave far to seaward that lifts in a line
And advances to shoreward in a slipping incline,

And climbs, and comes toppling, and advances in glory,
Mounting inwards, marching onwards, with his shoulders all hoary,
Sweeping shorewards with a shouting to burst on the sand,
So Right Royal sent meaning through the rein in each hand . . .

They went at Lost Lady's like Severn at flood,
With an urging of horses and a squelching of mud;
Then a whirl of urged horses thundered up, whipped and blown,
Soyland, Peterkinooks, and Red Ember the roan.
For an instant they challenged, then they drooped and were done;
Then the White Post shot backwards, Right Royal had won.

JOHN MASEFIELD (1878–1967)

VILE AMUSEMENTS IN NINETEENTH-CENTURY LONDON

I took leave of my kind hosts, and returned to London. The same evening L——— took me to a singular exhibition.

In a suburb, a good German mile from my lodging, we entered a sort of barn; dirty, with no other ceiling than the rough roof, through which the moon peeped here and there. In the middle was a boarded place, about twelve feet square, surrounded by a strong wooden breastwork: round this was a gallery filled with the lowest vulgar and with perilous-looking faces of both sexes. A ladder led up to a higher gallery, for the patrician part of the spectators, which was let out at three shillings a seat. There was a strange contrast between the 'local' and a crystal lustre hanging from one of the balks of the roof lighted with thick wax candles; as well as between the 'fashionables' and the populace among whom they were scattered, who – the latter I mean – were continually offering and taking bets of from twenty to fifty pounds. The subject of these was a fine terrier, the illustrious Billy, who pledged himself to the public to kill a hundred rats in ten minutes. As yet the arena was empty, and there was an anxious, fearful pause; while in the lower gallery huge pots of beer circulated from mouth to mouth, and tobacco-smoke ascended in dense clouds. At length appeared a strong man, bearing a sack looking like a sack of potatoes, but in fact containing the hundred live rats. These he set at liberty in one moment by untying the knot, scattered them about the place, and rapidly made his retreat into a corner. At a given signal Billy rushed in, and set about his murderous work with incredible fury. As soon as a rat lay lifeless, Billy's faithful esquire picked him up and put him in the sack; among these some might be only senseless, or perhaps there might be some old practitioners who feigned themselves dead at the first bite. However, be that as it may, Billy won in nine minutes and a quarter, according to all the watches; in which time a hundred dead, or apparently dead, rats were replaced in their old quarters – the sack.

The *amusements* ended with bear-baiting, in which the bear treated some dogs extremely ill, and seemed to suffer little himself. It was evident through the whole, that the managers were too chary of their animals to expose them in earnest; I therefore, as I said, suspected from the beginning some hidden talents for representation – even in the rats.

In a few months, cock-fights will be held in the same place. I shall send you a description of them.

PRINCE M. L. H. VON PÜCKLER-MUSKAU,
Tour in Germany, Holland and England,
in the Years 1826, 1827, and 1828, Vol. IV, 1832

Poor Percy is dead. I intend to have an old stone set up by his grave, with 'Cy gist li preux percie' and I hope future antiquarians will debate which hero of the house of Northumberland has left his bones in Tiviotdale.

LETTER TO ELLIS, 1808, FROM
SIR WALTER SCOTT (1771–1832)

EXTRAORDINARY AND DARING FEAT OF HORSEMANSHIP AT AYLESBURY

At the Steward's Ordinary at the White Hart Hotel, after the Aristocratic Steeplechase, Saturday Feb. 13th 1850, the conversation turned to the feat of bringing a horse up into the dining room in which the company were then assembled, which was once done by Lord Jocelyn and Mr Ricardo during the meeting of the Royal Hunt some few years ago. Mr Chas Symonds, of sporting notoriety at Oxford, offered to bring a grey horse of his up stairs, and lead him round the table. No sooner said than done – for off he started, fetched the animal from the stable, and very shortly announced his progress by a loud clattering on the old oak stair-case. In a few minutes the horse was gazing on the assembled company. His owner then led him over a flight of chairs which he jumped beautifully. Nothing then would satisfy the dear sons of Alma Mater but he must be jumped over the dining tables. Mr Flower, the proprietor of the hotel, fearing lest some serious accident might occur, as it is a room of great antiquity, having been built by the Earl of Rochester in the time of the Second Charles, made strong objections, but he was overruled and the horse was led over them, everything standing, the champagne glasses rattled, the plates quivered, the candlesticks shook, but nothing was displaced, back again he went, clearing everything at a bound, whereupon a most ardent sportsman and a very hard man across the vale Mr Manning of Wendover volunteered to ride him barebacked over, and to the astonishment of all present, he accomplished it without a bridle or saddle. The celebrated gentleman jock Captain Barlowe next assayed and managed to make a smash of one table and its contents. This was only a temporary check, for in the face of a tremendous fire, and the cheering of all present he achieved the feat gallantly. It was now time to desist and to get him downstairs. This was sooner said than done, for the stairs and passages being kept polished, the gallant grey slipped about dreadfully and was evidently afraid of the descent. At length at the suggestion of a worthy Baronet who was looking on at the fun he was blindfolded and thus descended into the entrance hall but managed to break about a dozen of the carved oak bannisters in his progress. After the horse was safely back in the stable the company returned and under the presidency of a noble earl, scion of a truly noble house, drank the health of the dining room Pet, and might he carry his owner safely through many a long run for many a long year to come.

Bucks Chronicle, SATURDAY 13 FEBRUARY, 1850

Hast thou given the horse strength? Hast thou clothed his neck with thunder?

Canst thou make him afraid as a grasshopper? The glory of his nostrils is terrible.

He paweth in the valley, and rejoiceth in his strength; he goeth on to meet the armed men.

He mocketh at fear, and is not affrighted; neither turneth he back from the sword.

The quiver rattleth against him, the glittering spear and the shield.

He swalloweth the ground with fierceness and rage; neither believeth he that it is the sound of the trumpet.

He sayeth among the trumpets, Ha, ha; and he smelleth the battle afar off, the thunder of the captains, and the shouting.

JOB 39: 19–25

I advanced up to the place slow and cautious. The snake was well concealed, but at last I made him out; it was a coulacanara, not poisonous, but large enough to have crushed any of us to death. On measuring him afterwards he was something more than fourteen feet long.... After skinning this snake I could easily get my head into his mouth, as the singular formation of the jaws admits of wonderful extension.

A Dutch friend of mine, by name Brouwer, killed a boa twenty-two feet long with a pair of stag's horns in his mouth. He had swallowed the stag, but could not get the horns down; so he had to wait in patience with that uncomfortable mouthful till his stomach digested the body, and then the horns would drop out. In this plight the Dutchman found him as he was going in his canoe up the river, and sent a ball through his head....

I had been at the siege of Troy for nine years, and it would not do now to carry back to Greece 'nil decimo nisi dedecus anno.' I mean I had been in search of a large serpent for years, and now having come up with one it did not become me to turn soft. So, taking a cutlass from one of the negroes, and then ranging both the sable slaves behind me, I told them to follow me, and that I would cut them down if they offered to fly....

On pinning him to the ground with the lance he gave a tremendous loud hiss.... We had a sharp fray in the den, the rotten sticks flying on all sides, and each party struggling for superiority. I called out to the second negro to throw himself upon me, as I found I was not heavy enough. He did so, and the additional weight was of great service. I had now got firm hold of his tail; and after a violent struggle or two he gave in, finding himself overpowered. This was the moment to secure him. So while the first negro continued to hold the lance firm to the ground, and the other was helping me, I contrived to unloose my braces and with them tied up the snake's mouth....

I had brought with me up into the forest a strong bag large enough to contain any animal that I should want to dissect. I considered this the best mode of keeping live wild animals when I was pressed for daylight; for the bag yielding in every direction to their efforts, they would have nothing solid or fixed to work on, and thus would be prevented from making a hole through it. I say fixed, for after the mouth of the bag was closed the bag itself was not fastened or tied to anything, but moved about wherever the animal inside caused it to roll. After securing afresh the mouth of the coula-canara, so that he could not open it, he was forced into this bag and left to his fate till morning.

I cannot say he allowed me to have a quiet night. My hammock was in the loft just above him, and the floor betwixt us half gone to decay, so that in parts of it no boards intervened betwixt his lodging-room and mine. He was very restless and fretful; and had Medusa been my wife, there could not have been more continued and disagreeable hissing in the bed-chamber that night.

CHARLES WATERTON (1782–1865),
Wanderings in South America, 1825

'Peace at any Price.'

It's a great day
It's a great day to be a husband
It's a great day to be a father
It's a great day to be a brother
It's a great day to be master of a fine dog
It's a great day to be a soldier
It's a great day to be an American.

GENERAL NORMAN SCHWARZKOPF, ON
ARRIVING AT MACGILL AIRFORCE BASE
AFTER THE GULF WAR, 28 APRIL 1991

I have two little monkeys at home of which I am exceedingly fond. They are really half educated in their way, and are almost fit to go up for a competitive examination. My monkeys' names are 'The Hag' and 'Tiny.' Hag's original name was 'Jenny,' but she has so much of the character of a disagreeable old woman about her that I call her 'The Hag,' and she 'answers to' that name. Tiny was originally a very little monkey indeed, not much bigger than a large rat. My friend Bartlett brought her to me from the Zoological Gardens as a dead monkey; she was 'as good as dead' – a perfect skeleton, and with but little hair on her. She arrived tied up in an old canvas bag. I put her into 'The Hag's' cage. The old Lady at once 'took to her,' and instantly began the office of nurse; she cuddled up poor Tiny in her arms, made faces, and showed her teeth, at anybody who attempted to touch her. Tiny had port-wine negus, quinine wine, beef-tea, egg and milk – in fact, anything she could eat; and 'The Hag' always allowed her to have 'first pull' at whatever was put into the cage. In time, Tiny, through Mrs Buckland's good nursing, stood up, then began to run, her hair all came again; and she is now one of the handsomest, most wicked, intelligent, funny little beasts that ever committed an act of theft. Steal? Why, her whole life is devoted to stealing, for the pure love of the thing. The Hag's Latin name is *Cercopithecus petaurista*, or the vaulting monkey. Tiny is a '*Mona*.' When pleased her cry is very like the word 'mona' prolonged.

Tiny and 'The Hag' are dressed like two sisters going to a ball, and it is difficult, for a person who does not know them well, to tell them apart. They are each a little larger than a big guinea-pig, with a long tail. 'The Hag' has a green head, a very handsome white beard, with a snow-white spot on her nose, and brilliant lustrous brown eyes; the cheeks are beautifully marked with silk-like black hairs; the ears are well turned, and very small. I put earrings once into 'The Hag's' ears, but Tiny pulled them out and crushed them up with her teeth. On the hair on the top of the head there are markings reminding us of the 'plate bonnets' once worn by ladies: the monkeys 'wear their own hair,' and not chignons. My monkeys are, summer and winter, dressed in seasonable garments: their wardrobe consists of three sets of dresses. 1st. Their common winter dress of thick white flannel, trimmed with red braid, and peg-top sleeves, with large capes: in these they look like the old-fashioned 'Charlies,' or night-watchmen. Their 'second best' dresses are of green baize without capes, made to fit quite tight, like a friar's frock, tied on round the waist by means of a girdle of ornamental ribbon or a patent leather strap. . . . Their best dress for summer evenings, at tea or dessert, when 'company is coming,' is a green velvet dress, trimmed with gold lace, like the huntsman of the Queen's staghounds.

Under their dresses, their chests are carefully wrapped round with warm flannel, sewed on. In very cold weather they have an extra thickness of flannel. I feel convinced that all valuable monkeys should be dressed in this way, and that this plan should *always* be adopted at the Zoological. . . .

When the fire is lighted in the mornings in my museum, the servant puts the monkeys in their night cage before it, and directly I come down to breakfast I let them out. . . .

Having poured out the tea, I open the *Times* newspaper quite wide, to take a general survey of its contents. If I do not watch her carefully, Tiny goes behind the chair on to the book-shelf, and comes crash, with a Léotard-like jump, into the middle of the *Times*, like a foxhunter charging at a five-barred gate. Of course, she cannot go *through* the *Times*; but she takes her chance of a fall somewhere, and her great aim seems to perform the double feat of knocking the *Times* out of my hand and upsetting the tea-cup and its contents, or, better still, the tea-pot on to the floor. Lately, I am glad to say, she did not calculate her fall quite right; for she put her foot into the hot tea and stung herself smartly, and this seems to have had the effect of making her more careful for the future. All the day of this misfortune she walked upon her heels, and not upon her toes as usual.

'The Hag' will also steal, but in a more quiet manner. She is especially fond of sardines in oil, and I generally let her steal them, because the oil does her good, though the servants complain of the marks of her oily feet upon the cloth. Sometimes the two make up a 'stealing party.' One morning I was in a particular hurry, having to go away on salmon inspection duty by train. I left the breakfast things for a moment, and in an instant Tiny snatched up a broiled leg of pheasant and bolted with it – carried it under her arm round and round the room, after the fashion of the clown in the pantomime. While I was hunting Tiny for my pheasant, 'The Hag' bolted with the toast: I could not find time to catch either of the thieves, and so had to go off without any breakfast.

Tiny and 'The Hag' sometimes go out stealing together. They climb up my coat and search all the pockets. I generally carry a great many cedar pencils: the monkeys take these out and bite off the cut ends. But the great treat is to pick and pick at the door of a glass cupboard till it is open, then to get in and drink the hair-oil which they know is there. . . .

Where I go to Herne Bay to attend to oyster cultivation, I take the monkeys with me for the benefit of the sea air. I always put up at Mr Walker's, the confectioner, in the Esplanade. Mrs Walker is very fond of the 'Coloured Ladies,' as she calls them, and allows them to take great liberties.

Mrs Walker is rather proud of the way she dresses her shop-window with cakes, buns, sweet-stuff, &c. One day 'The Hag' had crept very quietly into the shop, and was having a 'field day' all to herself in the shop-window among the sweets. Mrs Walker, sitting in the back parlour, was aroused by hearing a crowd of boys laughing outside the window. On coming into the front shop she found 'The Hag' all among the cakes, &c., in the window; both her cheek-pouches were as full as ever they could hold of lemon-peel, and she was still munching at a great lump of it. My lady was sitting on the top of a large cake like a figure on a twelfth-cake. Tiny was not in this bit of mischief for a wonder.

Mrs Walker declared she would send 'The Hag' before my friend, Captain Slark, the chief magistrate of the town, for stealing, and have her locked up for a fortnight; but the thief had first to be caught, and this was a difficult task, for she bolted out into the bakehouse, and upstairs into the loft where the flour is kept. There is a large wooden funnel through which the flour is passed into the bakehouse below. Trying to hide herself from Mrs Walker, 'The Hag' jumped into the open top of it, and, much to his astonishment, lighted on Mr Walker's head as he was making the bread below: she knew she was all right with Mr Walker, but she was one mass of flour. Her green baize coat was quite white, and she looked like a miller on a small scale, and the flour could not be brushed out of her for two or three days.

Mr Walker tied her up, and there she stayed by the warm oven, the rest of the day, chattering and telling him in monkey-language of all her troubles. . . .

When I come home in the evening tired from a long day's work, I let out the monkeys, and give them some sweet-stuff I bring home for them. By their affectionate greeting and amusing tricks they make me forget for a while the anxieties and bothers of a very active life. They know perfectly well when I am busy, and they remain quiet and do not teaze me. 'The Hag' sits on the top of my head, and 'looks fleas' in my hair, while Tiny tears up with her teeth a thick ball of crumpled paper, the nucleus of which she knows is a sugar-plum, one of a parcel sent by Mrs Owen, the kind-hearted wife of my friend, Mostyn Owen, of the Dee Salmon Board, and received through the post in due form, directed, 'Miss Tiny and Miss Jenny Buckland.'

I must now finish the 'Memoir,' though, if I had time, I could go on writing for a month describing my little pets. . . .

Although my monkeys do considerable mischief, yet I let them do it. I am amply rewarded by their funny and affectionate ways.

The reader may wonder that I like to keep my monkeys at all in my house; but I *do* like to keep them, and nothing whatever would induce me to part with them.

My monkeys love me, and I love my monkeys.

FRANK BUCKLAND (1826–80), *Logbook of a Fisherman and Zoologist*, 1875

For Wellington it was the worst, not forgetting Assaye. As he silently walked Copenhagen back to Waterloo, the moonlit fields on either side of the *chaussée* were littered with the dead and dying, with horses, weapons, helmets, caps, belts and feathers. Here and there a wounded man rose to his feet and staggered away or the sinister shadow of a robber bent over a corpse. . . .

[Wellington] was in a state of emotional shock. Nevertheless, life must go on. After dismounting at 11 p.m. outside his inn at Waterloo he gave Copenhagen an approving pat on the hind quarters, no doubt thinking that the little horse was as tired as himself. On the contrary, Copenhagen lashed out and nearly inflicted the wound which Wellington had miraculously escaped on the battlefield. Copenhagen had Arab blood from his dam and was a grandson of the famous race-horse Eclipse, whose son, Young Eclipse, was the second horse to win the Derby (1781). The Duke was never sentimental about Copenhagen, but the tribute he paid him after his death expressed, as always, exactly what he felt:

There may have been many faster horses, no doubt many handsomer, but for bottom and endurance I never saw his fellow.

ELIZABETH LONGFORD, *Wellington: The Years of the Sword*, 1969

"GOOD-BYE, OLD MAN"

THE HORSE

Where in this wide world can
man find nobility without pride,
friendship without envy or beauty
without vanity? Here, where
grace is laced with muscle,
and strength by gentleness confined.

He serves without servility; he has
fought without enmity. There is
nothing so powerful, nothing less
violent; there is nothing so quick,
nothing more patient.

England's past has been borne on
his back. All our history is his
industry; we are his heirs, he
our inheritance.

RONALD DUNCAN (b. 1914), 1962

HORSES ABROAD

Horses in horse cloths stand in a row
On board the huge ship that at last lets go:
Whither are they sailing? They do not know,
Nor what for, nor how. – They are horses of war,
And are going to where there is fighting afar;
But they gaze through their eye-holes, unwilling they are,
Their bones will bleach ere a year has passed,
And the item be as 'war waste' classed. –
And when the band booms, and the folk say 'Goodbye!'
And the shore slides astern, they appear awry
From the scheme Nature planned for them, – wondering why.

THOMAS HARDY (1840–1928),
Human Show Phantasies, 1925

7

Service to Man

Stop me not, but onward let me jog,
I'm Bob, the London fireman's dog.

REV. T. JACKSON,
Our Dumb Companions, c. 1863

I owe a great deal to rats. When a student at St George's Hospital I wrote an article on rats, which I sent to a magazine, and to my great amazement the publishers sent me a cheque for it. From that moment I have taken a great liking to my first patrons in literature, viz., 'Rats,' and I always somehow connect them in my memory with publishers.

FRANK BUCKLAND (1826–80), *Notes and Jottings from Animal Life*, 1882

How could man without the assistance of the dog, have been able to conquer, tame and reduce to servitude, every other animal? How could he discover, chase and destroy those that were noxious to him? In order to be secure, and to become master of all animated nature, it was necessary for him to begin by making a friend of part of them; to attach such of them to himself by kindness and caresses, as seemed fittest for obedience and active pursuit: thus the first art employed by man, was in conciliating the favour of the dog; and the fruits of this art were the conquest and peaceful possession of the earth.

OLIVER GOLDSMITH (1728–74), *History of the Earth and Animated Nature*, 1774

An officer in the Forty-fourth Regiment, who had occasion, when in Paris, to pass one of the bridges across the Seine, had his boots, which had been previously well polished, dirtied by a poodle dog rubbing against them. He in consequence went to a man who was stationed on the bridge, and had them cleaned. The same circumstance having occurred more than once, his curiosity was excited, and he watched the dog. He saw him roll himself in the mud of the river, and then watch for a person, against which he contrived to rub himself. Finding that the shoe-black was the owner of the dog, he taxed him with the artifice; and after a little hesitation he confessed that he had taught the dog the trick in order to procure customers for himself. The officer being much struck with the dog's sagacity, purchased him at a high price, and brought him to England. He kept him tied up in London some time, and then released him. The dog remained with him a day or two, and then made his escape. A fortnight afterwards he was found with his former master, pursuing his old trade on the bridge.

EDWARD JESSE (1780–1868), *Gleanings in Natural History*, 1832–5

Not long since there was a certain barber, who lived in Hampton Street, Walworth Road, who did a good business in bear's-grease, and this with *one* bear, which same bear he killed *three times a week*. He kept the bear in an area, where he could plainly be seen by the passers-by. He was a tame creature was this bear; the proprietor used to feed him with meat, placed on the end of a long stick, to make believe he was very savage; yet the little school-children, from the neighbouring 'penny a week school,' used to buy buns for him, and feed him with their fingers through the bars of the area, for the poor brute was half-starved. The children, sharp as children always are, were not taken in when the proprietor advertised, outside the shop, 'another bear just arrived.'

In order to carry out the trick, the barber simply took the poor old bear, and by means of flour and grease made him into a 'grey bear,' or he blackened his coat and made him into a 'black bear.'

Yet the children always knew their old friend, who had but one eye, and that a regular 'piercer;' nor would they fail to recognise his old moth-eaten coat, even when invited to come and see 'the pretty new bear,' who, strange to say, poked his mouth, 'with the broken tooth in it,' up to the bars of the area, as had been his wont for between five and six years. At the appointed day, when the bear was advertised to be killed, the poor beast was made to retire from the area; and shortly afterwards were heard, proceeding from under the barber's shop, the most dreadful yells and roarings, followed by groans as of the poor bear in the agonies of death. At last all was over, and the bear's cage was brought out (apparently empty) and taken off to the docks. A dried head and skin of a bear were duly hung up in the shop window, a plain proof that a bear had just been killed for grease.

The next morning another bear was brought back from the docks, and deposited in the area, in his turn to be killed, and so on.

But the truth was at last discovered. There was a certain Jew fishmonger, who went by the name of 'Leather-mouthed Jemmy,' on account of his tremendously powerful voice. This man was hired on bear-killing days to produce the roars and groans of the dying animal, which he did with a wonderful accuracy. On one unfortunate day the hair-dresser would not give the accustomed fee of five shillings. Leather-mouthed Jemmy immediately told the whole conceit, and the hairdresser was obliged to shut up his shop, and sell his only bear, that he used to kill three times a week, for what he would fetch, to the rival barber over the way.

FRANK BUCKLAND (1826–80),
Curiosities of Natural History, 1866

THE CHACE

...A diff'rent hound for ev'ry diff'rent chace
Select with judgment; nor the tim'rous hare
O'er-match'd destroy, but leave that vile offence
To the mean, murd'rous coursing crew, intent

On blood and spoil. O blast their hopes, just Heav'n!
And all their painful drudgeries repay
With disappointment and severe remorse.
But husband thou thy pleasures, and give scope
To all her subtle play: by nature led,
A thousand shifts she tries; t' unravel these
Th' industrious beagle twists his waving tail,
Thro' all her labyrinths pursues, and rings
Her doleful knell. See there with count'nance blithe,
And with a courtly grin, the fawning hound
Salutes thee cow'ring, his wide op'ning nose
Upward he curls, and his large sloe-black eyes
Melt in soft blandishments and humble joy;
His glossy skin, or yellow-pied, or blue,
In lights or shades by Nature's pencil drawn,
Reflects the various tints; his ears and legs,
Fleckt here and there, in gay enamel'd pride,
Rival the speckled pard; his rush grown tail
O'er his broad back bends in an ample arch;
On shoulders clean, upright and firm he stands;
His round cat-foot, straight hams, and wide-spread thighs,
And his low-dropping chest, confess his speed,
His strength, his wind, or on the steepy hill,
Or far extended plain; in ev'ry part
So well proportion'd, that the nicer skill
Of Phidias himself can't blame thy choice.
Of such compose thy pack....
But above all take heed, nor mix thy hounds
Of diff'rent kinds; discordant sounds shall grate
Thy ears offended, and a lagging line
Of babbling curs disgrace thy broken pack.
But if th' amphibious otter be thy chace,
Or stately stag, that o'er the woodland reigns;
Or if th' harmonious thunder of the field
Delight thy ravish'd ears; the deep-flew'd hound
Breed up with care, strong, heavy, slow, but sure;
Whose ears down-hanging from his thick round head
Shall sweep the morning dew, whose clanging voice
Awake the mountain echo in her cell,
And shake the forests: the bold talbot kind
Of these the prime, as white as Alpine snows;
And great their use of old....

WILLIAM SOMERVILLE (1675–1742), 1696

January 13, 1855

I am feeding up my few remaining dogs very carefully; but I have no meat for them except the carcasses of their late companions. These have to be boiled; for in their frozen state they act as caustics, and, to dogs famishing as ours have been, frozen food often proves fatal, abrading the stomach and oesophagus. One of these poor creatures had been a child's pet among the Esquimaux. Last night found her nearly dying at the mouth of our *tossut*, wistfully eyeing the crevices of the door as they emitted their forbidden treasures of light and heat. She could not move, but, completely subdued, licked my hand – the first time I ever had such a civilized greeting from an Esquimaux dog. I carried her in among the glories of the moderate paradise she aspired to, and cooked her a dead-puppy soup. She is now slowly gaining strength, but can barely stand.

ELISHA KENT KANE, *The Second Grimmel Expedition in Search of Sir John Franklin, 1853, 1854, 1855*

My dear Whuff,

You know how I dislike, chiefly on your behalf, attributing the thoughts and expressions of men to animals other than men. The comparison suggests patronage and a sense of superiority, whereas you and I know perfectly well that there is no question between us of better or worse, whether in intellect or in morals. We are just different. That you are imbued, as I shall show, with a greater humility than I gives me no excuse for conceit, but rather for as much more humility as I can muster, handicapped, as I am, by my species.

After such a profession of philosophy as this it may seem illogical to indite you a letter, since you cannot read it – though as a puppy you would have taken it in and duly digested it. But the human brain, so far as it issues in action, is 'saved by its want of logic,' as a great Frenchman said of the English people. I address the letter directly to you because in this way I can most easily make explicit to myself just what I feel about our friendship, and how the miracle is possible that you and I, as thousands of others in like case, can be friends and brothers....

You are, and have been these five years, as incapable of rancour as Mr Jarndyce. The old tag, of course, is a lying one:–

> A woman, a spaniel, and a walnut tree;
> The more you bash 'em, the better they be....

With a carelessness that I can now only regard as insensate, as definitely cruel, I allowed you to see me go forth with my gun, closing the door with furtive quickness between us. You did not bark or protest in any way; but when the sound of my footsteps died away you climbed upstairs to my bedroom, a place you had seldom entered, and, pulling from the cupboard an old shooting jacket, lay on it till I came back. You rejoiced in the return. There was no touch of rebuke in your welcome; but the mute protest as reported was almost unendurable, and next time I would rather offend a neighbour by taking out a dog too many than leave you to pine.

What was in your head all those hours as you lay with the smell of the shooting coat in your nostrils? Your immense joy in going out with a gun is inspired in part, I must believe, by the feeling that we are particularly *en rapport* when in pursuit of game. What is elemental in our two modes of life is in touch. We are primeval hunters, together seeking necessary food. We are in some sort equals; yet you are glad and proud to confess that I am the master.

SIR WILLIAM BEACH THOMAS,
The Way of a Dog, 1948

Sir Henry Wyat was imprisoned often; once in a cold and narrow tower, where he had neither bed to lie on, nor clothes sufficient to warm him, nor meat for his mouth. He had starved then, had not God, who sent a crow to feed his prophet, sent this and his country's martyr a cat both to feed and to warm him. It was his own relation unto them from whom I had it. A cat came one day down into the dungeon unto him, and as it were offered herself unto him. He was glad of her, laid her in his bosom to warm him, and by making much of her won her love. After this she would come every day unto him divers times, and, when she could get one, bring him a pigeon. He complained to his keeper of his cold and short fare. The answer was 'he durst not better it'. 'But', said Sir Henry, 'if I can provide any, will you promise to dress it for me?' 'I may well enough', said he, the keeper, 'you are safe for that matter'; and being urged again, promised him, and kept his promise, dressed for him, from time to time, such pigeons as his accator the cat provided for him. Sir Henry Wyat in his prosperity for this would ever make much of cats, as other men will of their spaniels or hounds; and perhaps you shall not find his picture anywhere but, like Sir Christopher Hatton with his dog, with a cat beside him. The prisoner had this faithful cat painted, with a pigeon in his paws, offering it through the grated windows of his dungeon.

JOHN TIMBS (1801–75), *Abbeys and Castles*, 1870

The exact dates of Nipper's birth and death can not, of course, now be given, but the best evidence available suggests that he was born in 1883 or 1884 and died in 1895. . . .

Floss, ship's mascot
HMS Cordelia,
1917.

He was not a thoroughbred, but we are told that he had a good deal of the bull-terrier in him. At any rate he seems to have inherited the characteristics of this breed, for it was said that he never hesitated to take on a fight with another dog and once he got a hold it was very difficult to make him let go. Ratting was another of his favourite pastimes and one that was responsible for his losing an eye when he ran into a thorn bush in the excitement of the chase. At times, too, he was reported to be not unpartial to an illicit pheasant from Richmond Park.

Francis Barraud in his *Strand Magazine* article recounted many stories of Nipper's escapades: 'Nipper was a splendid subject to play practical jokes on. One that never failed was to put a very realistic reproduction of a cat, which was cut out of cardboard, sitting up in his basket. He was always taken in and rushed madly at it, but, of course, it fell flat (I mean the cat, not the joke), and I suppose to him it disappeared as if by magic. He was taken in over and over again. It always interested me, because it proved to me that a realistic bit of painting does appear real to a dog. I have heard many people contend that a picture would only appear a flat surface to an animal, but I don't think, after this experiment, that this is the case. Another favourite joke was to give him some soda water in a saucer; he would go to drink it, when it would fizz. This annoyed him fearfully and he barked madly at it, but went on having sips, or rather laps, until he had finished it.' . . .

Some fifty-three years later ... the Gramophone Company, the owners of the 'His Master's Voice' trade mark, had decided to investigate the whereabouts of Nipper's grave, and, if they succeeded in discovering his remains, to bring them 'home' to Hayes, the Headquarters of 'His Master's Voice' which he had done so much to make famous.... and so, on August 4th 1950, a cavalcade of cars set out from their offices at Hayes, bearing representatives of the sales organization, pressmen and workmen from the Works Department of the Factory. After a formal lunch the workmen started digging. *The Hayes Gazette* takes up the story from here: 'They started at a spot indicated by Mark Barraud and were watched by him, by representatives, reporters, photographers, garage hands and by Miss Enid Barraud (great-niece of the artist). ... In brilliant sunshine they removed the oil-soaked crust of the garage courtyard and went down, down, down, putting the earth they removed on a square of wood. There it was sorted over by two students specially brought from the Royal College of Veterinaries by the Company to identify any bones that might be found. And bones indeed were found about three feet down. Some were identified out-of-hand as sheep bones – "Probably the remains of lunches" said Mark Barraud – and others could have been dog's bones. They were taken away for more positive identification. It was then looking at the brown encrusted pieces that you couldn't help thinking of the little dog who used to frisk about the lawn that now is a garage courtyard.'

Subsequent examination of the bones did not confirm that they could have been those of Nipper, and the old dog was thus left to rest where he had been buried all those years ago.

LEONARD PETTS, *The Story of Nipper and the His Master's Voice Picture Painted by Francis Barraud*, 1973

The Hedgehog may be rendered in a considerable degree domestic; and it has been frequently introduced into houses for the purpose of expelling those troublesome insects, the *Blattæ*, or Cock-roaches, which it pursues and devours with great avidity. In the huts of the Calmuc Tartars these animals are kept instead of cats. – There was a Hedgehog, in the year 1799, in the possession of a Mr Sample, of the Angel-inn at Felton in Northumberland, which performed the duty of a turn-spit, as well, in every respect, as the dog of that denomination. It ran about the house as familiarly as any other domestic quadruped, and displayed an obedience till then unknown in this species of animals. ...

The Hedgehog is occasionally an article of food, and is said to be very delicate eating. The skin was used by the ancients for the purpose of a clothes-brush.

REV. W. BINGLEY (1774–1823), *Animal Biography*, 1813

'Stop me not, but onward let me jog,
I'm Bob, the London fireman's dog.'

. . . At a fire in Lambeth, when the firemen were told that all the inmates were out of the burning premises, Bob was not satisfied with this testimony; he went to a side-door and listened, and there, by loud and continual barking, attracted the notice of the brigade. The men felt sure, from Bob's agitation, that some one was in the passage, and, on bursting open the door, a child was found nearly dead from suffocation. *Papa*: Bob saved this child's life. . . . 'He could all but speak,' said the men who loved him; and more than speak in the hour of danger; for his loud, sharp bark had a vast deal of meaning in it. But Bob was an orator in the sense of attending public meetings, and giving testimony. At the annual meeting of the Society for the Prevention of Cruelty to Animals, which was held in 1860, and on previous occasions, this brave dog went through a series of wonderful performances, to show how the fire-engines were pumped, and most kindly and effectually would he give his warning bark, and in his way tell the scenes that he had passed through. Fine, noble creature!

REV. T. JACKSON, *Our Dumb Companions, c.* 1863

The Emperor hath two Barons who are own brothers, one called Baian and the other Mingan; and these two are styled Chinuchi, which is as much as to say, 'The Keepers of the Mastiff Dogs.' Each of these brothers hath 10,000 men under his orders, each body of 10,000 being dressed alike, the one in red and the other in blue, and whenever they accompany the Lord to the chase, they wear this livery, in order to be recognized. Out of each body of 10,000 there are 2,000 men who are each in charge of one or more mastiffs, so that the whole number of these is very great. And when the Prince goes a-hunting, one of these Barons, with his 10,000 men and something like 5,000 dogs, goes towards the right, whilst the other goes towards the left with his party in like manner. They move along, all abreast of one another, so that the whole line extends over a full day's journey, and no animal can escape them. Truly it is a glorious sight to see the working of the dogs and the huntsmen on such an occasion. And as the Lord rides a-fowling across the plains, you will see these big hounds come tearing up, one pack after a bear, another pack after a stag, or some other beast, as it may hap, and running the game down now on this side and now on that, so that it is really a most delightful sport and spectacle.

The Book of Marco Polo,
TRANS. SIR HENRY YULE, 1871

Some sixty or seventy years ago a new breed of dog was started in Southern Austria for the express purpose of tracking animals which had been wounded and would otherwise have been lost in wooded and mountainous country. It was bred by means of a cross between a dachshund (one of the original dogs of the world and incidentally one of the first crosses in the creation of a Sealyham) and a small Bavarian bloodhound. . . . Such was Hippy, and he came into my life at a bad moment of it and shortly after I had been appointed His Majesty's Minister to Yugoslavia.

[One] of his marked characteristics was his undemonstrativeness. He never barked to show pleasure. He did wag his tail, but never for long and always very composedly. He never licked or used his tongue to indicate affection. The most he would ever do to show that he was pleased or glad to see me again, or wanted anything from me was to put a cold nose against my hand. And if there were strangers or other dogs about he would always jump up on to my knees; but that was merely to show that I belonged to him. In fact his convinced conception of our relationship was not that he was mine but that I was his property. I sometimes wondered how much I counted in comparison with his own pleasures. Not that it worried me, for I would not have changed him in any respect. And whether he did or not put sport above his master mattered not at all, for if ever there was one man's dog it was Hippy. He himself would have put it the other way round, namely that Nevile Henderson was a one dog's man.

To the whole of the rest of the world he was utterly and completely indifferent, though always perfectly courteous and polite to everybody, unless he thought they were hostile to me. If he thought that they were that, then he gave them unmistakable warning that he was there to defend me. Once in the summer of 1933 a very bumptious and offensive young Nazi tried to come up to speak to me when I was sitting in the park at Bled. It was a very hot day and my two dogs were lying concealed under the branches of the beech tree beside which I was sitting. I watched the man coming up with a smirk on his face with great distaste and my feelings must have been communicated by telepathy to Hippy, for just as he was about to say something to me out sprang Hippy, growling fiercely, followed by the other dog. The German fled for his life. On another occasion I was travelling by train down to Bosnia with Dimnik, the King's chief jäger [huntsman]. The conductor was not satisfied about the free railway pass which the Yugoslav Government had given me and began to argue about it in a whisper to Dimnik. Hippy was lying on the seat just opposite Dimnik and, realising that I was the subject of these suspicious whisperings, he slowly got up and very deliberately snapped his teeth with a terrifying click just an inch from the conductor's hand. The warning was so obvious and yet so scrupulously tactful that it was incredibly amusing, and the conductor troubled us no more. No, I was

Hippy's property and he would suffer nobody else to interfere with or to worry me. . . .

The curious part was that though Hippy treated the rest of the world with such complete and utter indifference, everyone who came into contact with him was devoted to him. It was his immense personality which made him so irresistible to all from the highest to the lowest. The late King Alexander of Yugoslavia had never in his life cared for dogs and there was not one in the Palace, but he, like all the others, fell a victim to Hippy's charm. Hippy had a standing invitation to the Palace, whether I went there for an audience or to dine, or for any other purpose, and if I failed to bring him with me, there was trouble. He was allowed to sit with impunity on any of the royal chairs or on the royal knees in the motor car. So much so that one day when the King himself rang me up on the telephone, as he sometimes used to do, to suggest I should go for a drive with him, I said I would with pleasure, though I fully realised that it was Hippy and not me that His Majesty wanted to accompany him.

SIR NEVILE HENDERSON (1882–1942),
Hippy, The Story of a Dog, 1943

A Farm-yard Guardian. – It is related by Meyer that a male and female crane, obtained when a few days old, soon became so tame as to follow their master wherever he went, and answered to their respective names. In the farm-yard they interposed in all quarrels and punished the offenders. Bulls, cows, foals, &c., were all subject to their control, but they declined to interfere with the pigs. They at all times showed the most marked affection for each other, and when the female died from an injury the male screamed piteously and tried to raise her up. After the body was removed, and all efforts on the part of the male to find it failed, he left the farm-yard for two or three days, and was found in the neighbourhood in a disconsolate and dejected condition. Some time after, for want of a companion, he took up with the bull, which he accompanied wherever he went, keeping off the flies while the animal grazed. In the morning, if the bull did not appear in time the crane went to fetch him. He ultimately became as serviceable as a shepherd's dog, and would not permit a single animal to stray from the rest. When horses were put to any carriage the crane, ostler-like, placed himself in front, and by blows from his bill and outspread wings prevented them from moving till the driver was ready to start. His attachment to the cook of the family, who usually fed him, was great, and he made it a rule not to go to bed till she took him under her arm and conveyed him to his sleeping department. On the other hand, he was unforgiving when punished. One day, being found in a neighbouring garden in search of food, the owner gave him a blow with a stick. In return for this, the crane soon after took his station on a bridge over which this person had to pass, and attacked him so fiercely that he had to fly for safety to his house. Now the victor, he remained the determined foe of this person. On one point this crane was a coward: he never could endure the sight of any black moving object, such as a dog, cat, or crow, and his greatest dread was the chimney-sweeper.

The Parlour Menagerie, DEDICATED TO THE
BARONESS BURDETT-COUTTS (1814–1906)

Man loves the dog, but how much more ought he to love it, if he considered, in the inflexible harmony of the laws of nature, the sole exception, which is that love which succeeds in piercing, in order to draw closer to us, the partitions, every elsewhere impermeable, that separate species! We are alone, absolutely alone, on this chance planet: and, amid all the forms of life that surround us, not one, excepting the dog, has made an alliance with us. A few creatures fear us, most are unaware of us and not one loves us. . . . This animal, our good familiar dog, simple and unsurprising as may today appear to us what he has done, in thus perceptibly drawing nearer to a world in which he was not born and for which he was not destined, has nevertheless performed one of the most unusual and improbable acts that we can find in the general history of life. When was this recognition of man by beast, this extraordinary passage from darkness to light effected? Did we seek out the poodle, the sheep-dog or the greyhound from among the wolves and the jackals, or did he come spontaneously to us? We cannot tell. So far as our human annals stretch, he is at our side, as at present; but what are human annals in comparison with the times of which we have no witness? The fact remains that he is there in our houses, as ancient, as rightly placed, as perfectly adapted to our habits as though he had appeared on this earth, such as he is now, at the same time as ourselves. We have not to gain his confidence or his friendship: he is born our friend: while his eyes are still closed, already he believes in us: even before his birth, he has given himself to man. But the word 'friend' does not exactly depict his affectionate worship. He loves us and reveres us as though we had drawn him out of nothing. He is, before all, our creature full of gratitude and more devoted than the apple of our eye. He is our intimate and impassioned slave, whom nothing discourages, whom nothing repels, whose ardent trust and love nothing can impair. He has solved, in an admirable and touching manner, the terrifying problem which human wisdom would have to solve if a divine race came to occupy our globe. He has loyally, religiously, irrevocably recognized the superiority of man and has surrendered himself to him body and soul, with not an afterthought, with no idea of drawing back, reserving of his independence, his instinct and his character only the small part indispensable to the continuation of the life prescribed by nature to his species. With a certainty, an unconstraint and a simplicity that surprises us a little, deeming us better and more powerful than all that exists, he betrays, for our benefit, the whole of the animal kingdom to which he belongs, and, without scruple, denies his race, his kin, his mother and even his young.

MAURICE MAETERLINCK (1862–1949, *My Dog*,
1906, TRANS. A. TEIXEIRA DE MATTOS

'PETER SPOTS' – NEW YORK FIREMAN

This is how Joe, the driver of the engine, told me the story of Peter Spots. . . .

'The alarm came in from a station that was in our half of the territory. A fire that ought to have been ours easily, but the harness got "jammed," – would not come down on the horses, – then when we started the horses shied, and we came near killing our lieutenant, who was opening the doors. This got the engine crooked, so that we could not get through the doorway, and we had to back her before we could get out, and I tell you, everything went wrong. We only lost a few seconds by these mishaps, but it was enough to lose us the station.

'When we finally got out and were going up the avenue, I tried to make up for lost time by giving the horses all the rein I could, and giving them the whip once in a while, but Peter was so excited by this time at the delay, that he began jumping at the horses' chests and biting at them, and they balked so they wouldn't go at all. I suppose he meant well enough, and wanted them to go faster, but he only made matters worse; and when I got to the fire there was our rival company at work, – line stretched in – and making all kinds of mean remarks as we pulled at a hydrant. Even the Chief was there, and he gave our Captain an awful lecture – wanted to know "if we were all asleep down at our quarters"; and "if we thought we were going to a funeral, that we took so much time!" This almost broke the old man's heart, and I tell you I never felt so cheap in all my life as I did when I found how late we were.

'When we got back to quarters again we all got a lecture from the Captain, and then he took me aside and said:

' "Joe, I don't like to do it, but we must get rid of Peter. He's bothering the horses a good deal, and I cannot take any more chances like that to-day. If I lose any more fires, you know what will happen.". . . And so Peter was suspended from active duty.

'It happened that I knew the very person to turn him over to. There was a baker who delivered bread to some of the houses around here, and whose shop was quite a way from here, – about thirty or forty blocks, – and in a street we were not apt to go through. He had taken a great liking to Peter, and had offered to buy him several times, and, of course, we had always refused. Peter had also come to like the baker very much, for he brought Peter, every once in a while, an odd kind of bread that Peter was very fond of. So that night, at my supper-hour, I took Peter down to his bake-shop, and transferred the smartest dog in the fire department from an engine-house to a bakery – a big come-down, I tell you.

Peter Spots – fireman.

'At first we missed him a good deal; but in a large fire department you get so used to changes and transfers from one company to another that in time you get so you don't miss anything or anybody.

'We had almost forgotten about Peter, and got used to not having him around, when one day a "third alarm" came in that took us out; and in getting to the station I had to drive through the street the baker's place was on. I never thought of it myself, but, on my word, Peter hadn't forgotten *us*; and when we made our appearance he showed up pretty quick. The baker told me all about it afterward, and this was the way it happened: Peter was lying asleep beside the stove in the center of the bake-shop, when all of a sudden he pricked up one ear, and then jumped on his feet and gave a bark. The baker was making out some bills behind the counter, and thought nothing of it until the next moment Peter gave one jump, and was in the show-window among the pies and cakes and such like. The baker hollered to him to get out; but Peter began to claw at the window, and bark and howl. You see he could hear our whistle and bell and had recognized us. Then the baker made up his mind that the dog had gone mad, and got frightened and got up on a chair, and began to holler himself; and what with the baker and Peter, there was a high old time in that bake-shop for a while. Every time Peter gave a kick he knocked a pie or a plate full of cakes out of the window until he had it clear of everything. Then we hove in sight; and through the side of the show-window he saw us and recognized me in the seat, and that settled it – no bake-shop would hold him then. He jumped back in the store, braced himself plumb in front of the pane of glass in the door, and when we were just about opposite he gave one last howl, and crash! out he came through glass and all!

'I heard the racket, and turned my head just in time to see him come flying out. I understood it all in a moment, and expected to see him roll over dead in the gutter; but not much! He came through so quick he scarcely got a scratch; and away he went, down the street ahead of us, barking at every one, and clearing the way just as he used to, and running around in a circle and jumping high in the air and cutting up gymnastics – and happy? – well, I just guess he was happy! Even the Captain heard him in all the racket behind the engine, and let up on the whistle long enough to holler ahead to me to look out and not run over him; but there was small fear of that, for he beat us by half a block all the way to the fire.

'When we got there we "stretched in and stood fast," as we call it, which means we stretched in the hose, and got ready to go to work when so ordered; but they didn't need us, for the fire was pretty well out then, and the third alarm had only been sent out as a sort of precaution; so in a few moments the Chief ordered us back to quarters.

'When we were "picking up," or putting the hose back in the wagon, Peter was around among us like old times, and every one of the "gang" had a kind word for him. He was cut a bit about the back with glass, so the Captain says: "Throw him in the wagon, boys, and we'll take him back to the house, and mend him up. I'll put him on probation; and if he acts right he can stay with us as long as he wants." And then he adds: "But you fellows will have to chip in and pay for that pane of glass." And we all laughed; for we were willing to pay for a whole show-window to get Peter back again. . . .'

St Nicholas Magazine, AUGUST 1897

*Philip Thicknesse was an eighteenth-century English
gentleman who took his family on the Grand Tour in somewhat
colourful style*

Our travellers themselves occasioned no small degree of surprise to the inhabitants; the Governor, dressed after the English manner, was seated on the forepart of a *cabriolet*, drawn by one horse, with a servant before, who acted in the original character of a *footman*, with his hair *en queue*, a monkey clothed after the French manner, in jack boots, and a red jacket faced with silver, acting the part of a postillion; his *belle esprit* wife, with two daughters, seated within; guiltars, bass-viols, together with a parrot, placed in proper order, and an English dog instead of a groom behind!

SYLVANUS URBAN, 'MEMOIRS OF MR THICKNESSE',
FROM *The Gentleman's Magazine*, 1829

8

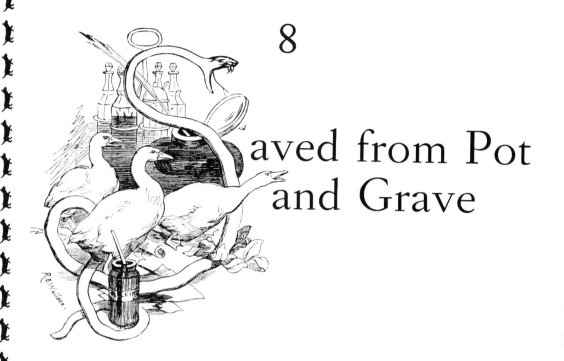

Saved from Pot and Grave

He heard bits chink as the horses shifted,
He heard hounds cast, then he heard hounds lifted,
But there came no cry from a new attack;
His heart grew steady, his breath came back.

JOHN MASEFIELD (1878–1967),
Reynard the Fox, 1919

In a voyage which I made to the East Indies with Captain Hamilton, I took a favourite pointer with me; he was, to use a common phrase, worth his weight in gold, for he never deceived me. One day when we were, by the best observations that we could make, at least 300 leagues from land, my dog pointed. I watched him for nearly an hour with astonishment, and mentioned the circumstance to the Captain, and to every officer on board, asserting that we must be near land, for my dog smelt game. This occasioned a general laugh; but that did not alter in the least the good opinion I had of my dog. After much conversation pro and con, I boldly told the Captain I placed more confidence in Tray's nose than I did in the eyes of every seaman on board, and therefore boldly proposed laying the sum I had agreed to pay for my passage (viz., 100 guineas) that we should find game within half an hour. The Captain (a good hearty fellow) laughed again, and desired Mr Crawford, the surgeon, who was prepared, to feel my pulse; he did so, and reported me in perfect health. The following dialogue between them took place; I overheard it, though spoken low and at some distance:

Captain.　　His brain is turned; I cannot with honour accept his wager.
Surgeon.　　I am of a different opinion; he is quite sane, and depends more upon the scent of his dog than he will upon the judgment of all the officers on board; he will certainly lose, and he richly merits it.
Captain.　　Such a wager cannot be fair on my side; however, I'll take him up, if I return his money afterwards.

During the above conversation, Tray continued in the same situation, and confirmed me still more in my former opinion. I proposed the wager a second time; it was then accepted.

Done! and done! were scarcely said on both sides when some sailors . . . harpooned an exceedingly large shark, which they brought on board and began to cut up for the purpose of barrelling the oil, when, behold, they found no less than six brace of live partridges in the creature's stomach.

They had been so long in that situation that one of the hens was sitting upon four eggs, and a fifth was hatching when the shark was opened! This young bird we brought up, by placing it with a litter of kittens that came into the world a few minutes before. The old cat was as fond of it as of any of her own four-legged progeny, and made herself very unhappy when it flew out of her reach until it returned again.

As to the other partridges, there were four hens amongst them; one or more were, during the voyage, constantly sitting, and consequently we had plenty of game at the Captain's table; and in gratitude to poor Tray (for being a means of winning 100 guineas) I ordered him the bones daily, and sometimes a whole bird.

RUDOLF ERICH RASPE (1737–94), *Original Travels and Surprising Adventures of Baron Munchausen,* 1785

Whilst serving on the cruiser HMS *Leander* in Malta during the summer of 1947, it was very hot, hence we had our awnings spread to keep the ship cool. I was petty officer then and lived on the POs' mess in the forward part of the ship and I was a boiler room watch keeper. I came off watch at midnight one night, went on deck for a breath of fresh air before I turned in, when I could hear a cat mewing. When I looked over the side, I could see our cat struggling at the waterline of the ship. I asked a couple of my mess mates to lower me over the side on a rope with a grapnel in the end, intending to be lowered just above the water, but because of my weight they could not hold me, so I had to swim for it. I managed to get the cat on my head, and then had to swim to the other end of the ship to the officers' gangway, and then explain to the officer of the watch how I came to be in the water with a cat on my head. Needless to say, I was still fully clothed, boots and all.

LETTER TO LUCINDA LAMBTON FROM A. D.
SAUNDERS, BEM, 9 FEBRUARY 1991

REYNARD THE FOX

... The fox swerved left and scrambled out,
Knocking crinked green shells from the brussels-sprout,
He scrambled out through the cobbler's paling,
And up Pill's orchard to Purton's Tailing,
Across the plough at the top of bent,
Through the heaped manure to kill his scent,
Over to Aldam's, up to Cappell's,
Past Nursery Lot with its whitewashed apples,
Past Colston's Broom, past Gaunt's, past Shere's,
Past Foxwhelps' Oasts with their hooded ears,
Past Monk's Ash Clerewell, past Beggars' Oak,
Past the great elms blue with the Hinton smoke ...

The pure clean air came sweet to his lungs,
Till he thought foul scorn of those crying tongues.
In a three mile more he would reach the haven
In the Wan Dyke croaked on by the raven.
In a three mile more he would make his berth
On the hard cool floor of a Wan Dyke earth,
Too deep for spade, too curved for terrier,
With the pride of the race to make rest the merrier.
In a three mile more he would reach his dream,
So his game heart gulped and he put on steam.

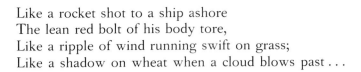

Like a rocket shot to a ship ashore
The lean red bolt of his body tore,
Like a ripple of wind running swift on grass;
Like a shadow on wheat when a cloud blows past . . .

On he went with a galloping rally
Past Maesbury Clump for Wan Brook Valley.
The blood in his veins went romping high,
'Get on, on, on to the earth or die.' . . .
He could make no sprint at a cry and cheer now,
He was past his perfect, his strength was failing,
His brush sag-sagged and his legs were ailing.
He felt, as he skirted Dead Men's Town,
That in one mile more they would have him down.

Through the withered oak's wind-crouching tops
He saw men's scarlet above the copse,
He heard men's oaths, yet he felt hounds slacken,
In the frondless stalks of the brittle bracken.
He felt that the unseen link which bound
His spine to the nose of the leading hound
Was snapped, that the hounds no longer knew
Which way to follow nor what to do;
That the threat of the hound's teeth left his neck,
They had ceased to run, they had come to check.
They were quartering wide on the Wan Hill's bent.

The terrier's chase had killed his scent.

He heard bits chink as the horsed shifted,
He heard hounds cast, then he heard hounds lifted,
But there came no cry from a new attack;
His heart grew steady, his breath came back.

He left the spinney and ran its edge
By the deep dry ditch of the blackthorn hedge;
Then out of the ditch and down the meadow,
Trotting at ease in the blackthorn shadow,
Over the track called Godsdown Road,
To the great grass heave of the gods' abode.
He was moving now upon land he knew:
Up Clench Royal and Morton Tew,
The Pol Brook, Cheddesdon, and East Stoke Church,
High Clench St Lawrence and Tinker's Birch.
Land he had roved on night by night,

For hot blood-suckage or furry bite.
The threat of the hounds behind was gone;
He breathed deep pleasure and trotted on.
JOHN MASEFIELD (1878–1967),
Reynard the Fox, 1919

By the middle of the eighteenth century equestrian exhibitions had become the rage in England. The wedding of the horse and circus took place on an open field in Lambeth in 1768. There Philip Astley ... featured Billy, 'the Little Military Learned Horse, three feet high from the deserts of Arabia.'

The fate of Billy is the stuff of which heartrending novels used to be made. William Davis took care of the horse after Astley's death, and in an act of kindness loaned him to Abraham Saunders, a talented but unfortunate trick rider and entrepreneur. Shortly thereafter Saunders' show was lost through debt and the little horse mistakenly auctioned with the rest of his stud. Billy was bought by a tradesman, who, though he called the horse 'Mountebank' because of his odd prancing, was ignorant of Billy's higher education. For three years the horse pulled the tradesman's cart until he was spotted by one of Astley's riders.

Thinking the horse might be Billy, the rider clicked his fingernails, the cue for the horse to tap his foreleg in a counting exhibition. When the horse perked up and began to count, he was purchased and taken home. 'Even in his old age he would ungirt his own saddle, wash his feet in a pail of water, fetch and carry a complete tea equipage, take a kettle of boiling water off the fire, and act like a waiter at a tea garden.'

Eventually losing his teeth and unable to eat corn, Billy was fed on soaked bread at considerable expense. Late in life he was still called upon to give an occasional performance. When he died at the age of forty-two his hide was fashioned into a special-effects thunderdrum used for many years in the amphitheater: a curious but laudatory gesture.
RICKY JAY, *Learned Pigs and*
Fireproof Women, 1986

U is for the Unnamed Soldier in the time of Charles II who constructed a wooden leg for a crane with such precision that the bird, according to John Evelyn, could use it as if it had been his own.
NEVILLE BRAYBROOKE, 'AN ANIMAL ALPHABET',
The Tablet, 21/28 DECEMBER 1991

I now subjoin a letter received from a lady who has always paid great attention to the experimental features of natural history, and delights in choosing for her pets the very creatures that would be thought most unfit for

such a purpose. I first heard the story of the two butterflies some years ago, but have asked her to relate the account in her own words, knowing that a narrative always gains spirit when from the pen of an eye-witness.

'Among the many pets that I have loved and lost, few have endeared themselves more to me than my butterflies, two of which I once kept for the space of a year and a half.

'They came into my possession when in their chrysalis state, and I, not knowing anything of entomology, shut them up for safety in a cabinet having glass doors. The cabinet stood near a small window in my bedroom. I was very unwell that winter, and therefore a fire was kept up in my room night and day. Therefore the room was very warm, and I suppose that the little butterflies were deceived thereby, and thought or dreamed that summer smiled upon the earth; for, a few days after Christmas, to my astonishment and delight, a little yellow butterfly was seen fluttering feebly within the cabinet. . . .

'I now became most anxious to feed the little thing; but how this was to be achieved I had not the slightest idea, nor could anyone in the house advise or help me in this important matter. Moreover, I was loudly ridiculed for the bare idea of trying to tame and feed butterflies.

'However, I remembered that the poets all agreed in saying that butterflies sipped nectar from the opening flowers, and therefore turned my attention to the manufacture of a substitute for nectar; so having obtained some honey, which I diluted with rose-water, I put one drop into the centre of the open blossoms of a fairy rose, and placed the little plant in the cabinet. I soon had the joy of seeing the little thing flutter around the rose, and finally settle upon it.

'Whether it really drank or not I cannot say. I thought that it must have done so, as it appeared to grow stronger and more lively every day. I fed it in this manner for a fortnight; and by the end of that time it became so tame that it would step off the flowers or anything else on which it might be standing, and appear quite happy and at rest upon my hand.

'It also appeared to understand that I wished it to come to me when I called it by the name of "Psyche," that being the name which I had given to the insect.

'About three weeks after the advent of Psyche, we were gladdened by the addition of another butterfly to our establishment – a peacock. He was strong and vigorous from the first, and flitted swiftly about, like a gleam of prismatic light. I used to fancy that they talked to each other, as he at once fell into the ways and habits of the other; and when I called Psyche, he too would come. I gave him another name; but he never seemed to understand that it belonged to him.

'They lived in this way until the earth had donned her glowing summer robe of lilies and roses, when I was told that their life-power could only extend over a month or two, and that it was cruel even to keep them as happy prisoners. I was therefore induced to give them their liberty. The

'The threads of insect life are seriously apportioned.'

cabinet was placed with opened doors before the window.

'It was many days before the butterflies ventured to leave the window-sill, and this much to my joy, for I thought that it might be affection for me that held them back. However, one day, with many bitter tears, I saw them depart and join some wild companions; but at night we found them again in the cabinet.

'On the following morning they left us, and came not back again until the cold and stormy September weather set in.

'Yet, when in the garden, they would come if I called them, and rest for a short time on my hair or hands. At length, on a cold windy day in September, we saw them on the window-sill, and on our opening the window they came in and resumed possession of their old quarters, and abode there for the winter.

'It is true they were but poor-looking objects to what they were when they went forth. The world seemed to have used them somewhat roughly, for the sheen had gone from the rich wings of the peacock butterfly, and the soft yellow bloom from Psyche's plumage. Nevertheless, they were welcome guests; and though ragged and wayworn, were not the less loved.

'We observed that during this winter they slept more than they did formerly. They also manifested pleasure when sung or talked to, and were very fond of being waved about and danced up and down in the air, while they would sit upon the hand quite calmly. I think that the movement must have reminded them of the nodding flowers and fresh breezes of their summer life.

'The sun and earth ran their appointed course, until they brought us to another bright June, and again I bestowed the boon of freedom on our fairy pets, who went forth gaily; but, alas! never to return. One day, after a heavy thunderstorm, we found the inanimate form of a yellow butterfly upon the window-sill. I took it up lovingly, and did my best to revive it; for I believed it to be the material form of my own beautiful Psyche, who had sought refuge from the storm, but found the window closed. Of this I cannot be sure, for all our efforts to restore her were in vain. The wondrous essence that had given it life, beauty, motion, affection, and memory, had returned to the hand of its mighty Creator, and with Him let it rest.

'The peacock butterfly never returned; and whether he fell a prey to that aerial shark, the dragon-fly, or died of age, sickness, or forgot his early friends, I know not.

'I have since tried to tame other butterflies, but never was so successful, although I have taught three or four to know me and to come at my call.'

REV. J. G. WOOD (1827–89),
Petland Revisited, 1890

Latimer Springfield was a rather cheerless, oldish young man, who went into politics somewhat in the spirit in which other people might go into half mourning. Without being an enthusiast, however, he was a fairly strenuous plodder, and Mrs Durmot had been reasonably near the mark in asserting that he was working at high pressure over this election. . . .

'I know he's going to sit up half the night working up points for his final speeches,' said Mrs Durmot regretfully; 'however, we've kept politics at arm's length all the afternoon and evening. More than that we cannot do.'

'That remains to be seen,' said Vera, but she said it to herself.

Latimer had scarcely shut his bedroom door before he was immersed in a sheaf of notes and pamphlets, while a fountain-pen and pocket-book were brought into play for the due marshalling of useful facts and discreet fictions. He had been at work for perhaps thirty-five minutes, and the house was seemingly consecrated to the healthy slumber of country life, when a stifled squealing and scuffling in the passage was followed by a loud tap at his door. Before he had time to answer, a much-encumbered Vera burst into the room with the question: 'I say, can I leave these here?'

'These' were a small black pig and a lusty specimen of black-red gamecock.

Latimer was moderately fond of animals . . . but he was pardonably unwilling to share even a commodious bedroom with samples of henroost and sty products.

'Wouldn't they be happier somewhere outside?' he asked, tactfully expressing his own preference in the matter in an apparent solicitude for theirs.

'There is no outside,' said Vera impressively, 'nothing but a waste of dark, swirling waters. The reservoir at Brinkley has burst.'

'I didn't know there was a reservoir at Brinkley,' said Latimer.

'Well, there isn't now, it's jolly well all over the place, and as we stand particularly low we're the centre of an inland sea just at present. You see the river has overflowed its banks as well.'

'Good gracious! Have any lives been lost?'

'Heaps, I should say. The second housemaid has already identified three bodies that have floated past the billiard-room window as being the young man she's engaged to. Either she's engaged to a large assortment of the population round here or else she's very careless at identification. Of course it may be the same body coming round again and again in a swirl; I hadn't thought of that.'

'But we ought to go out and do rescue work, oughtn't we?' said Latimer, with the instinct of a Parliamentary candidate for getting into the local limelight.

'We can't,' said Vera decidedly, 'we haven't any boats and we're cut off by a raging torrent from any human habitation. My aunt particularly hoped you would keep to your room and not add to the confusion, but she thought it would be so kind of you if you would take in Hartlepool's Wonder, the gamecock, you know, for the night. You see, there are eight other gamecocks, and they fight like furies if they get together, so we're putting one in each bedroom. The fowl-houses are all flooded out, you know. And then I thought perhaps you wouldn't mind taking in this wee piggie; he's rather a little love, but he has a vile temper. He gets that from his mother – not that I like to say things against her when she's lying dead and drowned in her sty, poor thing. What he really wants is a man's firm hand to keep him in order. I'd try and grapple with him myself, only I've got my chow in my room, you know, and he goes for pigs wherever he finds them.'

'Couldn't the pig go in the bathroom?' asked Latimer faintly, wishing that he had taken up as determined a stand on the subject of bedroom swine as the chow had.

'The bathroom?' Vera laughed shrilly. 'It'll be full of Boy Scouts till morning if the hot water holds out.'

'Boy Scouts?'

'Yes, thirty of them came to rescue us while the water was only waist-high; then it rose another three feet or so and we had to rescue them. We're giving them hot baths in batches and drying their clothes in the hot-air cupboard, but, of course, drenched clothes don't dry in a minute, and the corridor and staircase are beginning to look like a bit of coast scenery by Tuke. Two of the boys are wearing your Melton overcoat; I hope you don't mind.'

'It's a new overcoat,' said Latimer, with every indication of minding dreadfully.

'You'll take every care of Hartlepool's Wonder, won't you?' said Vera. 'His mother took three firsts at Birmingham, and he was second in the cockerel class last year at Gloucester. He'll probably roost on the rail at the bottom of your bed. I wonder if he'd feel more at home if some of his wives were up here with him? The hens are all in the pantry, and I think I could pick out Hartlepool Helen; she's his favourite.'

Latimer showed a belated firmness on the subject of Hartlepool Helen, and Vera withdrew without pressing the point, having first settled the gamecock on his extemporized perch and taken an affectionate farewell of the pigling. Latimer undressed and got into bed with all due speed, judging that the pig would abate its inquisitorial restlessness once the light was turned out. As a substitute for a cozy, straw-bedded sty the room offered, at first inspection, few attractions, but the disconsolate animal suddenly discovered an appliance in which the most luxuriously contrived piggeries were notably deficient. The sharp edge of the underneath part of the bed was pitched at exactly the right elevation to permit the pigling to scrape himself ecstatically backwards and forwards, with an artistic humping of the back at the crucial moment and an accompanying gurgle of long-drawn delight. The gamecock, who may have fancied that he was being rocked in the branches of a pine-tree, bore the motion with greater fortitude than Latimer was able to command. A series of slaps directed at the pig's body were accepted more as an additional and pleasing irritant than as a criticism of conduct or a hint to desist; evidently something more than a man's firm hand was needed to deal with the case. Latimer slipped out of bed in search of a weapon of dissuasion. There was sufficient light in the room to enable the pig to detect this manœuvre, and the vile temper, inherited from the drowned mother, found full play. Latimer bounded back into bed, and his conqueror, after a few threatening snorts and champings of its jaws, resumed its massive operations with renewed zeal. . . .

Towards dawn the pigling fell into a happy slumber, and Latimer might have followed its example, but at about the same time Stupor Hartlepooli gave a rousing crow, clattered down to the floor and forthwith commenced a spirited combat with his reflection in the wardrobe mirror. Remembering that the bird was more or less under his care Latimer performed Hague Tribunal offices by draping a bath-towel over the provocative mirror, but the ensuing peace was local and short-lived. The deflected energies of the gamecock found new outlet in a sudden and sustained attack on the sleeping and temporarily inoffensive pigling, and the duel which followed was desperate and embittered beyond any possibility of effective intervention. The feathered combatant had the advantage of being able, when hard pressed, to take refuge on the bed, and freely availed himself of this circumstance; the pigling never quite succeeded in hurling himself on to the same eminence, but it was not from want of trying.

Neither side could claim any decisive success, and the struggle had been practically fought to a standstill by the time that the maid appeared with the early morning tea.

'Lor, sir,' she exclaimed in undisguised astonishment, 'do you want those animals in your room?'

Want!

The pigling, as though aware that it might have outstayed its welcome, dashed out at the door, and the gamecock followed it at a more dignified pace.

'If Miss Vera's dog sees that pig –!' exclaimed the maid, and hurried off to avert such a catastrophe.

A cold suspicion was stealing over Latimer's mind; he went to the window and drew up the blind. A light, drizzling rain was falling, but there was not the faintest trace of any inundation.

Some half-hour later he met Vera on the way to the breakfast-room.

'I should not like to think of you as a deliberate liar,' he observed coldly, 'but one occasionally has to do things one does not like.'

'At any rate I kept your mind from dwelling on politics all the night,' said Vera.

Which was, of course, perfectly true.

'The Lull', SAKI (1870–1916)

Anyone who wishes to appreciate the manner in which a ship's company will treat an animal, should read Captain Basil Hall's account of the pig Jean and her treatment by the men. It is positively affecting, as well as extremely ludicrous, to read the account of that pig, and to see how the sailors took a fancy to her; how they actually begged her life when she was destined for the butcher's knife; how they fed her until she was so

'It is positively affecting to see how the sailors took a fancy to her.' Trotter, mascot of *HMS* Glasgow, *1909*

fat that she could not stand, much less walk; how they attended to all her wants, and actually put food into her mouth as she lay, huge and helpless, upon deck, a very mountain of a sow, with all her feet pointing to the sky through sheer exuberance of fat.

And then, after she had died the death naturally consequent on a life of such indulgence, they were as careful in disposing of her dead body and guarding it from being desecrated by Chinese cooking-pots, as they had been to pamper their mountainous pet into such a mass of obesity; rigging out a complicated apparatus of rods and ballast iron, which drove the huge body so deeply into the mud, that even the persevering Chinese could not recover it.

REV. J. G. WOOD (1827–89), *Petland Revisited*, 1890

THE RETIRED CAT

A poet's cat, sedate and grave
As poet well could wish to have,
Was much addicted to inquire
For nooks to which she might retire,
And where, secure as mouse in chink,
She might repose, or sit and think.
I know not where she caught the trick,
Nature perhaps herself had cast her
In such a mould philosophique,
Or else she learnt it of her master.
Sometimes ascending debonnair,
An apple tree, or lofty pear,
Lodged with convenience in the fork,
She watched the gardener at his work;
Sometimes her ease and solace sought
In an old empty watering-pot,
There, wanting nothing, save a fan,
To seem some nymph in her sedan
Apparelled in exactest sort,
And ready to be borne to court.
But love of change it seems has place
Not only in our wiser race:
Cats also feel, as well as we,
That passion's force, and so did she.
Her climbing, she began to find,
Exposed her too much to the wind,
And the old utensil of tin
Was cold and comfortless within;
She therefore wished instead of those

Some place of more serene repose,
Where neither cold might come, nor air
Too rudely wanton with her hair,
And sought it in the likeliest mode
Within her master's snug abode.
A drawer, it chanced, at bottom lined
With linen of the softest kind,
With such as merchants introduce
From India, for the ladies' use,
A drawer impending o'er the rest,
Half open in the topmost chest,
Of depth enough and none to spare,
Invited her to slumber there;
Puss with delight beyond expression
Surveyed the scene and took possession.
Recumbent at her ease ere long,
And lulled by her own humdrum song,
She left the cares of life behind,
And slept as she would sleep her last,
When in came, housewifely inclined,
The chambermaid, and shut it fast.
By no malignity impelled,
But all unconscious whom it held.
Awakened by the shock, cried Puss,
'Was ever cat attended thus!
The open drawer was left, I see,
Merely to prove a nest for me,
For soon as I was well composed,
Then came the maid and it was closed.
How smooth these kerchiefs and how sweet!
Oh what a delicate retreat!
I will resign myself to rest
Till Sol declining in the west,
Shall call to supper, when, no doubt,
Susan will come and let me out'.
The evening came, the sun descended,
And Puss remained still unattended.
The night rolled tardily away,
(With her indeed 'twas never day);
The sprightly morn her course renewed,
The evening gray again ensued,
And Puss came into mind no more
Than if entombed the day before.
With hunger pinched, and pinched for room,
She now presaged approaching doom,

Nor slept a single wink, or purred,
Conscious of jeopardy incurred.
That night, by chance, the poet watching,
Heard an inexplicable scratching;
His noble heart went pit-a-pat,
And to himself he said, 'What's that?'
He drew the curtain at his side,
And forth he peeped, but nothing spied.
Yet, by his ear directed, guessed
Something imprisoned in the chest,
And, doubtful what, with prudent care
Resolved it should continue there.
At length, a voice which well he knew,
A long and melancholy mew,
Saluting his poetic ears,
Consoled him, and dispelled his fears:
He left his bed, he trod the floor,
He 'gan in haste the drawers explore,
The lowest first, and without stop
The rest in order to the top.
For 'tis a truth well known to most,
That whatsoever thing is lost,
We seek it, ere it come to light,
In every cranny but the right.
Forth skipped the cat, not now replete
As erst with airy self-conceit,
Nor in her own fond apprehension
A theme for all the world's attention,
But modest, sober, cured of all
Her notions hyperbolical,
And wishing for a place of rest
Anything rather than a chest.
Then stepped the poet into bed
With this reflection in his head:

MORAL

Beware of too sublime a sense
Of your own worth and consequence.
The man who dreams himself so great,
And his importance of such weight,
That all around in all that's done,
Must move and act for him alone,
Will learn in school of tribulation
The folly of his expectation.

WILLIAM COWPER (1731–1800)

9

Strange as Can Be

Mr Hirst did not approve of horses excepting on the racecourse, and he went shooting mounted on the back of a bull of ample proportions and uncertain temper, whilst for pointers, he made use of the services of a crowd of vivacious and sagacious pigs.

EDITH SITWELL (1887–1964), *English Eccentrics*, 1933

The desire for equitation seems to be naturally implanted in the monkey mind. Not long ago, a gentleman who rather prided himself on a very fine stud of hunters, found that the horses did not appear properly refreshed by their nightly rest. One of the grooms, on being desired to keep a strict watch, discovered that a tame monkey belonging to the house was accustomed to ride on the horses' backs almost all night, preventing them from taking sufficient rest. His master, on discovering his penchant for riding, and being averse to killing the monkey on account of his horsemanship, succeeded in curing him effectually of his love for horses. The next time that the hounds met, he had the monkey put into a full hunting suit and secured by a strap to the saddle of his most spirited hunter, and took him away to the meet. When the fox was found the horse pricked up his ears at the well-known sound, and started off at once. The chase happened to be a particularly long and severe one, the monkey of course from his light weight being far ahead of the legitimate huntsmen. A countryman, who was coming from the direction which the fox had taken, was interrogated by some of the sportsmen who had been thrown out as to the position of the hunt, and told them that the fox was looking tired, but that none of the huntsmen was near, except a little gentleman in a yellow jacket, who took his leaps beautifully. Sure enough, Master Jacko was in at the death, but did not by any means appreciate the honour. After the fox had been killed, there was a long ride home again, by the end of which time the monkey seemed thoroughly wearied out. After the experience that he had of a day's hunting he was never known to mount a horse again.

REV. J. G. WOOD (1827–89), *Sketches and Anecdotes of Animal Life*, 1854

Animaux Célebres.

THE CRABS OF CAPTAIN DRAKE

The crab, a seafish in a shell, is an amphibious creature. It exists in every size: the large are carniverous and very dangerous. They live particularly in the Island of Cancres, in America. They have an horrible aspect and an astonishing strength. One reads, in the history of Captain Drake, of a French seaman who was eaten by the crabs when he was studying the islands of which we speak. Though he was well armed, though he defended himself with great courage, he nonetheless became the prey of this monster. Captain Marrion shared his fate: the moment he disembarked his vessel and set foot upon the shore, a crab of a dreadful size came suddenly out of the sea, threw itself at the captain, cut his body in two with his pincers, and ate him; there was never the least chance of him saving himself.

In 1627, when French colonists were disembarked one evening on the island of Saint Christophe, thirty ill people who were spending the night ashore were surprised and eaten by these animals, so that one found heaps of crabs as high as houses on the body of each victim.

THE GRATEFUL SWAN

There was a woman in Tarento who lost her husband, and suffered such a lively grief that she retired to weep by his grave. There she caught sight of a swan which had fallen to the ground and broken its leg. The woman took care of the swan, and nursed it back to health.

A year later the swan returned to the place, and showed, by the batting of its wings and its cries, its great joy at seeing the widow, which gave her much surprise.

The story goes that the swan brought a precious stone which allowed the widow to live out her days in an honourable comfort.

THE GOOD COW

In the month of Pluvoise in year 8, many people of the country round Auxonne were attacked by a famished she-wolf; one young girl had already perished in her murderous jaws. A young herdsman of about fourteen, whom she also intended for her victim, was saved in a very

extraordinary manner. This little boy, whose name was Fourcault, was watching a herd of cows in the fields of Villiers-les-Pots, a canton of Auxonne. It is known that these animals, moved by a communal sense of danger when the sight of a wolf threatens, come together and line up in a kind of circular phalanx, in such a way that they present to their enemy the defences nature gave them on their brows, hiding their un-defended and vulnerable flanks. The cows which Fourcault was watching were prompt to apply their instinctive tactic the instant they caught sight of the she-wolf; but it was not them the wolf was after; it was the young herdsman she was bearing down on, it was he her jaws menaced, it was the child she seized and shook violently as though to tear him in pieces. At this sight, one of the cows swiftly detached herself from the phalanx, ran at the wolf and by her attack succeeded in making her let go of her prey. The child took advantage of the battle which ensued between his enemy and his liberator to try to get himself to safety, but the she-wolf saw him, abandoned the cow and went to jump once again on Fourcault, whom she seized and shook like the first time. The cow hurried again in defence of the child, and so bothered the wolf that she was forced for the second time to release her prey. Villagers from Villiers-les-Pots arrived at this moment to finish the work of that courageous and well-meaning cow. They put the wolf to flight, for whom death soon waited in the forest of Long-Champ. The young Fourcault retired on account of his wounds, from which he was soon perfectly cured.

THE BALD MICE OF THE CARRIBBEAN

The bald mice of the Carribbean islands are fearsome. It is said that they choose one man in a hundred whom they bite once, that they may bite him again in the same place. It is also the case that the Carribbeans are powerfully afraid of them and honour them singularly. Though they dread them, yet they consider them as good angels who watch over their

cabins through the night. Any men who should kill them are considered to have committed a sacrilege.

A. ANTOINE, *Histoire des Animaux Célèbres*, 1813,
TRANS. KATE HARRIS 1992

A cat belonging to Mr Smith, the respectable bailiff and agent of the Earl of Lucan, at Lateham, is in the constant habit of taking her place on the rug before the parlour fire. She had been deprived of all her litter of kittens but one, and her milk probably incommoded her. I mention this in order to account in some degree for the following circumstance. One evening, as the family were seated round the fire, they observed a mouse make its way from the cupboard, which was near the fireplace, and lay itself down on the stomach of the cat, as a kitten would do when she is going to suck. Surprised at what they saw, and afraid of disturbing the mouse, which appeared to be full-grown, they did not immediately ascertain whether it was in the act of sucking or not. After remaining with the cat a considerable length of time, it returned to the cupboard. These visits were repeated on several other occasions, and were witnessed by many persons. The cat not only appeared to accept the mouse, but uttered that sort of greeting purr which the animal is so well known to make use of when she is visited by her kitten. The mouse had every appearance of being in the act of sucking the cat, but such was its vigilance, that it retreated as soon as a hand was put out to take it up. When the cat, after being absent, returned to the room, her greeting call was made, and the mouse came to her. The attachment which existed between these two incongruous animals could not be mistaken, and it lasted some time. The fate of the mouse, like that of most pets, was a melancholy one. During the absence of its nurse, a strange cat came into the room. The poor mouse, mistaking her for its old friend and protectress, ran out to meet her and was immediately seized and slain before it could be rescued from her clutches. The grief of the foster-mother was extreme. On returning to the parlour, she made her usual call; but no

mouse came to meet her. She was restless and uneasy, went mewing about the house, and showed her distress in the most marked manner. What rendered the anecdote I have been relating the more extraordinary, is the fact of the cat being an excellent mouser, and that during the time she was showing so much fondness for the mouse, she was preying upon others with the utmost avidity.

REV. T. JACKSON, *Our Dumb Companions*, c. 1863

An old lady had an Alderney cow, which she looked upon as a daughter. You could not pay the short quarter-of-an-hour call without being told of the wonderful milk or wonderful intelligence of this animal. The whole town knew and kindly regarded Miss Betsy Barker's Alderney; therefore great was the sympathy and regret when, in an unguarded moment, the poor cow tumbled into a lime-pit. She moaned so loudly that she was soon heard and rescued; but meanwhile the poor beast had lost most of her hair, and came out looking naked, cold, and miserable, in a bare skin. Everybody pitied the animal, though a few could not restrain their smiles at her droll appearance. Miss Betsy Barker absolutely cried with sorrow and dismay; and it was said she thought of trying a bath of oil. This remedy, perhaps, was recommended by some one of the number whose advice she asked; but the proposal, if ever it was made, was knocked on the head by Captain Brown's decided 'Get her a flannel waistcoat and flannel drawers, ma'am, if you wish to keep her alive. But my advice is, kill the poor creature at once.'

Miss Betsy Barker dried her eyes, and thanked the Captain heartily; she set to work, and by and by all the town turned out to see the Alderney meekly going to her pasture, clad in dark gray flannel. I have watched her myself many a time. Do you ever see cows dressed in gray flannel in London?

MRS GASKELL (1810–65), *Cranford*, 1891

I was greatly interested some months ago, when I received a letter from a young lady living at Brighton, announcing her possession of a veritable singing mouse. . . .

'In October, 1882, I was staying with a sister at Chiswick, in a villa adjoining market-gardens. . . . One morning my sister told me the maid had come running in the previous evening to call her to the scullery to listen to a mouse which sang like a bird. . . .

'Between 6 and 7 p.m., I was sent for, and, on entering the larder, heard a continuous twittering, not at all monotonous, but variably ascending and descending in short scales. I carefully moved out some pots, and came to a small piece of board, on which was a black spot, that proved to be the songster. Cautiously putting out my hand, I closed on it, and grasped air; it had vanished down a hole. . . . I then gently replaced everything, and set my trap baited with cheese. In a very short time we caught four or five mice, large and small, one so tiny and black that I thought it might be the singer, and kept it. I had to retire early, but begged that the trap should be examined the last thing.

'Later my sister came rushing up with the trap. The singer was in it making a great noise in the shrillest key. I heard her all that night without intermission; she was especially loud about 4 a.m. After that she became quiet. . . .

'When excited in carrying paper to and fro for a new bed, or in burrowing down a pot of mould, she always sang, on very high notes, a sort of undulating cadence impossible to describe. I think the swallow's song is the nearest approach to it I can imagine.

'As for its compass, she had rather low notes, about those of a deep contralto, which she used mostly when eating and perfectly undisturbed. She also uttered them during the day if she heard anyone enter the room where the cage hung. Her high notes I have no power of limiting, for they became so acute that my ear could scarcely distinguish them. Often when whirling her roundabout, I have heard her gradually getting higher and higher, then a short interval of silence, then she appeared to catch up the note on which I lost her ascent, and descend again, so that I felt sure she must have been singing all the time. . . .

'Some nights she would sing without intermission whilst running about the table, over the edge, and down the cloth, with surprising quickness, then up again into her cage before you could exclaim.'

REV. J. G. WOOD (1827–89),
Petland Revisited, 1890

We had in this village more than twenty years ago an idiot-boy, whom I well remember, who, from a child, showed a strong propensity to bees; they were his food, his amusement, his sole object. And as people of this cast have seldom more than one point in view, so this lad exerted all his few faculties on this one pursuit. In the winter he dosed away his time, within his father's house, by the fire-side, in a kind of torpid state, seldom departing from the chimney-corner; but in the summer he was all alert, and in quest of his game in the fields, and on sunny banks. Honey-bees, humble-bees, and wasps, were his prey wherever he found them: he had no apprehensions from their stings, but would seize them *nudis manibus*, and at once disarm them of their weapons, and suck their bodies for the sake of their honey-bags. Sometimes he would fill his bosom between his shirt and his skin with a number of these captives; and sometimes would confine them in bottles. He was a very *merops apiaster*, or bee-bird; and very injurious to men that kept bees; for he would slide into their bee-gardens, and, sitting down before the stools, would rap with his finger on the hives, and so take the bees as they came out. He had been known to overturn hives for the sake of honey, of which he was passionately fond. Where metheglin was making he would linger round the tubs and vessels, begging a draught of what he called bee-wine. As he ran about he used to make a humming noise with his lips, resembling the buzzing of bees. This lad was lean and sallow, and of a cadaverous complexion; and, except in his favourite pursuit, in which he was wonderfully adroit, discovered no manner of understanding. Had his capacity been better, and directed to the same object, he had perhaps abated much of our wonder at the feats of a more modern exhibitor of bees; and we may justly say of him now,

> '... Thou,
> Had thy presiding star propitious shone,
> Should'st Wildman be....'

When a tall youth he was removed from hence to a distant village, where he died, as I understand, before he arrived at manhood.

GILBERT WHITE (1720–93), *The Natural History and Antiquities of Selborne*, 1789

The rat conducting his blind companion.

The Rev. Mr Ferryman, walking out in some meadows one evening, observed a great number of rats in the act of migrating from one place to another, which it is known, they are in the habit of doing occasionally.

He stood perfectly still, and the whole assemblage passed close to him. His astonishment, however, was great, when he saw an old blind rat, which held a piece of stick at one end in its mouth, while another rat had hold of the other end of it, and thus conducted his blind companion.

MRS R. LEE, *Anecdotes of the Habits*
and Instincts of Animals, 1852

They gave us for dinner boiled ant-bear and red monkey, two dishes unknown even at Beauvilliers in Paris or a London City feast. The monkey was very good indeed, but the ant-bear had been kept beyond its time: it stunk as our venison does in England; and after tasting it, I preferred dining entirely on monkey.

CHARLES WATERTON (1782–1865),
Wanderings in South America, 1825

All was faery and yet simple among the fauna of my early home. You could never believe that a cat could eat strawberries? And yet because I have seen him so many times, I know that Babou, that black Satan, interminable and as sinuous as an eel, would carefully select in Madame Pomie's kitchen garden the ripest of the Royal Sovereigns or the Early Scarlets. He it was, too, who would be discovered poetically absorbed in smelling newly-opened violets.

Have you ever heard tell of Pelisson's spider that so passionately loved music? I for one am ready to believe it and also to add, as my slender contribution to the sum of human knowledge, the story of the spider that my mother kept – as my father expressed it – on her ceiling, in that year that ushered in my sixteenth spring. A handsome garden spider she was, her belly like a clove of garlic emblazoned with an ornate cross. In the daytime she slept, or hunted in the web that she had spun across the bedroom ceiling. But during the night, towards three o'clock in the morning, at the moment when her chronic insomnia caused my mother to relight the lamp and open her bedside book, the great spider would also wake, and after a careful survey would lower herself from the ceiling by a thread, directly above the little oil lamp upon which a bowl of chocolate simmered through the night. Slowly she would descend, swinging limply to and fro like a big bead, and grasping the edge of the cup with all her eight legs, she would bend over head foremost and drink to satiety. Then she would draw herself ceilingwards again, heavy with creamy chocolate, her ascent punctuated by the pauses and meditations imposed by an overloaded stomach, and would resume her post in the centre of her silken rigging.

COLETTE (1873–1954),
In My Mother's House, 1923

Not only is the rat fastidious in its choice of food, but it affords a dainty food itself, in connection with which fact I have had some amusing experiences.

Some years ago, I mentioned incidentally in the course of a lecture that the Chinese who eat the rat habitually, and the Parisians who did so under compulsion [during the siege of 1870], suffered no real hardship, as the flesh of the rat is not only nutritious, but really excellent and delicate food, far surpassing that of the rabbit.

Being pressed by the audience for further explanation, I told them that I spoke from practical experience, and that cold rat-pie (of course made from barn-rats) was a delicacy worthy of any epicure's table.

Reports of the lecture – mostly exaggerated – were published in the leading newspapers, and copied into nearly every journal in the land. An avalanche of correspondence poured on me, and I was greatly amused with the different views of the writers.

A few had summoned up courage to try the experiment, and were unanimous in their approbation. Many asked for details in the manufacture of the pie which would have taken up a whole cookery book if answered. Many more asked if I could kindly send them a ready-made pie, so that they might judge for themselves. Each pie, by the way, supposing it to contain four rats, cost at least 2s. 6d.

But the oddest view was that which was taken by many writers.

They first assumed that eating rats was an offence against morality, and then argued that as they knew that I would not offend against morality, I never had eaten a rat, and therefore was deliberately hoaxing the public. As the reader may not believe that such extraordinary beings could exist, I give, *verbatim*, one of the letters, which I found a few weeks ago, while sorting a mass of correspondence.

'April 21st, 1879

'Respected Sir. – I had occasion on the 13th of this month to look in our Weekly paper and in doing so I noticed your *lecture* on *rat pie* I do not know whether you talk for publicity or what but I can never believe for one moment that you have ever ate Rat pie I noticed in your address of the 20th ultimo about hundreds of people writing to you to know how it is made that I believe is a founded lie for this last two *Sunday Mornings* I have almost been turned sick on *observing* your lectures on the said *Rat pie* I think the lease you say about the matter the better you are very fond of saying so much about *Rat Pie* will you please make this public the next time you lecture that I the Undersigned write in the name of hundreds of people to protest against your assertion. All we eat now is not pure so I think you ought not to want us to eat *Vermin*.

'I am, Sir, A BELIEVER IN HUMAN FOOD.'
REV. J. G. WOOD (1827–89), *Petland Revisited*, 1890

On the French poet Gérard de Nerval (1808–55)

But soon his peculiarities showed themselves once more: it often became difficult to find excuses for him since his peculiarities left the realms of thought and showed themselves in his actions. It became necessary to watch him. This infuriated Gérard, because he could only suppose that the doctors were interested in his case because he went for walks in the gardens of the Palais Royal leading a live lobster on a blue ribbon.

'And why not,' he used to say. 'Is a lobster more ridiculous than a dog, a cat, a gazelle, a lion or any other animal that you take for walks? I have a taste for lobsters, who are quiet, serious beings who know the secrets of the seas; they don't bark and they don't monopolise the very souls of people like dogs do, which was so hateful to Goethe, and he wasn't mad.' And a thousand other reasons all more ingenious than the last.

THÉOPHILE GAUTIER (1811–72), *Portraits et Souvenirs Littéraires*, 1875

The following story of a strange animal friendship may be of interest to you. I am the owner of a pony and a small black pig, which, after the manner of the country, runs loose and finds its own food. The pony is kept stabled at night, but for part of the day is out grazing, and is fed with corn every morning near the dining-room window. The corn is put in a box which rests on the ground. When the pony comes he is always accompanied by the pig, who sits between his forefeet and occasionally takes a little corn himself. If he takes too much, the pony gives him a gentle bite, as a reminder not to be greedy. The pig sleeps against the stable door (it is not allowed to go inside) so as to be near its friend, and when the pony is grazing the pig is always just alongside. When I go out for a drive or a ride, and the pony has to wait, friend pig lies down between his feet until we start. On returning from the drive, the pig jumps about, making the most absurd antics to greet his friend, and grunts a queer sharp grunt, looking for all the world like a fat and clumsy old spaniel greeting his master. He then trots off to the stable with his friend.

P.S. – No other pig dare go near the pony, as he has a great dislike for the animals as a whole, and bites them savagely.

LETTER TO THE *Spectator*, 12 MARCH 1904

'A guinea fowl always used to go out with Mr Allgood's hounds of Northumberland, whenever the hounds were let out of kennel, nothing but shutting up the bird could restrain it. The manner of its following them was running and flying. Once it got the distance of seventy miles from Mr Allgood's house, and it was so fatigued with the exertion that it was four days in returning home. In a general way it kept pace with the leading hound.'

The above cutting was sent by Mr Kearney of Lanchester to Mr or Mrs Hunter Allgood, letter dated March 6th 1868, and said the cutting had belonged to an uncle of his wife who died 50 years before which would be 1818.

Mrs M. E. Bell (*née* Allgood 1871–1941) said that the guinea fowl met its death through having its head bitten off by a young hound which did not know it.

LETTER TO LUCINDA LAMBTON FROM
LANCELOT ALLGOOD, 1991

BREAKFAST FOR THREE

At a farmhouse at which we have been staying a terrier, Rough, shares always his master's first breakfast – the bread-and-cream accompanying a cup of tea. Three corners he breaks off and gives to Rough, who eats the first two. Off the third he licks the cream, then carries the crust to a hen who each morning comes across the field where the fowls are kept, and at the gate awaits her friend's arrival. Should other hens appear Rough 'barks them off' while his favourite devours her portion.

E. PARKER AND A. CROXTON-SMITH,
The Dog Lover's Week-End Book, 1950

In the flat countryside near Doncaster, stout Mr Jemmy Hirst, the Rawcliffe tanner, who had retired from business with a large fortune, might have been seen any autumn day about the year 1840, leaving his house to go shooting. His jolly, if coarse, face was as round as the autumn sun, and shone like brightly polished leather, whilst everything about him had a strong, horsy, leathery smell. But Mr Hirst did not approve of horses excepting on the racecourse, and he went shooting mounted on the back of a bull of ample proportions and uncertain temper, whilst for pointers, he made use of the services of a crowd of vivacious and sagacious pigs, all of whom answered to their names, and did their duty irreproachably. It is said that Mr Hirst rode the bull when hunting with the Badsworth hounds. If he did, his presence must, I imagine, have lent animation to the scene, and speed to the chase.

EDITH SITWELL (1887–1964),
English Eccentrics, 1933

April [1829] . . . At Paris, aged 72, the Right Hon. and Rev. Francis Henry Egerton, eighth Earl of Bridgewater, ninth Viscount Brackley and Baron Ellesmere, and a Prince of the Holy Roman Empire, senior Prebendary of Durham, Rector of Whitchurch and Middle in Shropshire, MA, FRS and FSA. . . .

The Earl's singularities were a general topic for conversation at Paris. He had, at the time of his death, his home nearly filled with dogs and cats, which he had picked up at different places. Of the fifteen dogs which he kept, two were admitted to the honours of his table, and the whole of them were frequently dressed up in clothes like human beings. Sometimes a fine carriage, containing half a dozen of them, was seen in the streets drawn by four horses, and accompanied by two footmen. In his last days, when so debilitated as to be unable to leave his own grounds, he is said to have adopted a strange substitute for the sports of the field, to which he had been addicted. Into the garden at the back of his house, there were placed about 300 rabbits, and as many pigeons and partridges, whose wings had been cut. Provided with a gun, and supported by servants, he would enter the garden and shoot two or three head of game, to be afterwards put upon the table as his sporting trophies!

SYLVANUS URBAN, OBITUARY OF THE
8TH EARL OF BRIDGEWATER (1756–1829)
IN *The Gentleman's Magazine*, 1829

*Thomas Henry and
Mary Ann.*

At that time I possessed a very pretty small tortoise, called 'Mary Ann.' Thomas Henry [a cat] was devoted to her. They used to drink milk out of the same saucer, and when they had finished, Thomas Henry would lick the milk off Mary Ann's hand and neck, and tidy her up generally. She was so used to the process that she only used to blink her eyes, and did not even trouble herself to draw in her head.

When evening came, and Mary Ann was too sleepy to toddle about with him, he used to have a game of his own invention.

He used to pick up Mary Ann in his arms, and see how far he could run on his hind-legs before letting her fall. I have often seen him run six or seven yards before letting her down. This absurd game always went on in a passage with an oil-cloth floor, so that the quick scurrying footsteps could be heard at some distance, and every tumble made a great bang.

REV. J. G. WOOD (1827–89), *Petland Revisited,*

It is unusual to see a Cow playing the games of a puppy; but the gentleman had among his Cows, one which was called *Cherry*, with which his sons, then stout lads, would play as though she had been a dog. When they called her, she would gallop across the field after them, first chasing one and then another; and as each fell, or seemed to be knocked down by her head, (which had no horns), *Cherry* fell too, rolling about with them in a way which no stranger to her education would have envied. Yet, so careful was this gentle creature, that she would not even hurt an infant. One of the young men assured the writer, that at her fastest speed when running after them, 'rather than run against any one to do an injury, he was sure she would drop short and break her neck.' And it seemed that she was no less careful in using her legs when rolling with her young biped friends upon the grass.

W. R. MACDONALD, *The Book of Quadrupeds for
the Instruction of Young People*

Once when I was at Dorchester he showed me a letter from a firm which had presented him with a broadcasting set. They said they were delighted to hear from him that it gave pleasure, but that they were rather damped to learn from another source that it was not he who listened, but his dog.

This was quite true.

We went that afternoon to a local rehearsal of the play of *Tess*, and the dog [a wire-haired terrier], who was with us, behaved beautifully until the time came when he knew the wireless would be putting on 'the Children's Hour'. It was his favourite item. He howled for it so that even Tess's champion had to desert her and hurry home with him.

The dog afterwards discovered that a weather report, or something of the kind, was issued in the early morning, and I understand his master used to go downstairs in the cold and turn it on for him.

J. M. BARRIE (1860–1937) ON THOMAS
HARDY'S DOG WESSEX, IN A SPEECH TO
THE SOCIETY OF AUTHORS,
28 NOVEMBER 1928

A most singular instance of attachment between two animals, whose natures and habits were most opposite, was related to me by a person on whose veracity I can place the greatest reliance. He had resided for nine years in the American States, where he superintended the execution of some extensive works for the American Government. One of these works consisted in the erection of a beacon in a swamp in one of the rivers, where he caught a young alligator. This animal he made so perfectly tame that it followed him about the house like a dog, scrambling up the stairs after him, and showing much affection and docility. Its great favourite, however, was a cat, and the friendship was mutual. When the cat was reposing herself before the fire (this was at New York), the alligator would lay himself down, place his head upon the cat, and in this attitude go to sleep. If the cat was absent the alligator was restless; but he always appeared happy when the cat was near him. The only instance in which he showed any ferocity was in attacking a fox, which was tied up in the yard. Probably, however, the fox resented some playful advances which the other had made, and thus called forth the anger of the alligator. In attacking the fox he did not make use of his mouth, but beat him with so much severity with his tail, that, had not the chain which confined the fox broke, he would probably have killed him. The alligator was fed on raw flesh, and sometimes with milk, for which he showed a great fondness. In cold weather he was shut up in a box, with wool in it; but, having been forgotten one frosty night, he was found dead in the morning.

Romanes' Animal Intelligence, QUOTED IN EDWARD
JESSE (1780–1868), *Gleanings in
Natural History*, 1832–5

He was thrice married, the third marriage taking place in a pets' cemetery with Addams's dog, Alice B. Cur, as the only attendant. The bride wore black.

OBITUARY OF *New Yorker* CARTOONIST
CHARLES ADDAMS (1912–88),
Daily Telegraph, 1 OCTOBER 1992

A gentleman residing in Northumberland assured Mr Jesse that he had a tame fox who was so much attached to his harriers, and they to him, that they lived together, and that the fox always went out hunting with the pack. This fox was never tied up, and was as tame, playful, and harmless as any dog could be. He hunted with the pack for four years, and was at last killed by an accident.

REV. T. JACKSON M. A., *Our Dumb Neighbours*, 1868

Even great disparity of kind and size does not always prevent social advances and mutual fellowship. For a very intelligent and observant person has assured me that, in the former part of his life, keeping but one horse, he happened also on a time to have but one solitary hen. These two incongruous animals spent much of their time together in a lonely orchard, where they saw no creature but each other. By degrees an apparent regard began to take place between these two sequestered individuals. The fowl would approach the quadruped with notes of complacency, rubbing herself gently against his legs: while the horse would look down with satisfaction, and move with the greatest caution and circumspection, lest he should trample on his diminutive companion. Thus, by mutual good offices, each seemed to console the vacant hours of the other: so that *Milton*, when he puts the following sentiment in the mouth of *Adam*, seems to be somewhat mistaken:

'Much less can *bird* with *beast*, or fish with fowl,
'So well converse, nor with the ox the ape.'

GILBERT WHITE (1720–93), *Natural History
and Antiquities of Selborne*, 1789

. . . carriage accidents were Squire Mytton's strongest point. Having bought some carriage horses from a horse-dealer named Clarke, he put one of them into a gig, tandem, to see if it would make a good leader. 'Do you think he is a good timber-jumper?' he inquired of the alarmed Mr Clarke, who sat beside him. Not waiting for that unhappy gentleman's reply, the Squire exclaimed: 'We'll try him.' And a closed turnpike being before him, he gave the horse his head. The horse did himself credit, leaving Squire Mytton, the other horse, Mr Clarke, and the gig at the other side of the gate in grand style and almost inextricable confusion. But once again, nobody was hurt. The Squire had, too, a horse that would rear up in his gig at the word of command, 'until the hinder part of it absolutely touched the ground'. But in spite of this talented animal's frequently repeated achievement, the Squire remained alive.

Master and horses were so friendly with the country people that they would help themselves to anything that took their fancy on their way home from hunting, and Squire Mytton, if his coat was wet, would think nothing of taking a country woman's red flannel petticoat from a hedge, slipping it over his head, and leaving his coat drying in its place. It was, too, not in the least unusual for Squire Mytton, if he felt cold when out hunting, to go into the house of a cottager, accompanied by his favourite horse Baronet, and ask her to light a good fire to warm Baronet and himself, for he did not believe in a heaven from which animals were excluded. Baronet and he would then lounge by the fire, side by side, until they were warm again, and then they would start for home. But

alas, there was one moment when disaster came from the master's habit of sharing all good things with the subject beast, for a horse named Sportsman dropped dead because John Mytton, out of kindness of heart, had given him a bottle of mulled port.

EDITH SITWELL (1887–1964),
English Eccentrics, 1933

'I wonder whether he is a good timber jumper!'

We have had a fine day's hunting here. The weather was remarkably clear and sunny, and at least a hundred red coats took the field. Such a sight is certainly full of interest; the many fine horses; the elegantly dressed huntsmen; fifty or sixty beautiful hounds following Reynard over stock and stone; the wild mounted troop behind; the rapid change of wood and hill and valley; the cries and shouts – it is a miniature war. . . .

The most striking thing, however, in the whole business, to German eyes, is the sight of the black-coated parsons, flying over hedge and ditch. I am told they often go to the church, ready booted and spurred, with the hunting-whip in their hands, throw on the surplice, marry, christen, or bury, with all conceivable velocity, jump on their horses at the church-door, and off – tally-ho! They told me of a famous clerical fox-hunter, who always carried a tame fox in his pocket, that if they did not happen to find one, they might be sure of a run. The animal was so well trained that he amused the hounds for a time; and when he was tired of running, took refuge in his inviolable retreat – which was no other than the altar of the parish church. There was a hole broken for him in the church-wall, and a comfortable bed made under the steps. This is right English religion.

PRINCE M. L. H. VON PÜCKLER-MUSKAU,
Tour in Germany, Holland and England,
in the Years 1826, 1827, and 1828, Vol. IV, 1832

10

Out of the Run of Things

'Party at the Deanery,' one guest wrote; 'tripe for dinner; don't like crocodile for breakfast.'

GEORGE C. BOMPAS,
Life of Frank Buckland, 1885

GILDING *of live-fish*, as craw-fish, carps, &c. may be performed without injuring the fish, by means of a cement; which Mr Hooke, in his post-humous papers, directs to be prepared in the following manner: Put some Burgundy pitch into a new earthen pot, and warm the vessel till it receives so much of the pitch as will stick round it; then strew some finely powdered amber over the pitch when growing cold; add a mixture of three pounds of linseed oil, and one of oil of turpentine: cover the vessel, and boil the contained ingredients over a gentle fire; grind the mixture as it is wanted, with so much pumice-stone in fine powder as will reduce it to the consistence of paint. When the fish has been wiped dry, this mixture is spread upon it, and the gold leaf laid over it, and gently pressed down; after which, the fish may be immediately put into water, and the cement will harden, and be in no danger of falling off.

ABRAHAM REES, The Cyclopaedia; or, Universal
Dictionary of Arts, Sciences,
and Literature, 1 8 1 9

We found a cayman [a kind of alligator] ten feet and a half long fast to the end of the rope. Nothing now remained but to get him out of the water without injuring his scales: 'hoc opus, hic labor'. . . .

I now walked up and down the sand, revolving a dozen projects in my head. The canoe was at a considerable distance, and I ordered the people to bring it round to the place where we were. The mast was eight feet long, and not much thicker than my wrist. I took it out of the canoe and wrapped the sail round the end of it. Now it appeared clear to me that, if I went down upon one knee and held the mast in the same position as the soldier holds his bayonet when rushing to the charge, I could force it down the cayman's throat should he come open-mouthed at me. When this was told to the Indians they brightened up, and said they would help me to pull him out of the river. . . .

I now took the mast of the canoe in my hand (the sail being tied round the end of the mast) and sunk down upon one knee, about four yards from the water's edge, determining to thrust it down his throat in case he gave me an opportunity. I certainly felt somewhat uncomfortable in this situation, and I thought of Cerberus on the other side of the Styx ferry. The people pulled the cayman to the surface; he plunged furiously as soon as he arrived in these upper regions, and immediately went below again on their slackening the rope. I saw enough not to fall in love at first sight. I now told them we would run all risks and have him on land immediately. They pulled again, and out he came – 'monstrum horrendum, informe.' This was an interesting moment. I kept my position firmly, with my eye fixed steadfast on him.

By the time the cayman was within two yards of me I saw he was in a state of fear and perturbation. I instantly dropped the mast, sprung up and jumped on his back, turning half round as I vaulted, so that I gained my seat with my face in a right position. I immediately seized his fore-legs, and by main force twisted them on his back; thus they served me for a bridle.

He now seemed to have recovered from his surprise, and probably fancying himself in hostile company he began to plunge furiously, and lashed the sand with his long and powerful tail. I was out of reach of the strokes of it by being near his head. He continued to plunge and strike and made my seat very uncomfortable. It must have been a fine sight for an unoccupied spectator.

The people roared out in triumph, and were so vociferous that it was some time before they heard me tell them to pull me and my beast of burden farther inland. I was apprehensive the rope might break, and then there would have been every chance of going down to the regions under water with the cayman. That would have been more perilous than Arion's marine morning ride:

Delphini insidens vada cærula sulcat Arion.

The people now dragged us above forty yards on the sand: it was the first and last time I was ever on a cayman's back. Should it be asked how I managed to keep my seat, I would answer, I hunted some years with Lord Darlington's fox-hounds.

CHARLES WATERTON (1782–1865),
Wanderings in South America, 1825

A very little canary, who was so tame that he was brought down by Mr Boythorn's man, on his forefinger, and after taking a gentle flight round the room, alighted on his master's head. To hear Mr Boythorn presently expressing the most implacable and passionate sentiments, with this fragile mite of a creature quietly perched on his forehead, was to have a good illustration of his character, I thought.

'By my soul, Jarndyce,' he said, very gently holding up a bit of bread for the canary to peck at, 'if I were in your place, I would seize every Master in Chancery by the throat to-morrow morning, and shake him until his money rolled out of his pockets, and his bones rattled in his skin. I would have a settlement out of somebody, by fair means or by foul. If you would empower me to do it, I could do it for you with the greatest satisfaction!' (All this time the very small canary was eating out of his hand).

CHARLES DICKENS (1812–70),
Bleak House, 1852–3

Sage-brush is very fair fuel, but as a vegetable it is a distinguished failure. Nothing can abide the taste of it but the jackass and his illegitimate child the mule. But their testimony to its nutritiousness is worth nothing, for they will eat pine knots, or anthracite coal, or brass filings, or lead pipe, or old bottles, or anything that comes handy, and then go off looking as grateful as if they had had oysters for dinner. Mules and donkeys and camels have appetites that anything will relieve temporarily, but nothing satisfy. In Syria, once, at the head-waters of the Jordan, a camel took charge of my overcoat while the tents were being pitched, and examined it with a critical eye, all over, with as much interest as if he had an idea of getting one made like it; and then, after he was done figuring on it as an article of apparel, he began to contemplate it as an article of diet. He put his foot on it, and lifted one of the sleeves out with his teeth, and chewed and chewed at it, gradually taking it in, and all the while opening and closing his eyes in a kind of religious ecstasy, as if he had never tasted anything as good as an overcoat before, in his life. Then he smacked his lips once or twice, and reached after the other sleeve. Next he tried the velvet collar, and smiled a smile of such contentment that it was plain to see that he regarded that as the daintiest thing about an overcoat. The tails went next, along with some percussion caps and cough candy, and some fig-paste from Constantinople. And then my newspaper correspondence dropped out, and he took a chance in that – manuscript letters written for the home papers. But he was treading on dangerous ground, now. He began to come across solid wisdom in those documents that was rather weighty on his stomach; and occasionally he would take a joke that would shake him up till it loosened his teeth; it was getting to be perilous times with him, but he held his grip with good courage and hopefully, till at last he began to stumble on statements that not even a camel could swallow with impunity. He began to gag and gasp, and his eyes to stand out, and his forelegs to spread, and in about a quarter of a minute he fell over as stiff as a carpenter's work-bench, and died a death of indescribable agony. I went and pulled the manuscript out of his mouth, and found that the sensitive creature had choked to death on one of the mildest and gentlest statements of fact that I ever laid before a trusting public.

MARK TWAIN (1835–1910),
Roughing It, 1875

*In 1786 the novelist and diarist Fanny Burney (1752–1840)
was appointed Second Keeper of the Robes to Queen Charlotte,
consort of George III*

Fanny was not a happy courtier. She was easily insulted, but easily gave insults herself, often because she was short-sighted and prohibited by court etiquette from wearing glasses. The work she did was menial and lasted from six in the morning till midnight. She disliked her superior, 'Mrs' Schwellenberg, a bad-tempered German woman, who forbade Fanny to entertain friends, lived in greater style than the Queen and had her rooms so placed that she could bar the way to the royal apartments.

Fanny was forced to spend long evening hours with this lady, submitting to insults or playing piquet with her, according to her whim. Here she met the two pet frogs which Mrs Schwellenberg kept in glasses for fondling, and heard her describe them to Mrs Delany. 'A commendation ensued, almost ecstatic, of their most recreative and dulcet croaking and of their amiable ways of snapping live flies.' They were trained to croak, Mrs Schwellenberg explained on another occasion. ' "I only go so to my snuff-box, knock knock knock, they croak all I please." '

' "Very pretty, indeed!" exclaimed Colonel Goldsworthy.

' "I thought to have some spawn," she continued; "but Lady Maria Carlton . . . came and frightened them. I was never so angry!"

' "I am sorry for that," cried the Major, very seriously, "for else I should have begged a pair." '

THOMAS HINDE, *Courtiers: 900 Years of Court Life*, 1986

At the corner of the great Quadrangle of Christ Church lived Dr Buckland, always ready to help me. . . . At his breakfast-table I met the leading scientific men of the day, from Herschel downwards, and often intelligent and courteous foreigners. . . . Every one was at ease and amused at that breakfast-table, – the menu and science of it usually in themselves interesting. I have always regretted a day of unlucky engagement on which I missed a delicate toast of mice; and remembered with delight being waited upon one hot summer morning by two graceful and polite little Carolina lizards, who kept off the flies.

JOHN RUSKIN (1819–1900), *Praeterita*, 1885–9

The most enjoyable purely literary event that she attended was a literary lunch at the Dorchester Hotel in March 1939. Elinor was a guest speaker and she wore, with fine panache, her huge Persian cat, Candide, asleep round her neck instead of a fox fur. During the speeches Candide would open a baleful and somewhat desiccating eye upon the speaker and then go to sleep again. 'Great Success!' commented Elinor in her diary.

ANTHONY GLYN, *Elinor Glyn: A Biography*, 1955

[Dante Gabriel Rossetti's] famous menagerie with its varied assortment of animals was a caliph's whim.

He collected peculiar creatures as he collected peculiar people. It was another alternative world. Detached from it as he was detached, in fact, from Howell and Swinburne, he took pleasure in the imaginative fancy of creation. He understood animals, their antics and their humour . . . ; and he found in them, with a sort of irony, the equivalents of human beings. His purchases were made principally at Jamrach's in the Ratcliffe Highway, now a dreary little row of empty houses and shops, marked down for demolition and renamed St George's Street (subsequently *The Highway*); then a lurid quarter of sing-song caves, opium dens, sailors' orgies and savage misery.

It was not a phase that outlasted the 'sixties, but during these years he possessed kangaroos, a wallaby, a chameleon, some salamanders, wombats, an armadillo, a marmot, a woodchuck, a deer, a jackass, a racoon and smaller animals galore. The birds included peacocks, parakeets, Chinese horned owls and a raven. There was a marquee and cages in the desolate garden to house them. He bought a Brahmin bull because he said it had eyes like Janey Morris. He would have bought a lion only this would have entailed a special arrangement of hot-water pipes. He wanted to buy an elephant. Browning, it is said, asked him what he wanted an elephant for. He replied that he meant to teach it to clean the windows so that people, seeing it, would ask who lived there and then come and buy his pictures.

WILLIAM GAUNT (1900–1980),
The Pre-Raphaelite Tragedy, 1943

There were two other curiosities – a pair of armadillos which, under the idea that they were harmless, had the run of the garden. They, too, seemed to have caught the contagion for mischief. Now and then our neighbour's garden would be found to have large heaps of earth thrown up, and some of his choicest plants lying waste over the beds. This was the work of the armadillos.

As in the racoon escapades, letters of complaint were received, and so baits were laid for the pests in the form of bits of beef saturated with prussic acid. The beef disappeared, and so, it was hoped, had the armadillos; but no – after about three months they re-appeared in a sadly mangy and out-at-elbows state; they had evidently shed their scales during their absence, and new ones were forming. I suppose that after taking the dose of poison, feeling the worse for it, they must have betaken themselves to a hospital, and were just discharged as convalescent. Very soon after their return, I am sorry to say they slid back into their old mischievous habits, and at last had to be made over to the Zoological Gardens, where no doubt they were better guarded.

Amongst this curious collection of odd animals were a couple of kanga-roos – mother and son. As far as my observation went, I do not think they lived on very good terms with each other. At any rate, the mother was found dead one morning, murdered by her bloodthirsty son. There must have been an unusually fierce quarrel over family matters in the night, with this as a consequence. Nemesis, however, overtook the wicked son, for he also was found dead in his cage some few days after, but whether he committed suicide through remorse, or whether the racoon, who was strongly suspected, polished him off, was an open verdict.

HARRY DUNN, QUOTED IN
GALE PEDRICK, *Life with Rossetti*, 1964

As a punishment for adultery Mazeppa, a seventeenth-century
Polish grandee, was tied naked to the back of a wild horse which
was then set loose

MAZEPPA

'Bring forth the horse!' – the horse was brought;
In truth, he was a noble steed,
A Tartar of the Ukraine breed,
Who looked as though the speed of thought
Were in his limbs; but he was wild,
Wild as the wild deer, and untaught,
With spur and bridle undefiled –
'Twas but a day he had been caught;
And snorting, with erected mane,
And struggling fiercely, but in vain,
In the full foam of wrath and dread
To me the desert-born was led:
They bound me on, that menial throng,
Upon his back with many a thong;
They loosed him with a sudden lash –
Away! – away! – and on we dash! –
Torrents less rapid and less rash . . .
Away, away, my steed and I,
Upon the pinions of the wind.
All human dwellings left behind,
We sped like meteors through the sky,
When with its crackling sound the night
Is chequered with the northern light: . . .
At times I almost thought, indeed,
He must have slackened in his speed;
But no – my bound and slender frame

Was nothing to his angry might,
And merely like a spur became:
Each motion which I made to free
My swoln limbs from their agony
Increased his fury and affright:
I tried my voice, – 'twas faint and low,
But yet he swerved as from a blow;
And, starting to each accent, sprang
As from a sudden trumpet's clang:
Meantime my cords were wet with gore,
Which, oozing through my limbs, ran o'er;
And in my tongue the thirst became
A something fierier far than flame.

LORD BYRON (1788–1824)

SHERRY of Culcheth, age 24, 1989

Never before has a small, sweet sherry warranted a certificate from the Queen, but Sherry, a Jack Russell belonging to Mrs. Ada Yates, of Thames Road, Culcheth, has just celebrated her 23rd birthday – 161 years old in canine years! . . .

A connoisseur of good eating, Sherry shuns dog foods, preferring lambs' liver or stewed meat lightly fried, although she also has a soft spot for Lancashire hot-pot.

Although Sherry has never been washed, it would not appear she needs to be and, despite the onset of arthritis and a need to sleep up to 16 hours a day, is still enjoying her dog's life. . . .

Ada believes that Sherry's recipe for a long and cold-nosed life would be love, kindness and a good deal of good food.

Warrington Guardian, 10 JUNE 1988
(Sherry died the next year aged 24.)

Some time in 1913, at this address, my wife and I acquired a young fox-terrier. We debated as to what to call him and as Henry James had just been having his seventieth birthday, and as his books had given me more pleasure than those of any other living man, I, rather priggishly perhaps, insisted that the dog should be known as James. But this was a name which Italian peasants, who are the only neighbours we have, of course would not be able to pronounce at all. So we were phonetic and called the name of the dog *Yah-mès*. And this did very well. By this name he was known far and wide – but not long; for alas, he died of distemper.

Now we are re-established here, we haven't another dog; dogs aren't so necessary to one as they seem to be in England, and they have an odd and tactless way of making one feel that one *is* in England – perhaps because they don't gesticulate and don't speak one word of Italian and seem to expect to find rabbits among the olive-groves and to have bones of Welsh mutton thrown to them from the luncheon table.

But the other day we were given a small kitten – charming in itself and somehow not destructive of local colour. The old question arose; what shall we call it? Again I laid myself open to the charge of priggishness, perhaps. And again you will perhaps think I have taken a liberty. But – well, there it is: no book by a living man has given me so much pleasure – so much lasting pleasure in dipping and in re-reading since I wrote to you – as your *Eminent Victorians*. And the name of that kitten is, and the name of that cat will be, *Stré-chi* (or rather, *Stré-ici*). I hope you don't mind. I am sure you would be amused if you heard the passing-by of peasants enticing it by your hardly-recognisable name. We will re-christen it if you like.

LETTER TO LYTTON STRACHEY (1880–1932),
FROM MAX BEERBOHM (1872–1956),
7 JULY 1920, WRITTEN FROM RAPALLO IN ITALY

Dr Merriman, his praefect at Winchester, renewed his friendship with him at Oxford. . . .

'He came down to me one day for the purpose of telling me what he had for dinner the day before – namely, panther chops! He was a great friend of the curator of the then existing Surrey Zoological Gardens. From him Frank heard one day that the panther was dead. "I wrote up at once," he said, "to tell him to send me down some chops. It had, however, been buried a couple of days, but I got them to dig it up and send me some. *It was not very good.*"'

GEORGE C. BOMPAS,
Life of Frank Buckland, 1885

I carried the indignant owlet back home in my pocket, and introduced him to the family with a certain trepidation. To my surprise, he was greeted with unqualified approval, and no objection was raised to my keeping him. He took up residence in a basket kept in my study and, after much argument, he was christened Ulysses. From the first he showed that he was a bird of great strength of character, and not to be trifled with. Although he would have fitted comfortably into a tea-cup, he showed no fear and would unhesitatingly attack anything and everyone, regardless of size. As we all had to share the room, I felt it would be a good idea if he and Roger got on intimate terms, so, as soon as the owl had settled down, I performed the introductions by placing Ulysses on the floor, and telling Roger to approach and make friends. Roger had become very philosophical about having to make friends with the various creatures that I adopted, and he took the appearance of an owl in his stride. Wagging his tail briskly, in an ingratiating manner, he approached Ulysses, who squatted on the floor with anything but a friendly expression on his face. He watched Roger's approach in an unwinking stare of ferocity. Roger's advance became less confident. Ulysses continued to glare as though trying to hypnotize the dog. Roger stopped, his ears drooped, his tail wagging only feebly, and he glanced at me for inspiration. I ordered him sternly to continue his overtures of friendship. Roger looked nervously at the owl, and then with great nonchalance walked round him, in an effort to approach him from the back. Ulysses, however, let his head revolve too, and kept his eyes still fixed on the dog. Roger, never having met a creature that could look behind itself without turning round, seemed a trifle nonplussed. After a moment's thought he decided to try the skittish, let's-all-have-a-jolly-game approach. He lay down on his stomach, put his head between his paws, and crept slowly towards the bird, whining gently and wagging his tail with abandon. Ulysses continued to look as though he were stuffed. Roger, still progressing on his stomach, managed to get quite close, but then he made a fatal mistake. He pushed his woolly face forward and sniffed loudly and interestedly at the bird. Now Ulysses would stand a lot, but he was not going to be sniffed at by a mountainous dog covered with black curls. He decided that he would have to show this ungainly and wingless beast exactly where he got off. He lowered his eyelids, clicked his beak, hopped up into the air, and landed squarely on the dog's muzzle, burying his razor-sharp claws in the black nose. Roger, with a stricken yelp, shook the bird off and retired beneath the table; no amount of coaxing would get him to come out until Ulysses was safely back in his basket.

GERALD DURRELL (b. 1925), *My Family and Other Animals*, 1956

I may ... mention another *marvellous* instance of cool courage. I allude to an occurrence in the Zoological Gardens, in London, in the year 1861, when, after much entreaty, on the part of Mr Waterton, he was permitted, by the then curator, Mr Mitchell, now deceased, to pay his personal respects to a large orang-outang, from Borneo, which was reputed to be very savage. Indeed, the keepers, one and all, declared that 'he would worry the Squire, and make short work of it,' if he should enter his den, especially as he was just then in a horrid temper, having been recently teased by some mischievous boys. The late Mr Mitchell, even at last, yielded to Mr Waterton's urgent request, with great reluctance. Nothing daunted by all this badinage of the keepers, the Squire, to the very great horror of the numerous spectators, entered the palisaded enclosure with a light heart. The meeting of these two celebrities was clearly a case of

'love at first sight,' as the strangers embraced each other most affectionately; nay, they positively hugged each other, and in their apparently
uncontrollable joy, they kissed one another many times, to the great
amusement of the numerous spectators.

Mr Waterton, who has written specially on the monkey tribe, had long
been anxious to minutely inspect the palm of its hand during life, and
was then also wishful to examine the teeth of his newly-acquired friend,
both of which investigations were graciously conceded to the Squire,
without a murmur, his fingers being freely admitted within its jaws.
These little ceremonies having been accomplished on the part of Mr
Waterton, his apeship claimed a similar privilege.

RICHARD HOBSON, *Charles Waterton:*
His Home, Habits and Handiwork, 1866

. . . . Sir Michael Howard, the historian, was also at Abinger at the same
time and recounts how Edward [Boyle] had 'a revolting but hugely
popular parlour trick which consisted of folding his capacious belly into
a kind of marsupial pouch and lapping out of it the milk we had at
elevenses.'

I remember the parlour trick as being slightly different and much more
revolting. First, he would pour the milk into the great creases of his
tummy and then float dead flies down a kind of fatty Cresta run which
started under his chin and ended in his private parts. His admiring friends
were asked to bet on which fly would reach the finishing post first.

PEREGRINE WORSTHORNE, *Daily Telegraph,*
24 APRIL 1991

My big hit . . . was made by my tame oyster. This was quite my own idea,
and puzzled and amazed not only the public, but at its start the other show-
men, who became quite envious of the novel attraction. It was a daring
thing to do, but the result quite justified my audacity, though it was not
without some qualms that I prepared a piece of calico, four yards long by
a yard wide, and painted on it the legend: 'The Only Novelty in the Fair.
The Wonderful Performing Fish and a Tame Oyster that sits by the fire
and smokes his yard of clay.' How they did bite at this! My show was
crowded as it had never been before. . . .

I had prepared a fine big oyster shell, the two halves fitting closely
together as though it was a nice fresh bivalve. Inside were fixed two little
pieces of piping opening to two holes in the lip of the oyster shell. These
pipes were connected with two pieces of black rubber tubing that ran down
under my conjuring table on the raised platform.

I had handy two or three nice fresh oysters that in appearance and size
closely matched the dummy shell, which, laid on a dark grey cloth, was not

visible to the audience. When all was ready I took one of my good oysters, and introducing it as 'The Tame Oyster, the only one in the world!' handed it round, so that the company could see there was 'no deception.' When I got the oyster back I pretended to place it in the grey cloth, but really dropped it at the back of the table, while I pulled forward the prepared shell. This I lifted up on to a black bottle, with the two tubes running down behind well out of sight. Then, calling attention to the fact that everybody could see the oyster, I would get a clay pipe, put some tobacco in it, and then with a request to the oyster: 'Now, sir, let the company see that you really are trained and intelligent by showing how you can smoke a pipe!' I would insert the stem of the pipe into the hole made for it in the shell, call a boy from the audience, give him a spill, and ask him to light the tobacco. As he did so, my boy who used to do the 'Suspension by Ether,' and who was concealed under the table, would draw the smoke down one tube and blow it back through the other, so that it really looked as though the oyster was puffing away at his pipe.

The trick never failed to amaze as well as amuse, more especially when, as if considering the oyster had smoked enough, I would say: 'That will do, sir! You will make your head ache if you smoke too much!' With this I took the pipe away, and threw the corner of the grey cloth over the prepared oyster, at the same moment pulling the latter off the bottle and dropping it, tubes and all, into the drawer at the table back while I deftly substituted a real oyster. Then, throwing the corner of the cloth back, I would say: 'There he is, ladies and gentlemen! Looks none the worst for his smoke, I think, but see for yourselves!' And the real oyster was handed round again, all believing it to be the one that had just been smoking. They used to go away quite convinced that they had seen an oyster enjoying a smoke and that there was 'no deception.'

'LORD' GEORGE SANGER (1825–1911),
Seventy Years a Showman, 1910

'We are much taken with the old church,' wrote a well-known public man a few years ago to a friend, 'to say nothing of the vicar thereof, who reminds me immensely of Cardinal Wiseman. He is a sight to see, as well as a preacher to hear, as he stands in his quaint garb and quaint pulpit, and looks as if he belonged to the days of Morwenna Abbatissa herself.'

He was usually followed to church by nine or ten cats, which entered the chancel with him, and careered about it during service. Whilst saying prayers Mr Hawker would pat his cats, or scratch them under their chins. Originally ten cats accompanied him to church; but one, having caught, killed, and eaten a mouse on a Sunday, was excommunicated, and from that day was not allowed again within the sanctuary.

A friend tells me that on attending Morwenstow church one Sunday morning, nothing amazed him more than to see a little dog sitting upon the altar step behind the celebrant, in the position which, in many churches, is occupied by a deacon or a server. He afterwards spoke to Mr Hawker on the subject, and asked him why he did not turn the dog out of the chancel and church.

'Turn the dog out of the ark!' he exclaimed: 'all animals, clean and unclean, should find there a refuge.'

SABINE BARING-GOULD (1834–1924),
The Vicar of Morwenstow, 1886

*An episode from the childhood of the American-born artist
Benjamin West (1738–1820)*

The want of pencils soon became a matter of thoughtful regret – he was told they were small brushes of camel's hair; they could not be caught or killed in the forest, and he had never seen them sold. This was a serious dilemma, from which the bushy tail of a large black cat, which was an especial favorite of his father, ere long relieved him. Armed with his mother's scissars, he made his first pencil from the fur of Tom's tail. But the tail only furnished one pencil, and that was soon worn out; so that he was next obliged to have recourse to the back, which he soon clipped so bare, that his father observed the sad reduction of sleekness and rotundity in the person of his favorite, and became so uneasy with his fears for Tom's decay, that the aspiring artist could not avoid confessing the true cause: forgiveness was the reward of his ingeniousness.

GEORGE LEWIS SMYTH, *The Monuments of Genii of
St Paul's and Westminster Abbey*, 1826

Frank's training for the acclimatisation of animals for food, to which he afterwards devoted much energy, began early.

At his father's table at Christ Church the viands were varied. A horse belonging to his brother-in-law having been shot, Dr Buckland had the tongue pickled and served up at a large luncheon party, and the guests enjoyed it much, until told what they had eaten.

Alligator was a rare delicacy, as told in the first volume of 'Curiosities,' but puppies were occasionally, and mice frequently eaten. So also at the Deanery, hedgehogs, tortoise, potted ostrich, and occasionally rats, frogs, and snails, were served up for the delectation of favoured guests. 'Party at the Deanery,' one guest notes; 'tripe for dinner; don't like crocodile for breakfast.'

GEORGE C. BOMPAS,
Life of Frank Buckland, 1885

When I was down in the Masai Reserve, doing transport for the Government, I one day saw a strange thing, such as no one I know has ever seen. It took place in the middle of the day, while we were trekking over grass-country.

The air in Africa is more significant in the landscape than in Europe, it is filled with loomings and mirages, and is in a way the real stage of activities. In the heat of the midday the air oscillates and vibrates like the string of a violin, lifts up long layers of grass-land with thorn-trees and hills on it, and creates vast silvery expanses of water in the dry grass.

We were walking along in this burning live air, and I was, against my habit, a long way in front of the waggons, with Farah, my dog Dusk and the Toto who looked after Dusk. We were silent, for it was too hot to talk. All at once the plain at the horizon began to move and gallop with more than the atmosphere, a big herd of game was bearing down upon us from the right, diagonally across the stage.

I said to Farah: 'Look at all these Wildebeests.' But a little after, I was not sure that they were Wildebeests; I took up my field-glasses and looked at them, but that too is difficult in the middle of the day. 'Are they Wildebeests, Farah, do you think?' I asked him.

I now saw that Dusk had all his attention upon the animals, his ears up in the air, his far-seeing eyes following their advance. I often used to let him have a run after the gazelles and antelopes on the plains, but to-day I thought that it would be too hot, and told the Toto to fasten his lead to his collar. At that same moment, Dusk gave a short wild yell and jumped forward so that the Toto was thrown over, and I snatched the lead myself and had to hold him with all my might. I looked at the game. 'What are they?' I asked Farah.

It is very difficult to judge distances on the plains. The quivering air and the monotony of the scenery make it so, also the character of the scattered thorn-trees, which have the exact shape of mighty old forest trees, but are in reality only twelve feet high, so that the Giraffes raise their heads and necks above them. You are continually deceived as to the size of the game that you see at a distance and may, in the middle of the day, mistake a jackal for an Eland, and an ostrich for a Buffalo. A minute later Farah said: 'Memsahib, these are wild dogs.'

The wild dogs are generally seen three or four at a time, but it happens that you meet a dozen of them together. The Natives are afraid of them, and will tell you that they are very murderous. Once as I was riding in the Reserve close to the farm I came upon four wild dogs, which followed me at a distance of fifteen yards. The two small terriers that I had with me then kept as close to me as possible, actually under the belly of the pony, until we came across the river and on to the farm. The wild dogs are not as big as a Hyena. They are about the size of a big Alsatian dog. They are black, with a white tuft at the tip of the tail and of the pointed ears. The skin is no good, it has rough uneven hair and smells badly.

Here there must have been five hundred wild dogs. They came along in a slow canter, in the strangest way, looking neither right nor left, as if they had been frightened by something, or as if they were travelling fast with a fixed purpose on a track. They just swerved a bit as they came nearer to us; all the same they hardly seemed to see us, and went on at the same pace. When they were closest to us, they were fifty yards away. They were running in a long file, two or three or four side by side, it took time before the whole procession had passed us. In the middle of it, Farah said: 'These dogs are very tired, they have run a long way.'

When they had all gone by, and were disappearing again, we looked round for the Safari. It was still some way behind us, and exhausted by our agitation of mind we sat down where we stood in the grass, until it came up to us. Dusk was terribly upset, jerking his lead to run after the wild dogs. I took him round the neck, if I had not tied him up in time, I thought, he would by now have been eaten up.

The drivers of the waggons detached themselves from the Safari and came running up to us, to ask us what it had all been. I could not explain to them, or to myself, what had made the wild dogs come along in so great a number in such a way. The Natives all took it as a very bad omen, – an omen of the war, for the wild dogs are carrion-eaters. They did not afterwards discuss the happening much among themselves, as they used to discuss all the other events of the Safari.

I have told this tale to many people and not one of them has believed it. All the same it is true, and my boys can bear me witness.

KAREN BLIXEN (1883–1962),
Out of Africa, 1937

Like many childless women, the Duchess of York lavished her affections on her pets. Some of these were indeed exotic but her especial favourites were her dogs of which there seem to have been a large number. According to Thomas Raikes: 'There were sometimes twenty or thirty of different sorts in the house and many a morning have I, to my annoyance, been awakened from an incipient slumber, after a long sitting at whist, by the noisy pack rushing along the gallery next to my bedroom, at the call of old Dawe the footman to their morning meal'.

J. W. LINDUS FORGE, RIBA,
The Dogs' Cemetery, Oatlands,
WALTON AND WEYBRIDGE
LOCAL HISTORY SOCIETY, 1974

In September 1846, after a short tour in Switzerland, Frank Buckland returned to Oxford. He brought with him this time a jar full of the red slugs he wished to introduce into England. . . . In the opposite corner of the diligence placidly slumbered a traveller with ample bald head; Frank also slept, but, waking at midnight, he saw with horror that two of his red slugs had escaped and were crawling over the traveller's bald pate. What was to be done? To remove them might waken the sleeper. Frank sat as it were on tenterhooks until the diligence stopped at the next stage, when firmly covering up the jar and what remained of the slugs, he slipped quietly out of the diligence, resolved to proceed on his journey by another conveyance next morning, rather than face that man's awakening.

GEORGE C. BOMPAS, *Life of*
Frank Buckland, 1885

Lord B.'s establishment consists, besides servants, of ten horses, eight enormous dogs, five cats, an eagle, a crow, and a falcon; and all these, except the horses, walk about the house, which every now and then resounds with their unarbitrated quarrels, as if they were the masters of it. . . .

After I have sealed my letter, I find that my enumeration of the animals in this Circaean palace was defective, and that in a material point. I have just met on the grand stair-case five peacocks, two guinea-hens, and an Egyptian crane. I wonder who all these animals were, before they were changed into these shapes.

PERCY BYSSHE SHELLEY (1792–1822),
DESCRIBING BYRON'S MENAGERIE
IN A LETTER TO THOMAS LOVE
PEACOCK, 1821

Chestertown, Maryland

DEAR JACK-IN-THE-PULPIT: One day a circus and menagerie train halted at the railway station on its way through this town. Of course there was great curiosity among the railroad men to inspect this queer

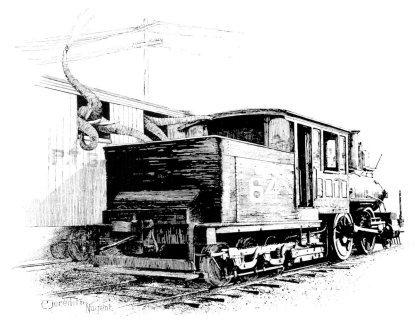

special train; and with the others the engineer and the fireman of one of the locomotives in the yard left their posts for a short time to see the different menagerie cars.

When they came back and were ready to move their locomotive, they noticed that the cover of the water-tank was open! Further, they luckily discovered that the tank was nearly empty – although it had been full to the brim when they left it.

Such an extraordinary thing had never happened before! No wonder there was great surprise on all sides; every one knew the tank was full when the men had left it; in fact some of the 'hands' had seen it filled, neither was there a leak in it, and yet, the tank was empty. The question was, where had the water gone?

Seven thirsty elephants, shut up all day and all night in a car that gave them hardly room to move; their warm bodies fairly touching one another, a paltry allowance of water to quench their thirst, and, then, to be left standing on the hot railroad-track, the sun's rays pouring down upon the roof of the car, and with only such air as could come through the small open windows! Was it any wonder, when their keen scent told them water was near, that they should search for it? How were they to know that it was not there for their convenience. At any rate, no sooner were the men gone, than through a small window of the elephant car, the dusky trunk of an elephant made its way sinuously out. Another followed its example, then another, until seven trunks had felt and snuffed around, over engine, tender, and coal. What they sought was not there; but they still kept moving about, and, coming to the water-tank, one of them stopped, felt all over the cover, and at last managed to get the finger-like end under the edge of the cover. Then slowly and carefully it was opened; when, behold! there was what the elephants wanted – water, and plenty of it. The owner of that particular trunk took a long draught, its companions meanwhile shoving and pushing one another, in their anxiety to drink. One after another they filled their trunks with the cool water, and poured it down their dry parched throats.

How grateful! How refreshing! After the long dusty ride, with what keen enjoyment they squirted the water over their tired, hot bodies, until they were cool and comfortable.

The mystery of the empty tank was a mystery but a short time. The keeper of the elephants on visiting the car had found it and the elephants deluged with water. A few inquiries, and the matter was explained to everyone's satisfaction.

Yours truly,
M.B.D.

LETTER TO *Jack-in-the-Pulpit*,
St Nicholas Magazine, JANUARY 1891

It was a grand year, for the marvellous collection of Humming Birds was added to the museum, and in November 1894 the live zebras arrived at Tring and were, amid great excitement, put into loose boxes in the carriage stables and Walter began the business of breaking them in. This proved no mean task, for the zebras objected strongly to harness and bridles and he had to devise a method of letting down their collars from the ceiling. Loulou Harcourt mentions in his diary that after other guests had departed from Tring he stayed behind to play croquet with Evelina and help Walter drive one of the zebras which was 'beginning to go very well'. Walter first accustomed them singly to a small trap, but he was eventually able to drive them down Piccadilly as a four-in-hand – three zebras and a small pony was the usual combination – and into the fore-court of Buckingham Palace. Charles told the children that the zebra camouflage was so good that half way down the street they seemed to vanish, leaving Walter bowling along Piccadilly in a horseless carriage.... Natty was not pleased about the expedition to the Palace which, in his opinion, invited disaster. Walter himself admitted it was rather risky – his heart was in his mouth when Princess Alexandra tried to pat the leading zebra.

MIRIAM ROTHSCHILD, *Dear Lord Rothschild*,
1983

11

Tricks and Treats

At one end of the box sits an orchestra composed of fleas, each tied to its seat . . . the exhibitor touches each of the musicians with a bit of stick, and they all begin waving their hands about.

FRANK BUCKLAND (1826–80),
Land and Water, 1869

We are told that, in the thirteenth Century, a *horse* was exhibited by the joculators, which danced upon a rope; and *oxen* were rendered so docile as to ride upon horses, holding trumpets in their mouths as though they were sounding them.

JOSEPH STRUTT, *Sports and Pastimes of the People of England*, 1805

[In 1852] came out Mons. Auguste Reinham's 'Industrious Fleas, whose extraordinary performances have received the distinguished patronage of the continental sovereigns. These surprising little creatures consist of a troupe of one hundred, who, after the most unwearied perseverance, have been taught to go through a variety of performances truly wonderful, of which the following form the principal features:–

1. The BALL ROOM, in which two ladies and two gentlemen dance a polka. The orchestra is composed of fifteen musicians, playing on different instruments of proportionate size. Four having a game at whist. A little brunette on a sofa is flirting with a fashionable beau, while her mamma's mind is intensely engaged in the politics of a newspaper. The saloon is lighted by three elegant chandeliers. The performers in this as in all the following pieces are fleas, dressed and instructed according to their respective tasks, &c., &c., &c.' . . .

TOM TAYLOR, *Leicester Square: Its Associations and Its Worthies*, 1874

In the month of July, 1856, I discovered an individual who for twenty years had devoted his life to the intellectual training of fleas. He carried on his operations in a little room in Marylebone-street, London. I entered, and saw fleas here, fleas there, fleas everywhere; no less than sixty fleas imprisoned and sentenced to hard labour for life. All of them are luckily chained, or fastened in some way or other, so that escape and subsequent feasting upon visitors is impossible. A little black speck jumps up suddenly off the table whereon the performance takes place – I walk up to inspect, and find that it is a monster flea attired 'à la convict;' he is free to move about, but, wherever he goes, a long gilt chain, tightly fastened round his neck, accompanies him.

Occasionally he tries to jump; the chain instantly brings him down again, strong as he is. If a flea be fastened to the end of an unbroken wheat straw, he will be strong enough to lift it right off the table on which it is placed. This discovery was first made by the flea proprietor, and made him turn his attention towards utilising the race. One would think it were easy enough to procure troops of fleas, and to train them to perform; but it appears that neither is an easy matter. It is not easy to procure a lot of able-bodied fleas, and it is not every sort of flea that will do. They must be human fleas: dog fleas, cat fleas, and bird fleas, are of no use – they are not lively enough nor strong enough, and soon break down in their training. Human fleas, therefore, must be obtained, and our friend has created a market for them. The dealers are principally elderly females, who supply the raw material; the trade price of fleas, moreover (like the trade price of everything else), varies, but the average price is threepence a dozen. In the winter-time it is sixpence; and, on one occasion, the trainer was obliged to give the large sum of sixpence for one single flea. He had arranged to give a performance; the time arrived; he unpacked the fleas; one, whose presence was necessary to make up a certain number with, was gone. What was to be done? the vacancy must be filled. At last, an ostler, pitying the manager's distress, supplied the needful animal; but he required sixpence for it, and sixpence he got.

While I was looking at the performance there came in a fresh supply of fleas; a swarm of them, in a vial bottle, huddled all together at the bottom. The flea trainer gave them a shake, and immediately they all began hopping about, hitting their little horny heads against the sides of the bottle (which was held sideways) with such force that there was a distinct noise, as if one had gently tapped the bottle with the nail. These fleas were not very good friends, for they were perpetually getting entangled in masses, and fighting with their tiny but powerful legs, and rolling over and over as if in mortal combat. It was not, however, a case of life and death; for I did not see one that was looking injured or tired after the mêlée.

I then observed one fact, which gave me great pleasure; namely, that fleas are at enmity with bugs. There was one bug in the bottle surrounded by many fleas; the poor bug rushed continually from one end of the bottle to the other, running the gauntlet of the assembled fleas; every flea he came near attacked him, and retreated immediately as though half afraid of him; the bug, overwhelmed by numbers, had the worst of it, and beat an ignoble retreat into a bit of flannel.

Fleas are not always brought to market in vial bottles. The best fleas are imported from Russia, and come over in pill-boxes packed in the finest cotton-wool. These fleas are big, powerful, and good workers. I wonder whether the Custom House authorities think it worth while to examine the contents of these pill-boxes. When our friend in Marylebone makes his annual tour into the provinces, his wife sends him weekly a supply of fleas in the corner of an envelope, packed in tissue-paper. She is careful not to put them in the corner where the stamp goes, as the post-office clerk would, with his stamp-marker, at one blow, smash the whole of the stock.

A flea cannot be taken up from its wild state and made to work at once; like a colt or a puppy, it must undergo a course of training and discipline. The training is brought about as follows:– The flea is taken up gently, and a noose of the finest 'glass-silk' is passed round his neck, and there tied with a peculiar knot. The flea, unfortunately for himself, has a groove or depression between his neck and his body, which serves as a capital hold-fast for the bit of silk; it can slip neither up nor down, and he cannot push it off with his legs; he is a prisoner, and is thus tied to his work. This delicate operation is generally performed under a magnifying glass; but, after a time, the eye gets so accustomed to the work that the glass is not always used. In no way is the performing flea mutilated; his kangaroo-like springing legs are not cut off, nor are his lobster-like walking legs interfered with, – a flea must be in perfect health to perform well.

The first lesson given to the novice is the same as that given to a child, namely, to walk. To effect this he is fastened to the end of a slip of card-board, which works on a pin as on a pivot; the moment he feels himself free from the hands, or rather forceps, of the harnesser, he gives a tremendous spring forward: what is the consequence? he advances in a circle, and the weight of the card-board keeps him down at the same time. He tries it again with the same result; finally, he finds the progress he makes in no way equal to his exertions; he therefore, like a wise flea, gives it up, and walks round and round with his card-board as quietly as an old blind horse does in a mill. To arrive at this state of training requires about a fortnight; some fleas have more genius for learning than others, but a fortnight is the average time.

There is another mode of training fleas: it is to shut them up in a small glass box, which turns easily between two upright supporters. The flea, when first put in, hops wildly about, but he only hits his head against the top of the box, and at the same time is supposed to get giddy with the turning round of his prison. I am not aware which system of training has proved the more successful.

Among the trained fleas already at work, I noticed the following: there was a coach with four fleas harnessed to it, who draw it along at a pretty good pace; and I should feel inclined to back the coach in a race against a common garden snail. It is very heavy for the little creatures to drag along, for one pane of glass in the coach is equal to the weight of one hundred fleas. There is a large flea, whose daily task is to drag along a little model of a man-of-war; it is amusing to see him push and struggle to get it along; but get it along he does, although it is two hundred and forty times his own weight. Again, there are two fleas secured, one at each end of a very little bit of gold-coloured paper. They are placed in a reversed position to each other – one looking one way, the other another way. Thus tied, they are placed in a sort of arena on the top of a musical box; at one end of the box sits an orchestra composed of fleas, each tied to its seat, and having the resemblance of some musical instrument tied on to the foremost of their legs. The box is made to play, the exhibitor touches each of the musicians with a bit of stick, and they all begin waving their hands about, as performing an elaborate piece of

music. The fleas tied to the gold paper feel the jarring of the box below them, and begin to run round and round as fast as their little legs will carry them. This is called the Flea's Waltz.

Tightly secured in a tiny chair sits a flea facing a tiny cannon. Several times a-day this unfortunate insect fires this cannon, and in this wise:– One of the little slips which form the feather of a quill-pen is fastened on to one of his legs, and a little detonating powder placed on its tip; the exhibitor then presses the wand down on to the cannon, and scratches the detonating powder; it goes off with a sharp report, making the look-ers-on jump, but it astonishes nobody more than the flea himself; he flourishes the burnt remains of his firing wand madly about in the air, his numerous legs kick about violently, his little head bobs up and down, and altogether he shows as many symptoms of alarm as it is possible for a flea to exhibit. The individual flea that we saw in this state of trepidation did not seem to have got used to his work, though the poor thing had been firing his cannon about thirty times a-day for a month.

The fleas are not kept always in harness; every night each flea is taken out of his harness, is fed, and placed in a private compartment in a box for the night; before they go to bed they have their supper, and in the morning also their breakfasts, upon the hand of their owner – sometimes he has nearly all his fleas on the backs of his hands at the same moment, all biting and sucking away. For more than twenty years has he thus daily fed his fleas without any detriment to his health: the quantity of blood each flea takes away being imperceptibly small – one drop of blood, he considers, would feed a flea for many weeks; but it is the itching sensation caused by the flea cutting the skin which is unpleasant. This feeling of itching he felt painfully when he first began to submit himself to the tender mercies of his little performers: now he is so hard-ened that he feels them not at all, whether biting or sucking. When, however, there are many on his hands at the same time, he suffers from a sensation of great irritation all over his body, which passes away when their supper is over. He has remarked that fleas will not feed if his hand be not kept perfectly motionless; the act, therefore, of feeding and harnessing is troublesome, and he is obliged to devote two hours in the morning and two in the afternoon to it. His fleas generally live a long time, provided they are properly fed and taken care of. He once had a flea, a patriarch, who for eighteen months was occupied in pulling up a little bucket from a well: this flea lived longer than any other flea he ever had, and he believes he died finally from pure old age; for he was found dead one day, faithful to his post, with his bucket drawn half-way up the well.

FRANK BUCKLAND (1826–80),
Land and Water, MAY 1869

Les Serins Savants.
The escapades of
trained canaries.

OISEAUX MERVEILLEUX
Willis' Rooms
King Street
St James'

Mlle Emilie VANDERMEERSCHT has the honour to announce that she
will give THREE MORNING PERFORMANCES JULY at the rooms
above, on THURSDAY July 15; Friday July 16; and Saturday July 17,
with her celebrated troupe of LEARNED BIRDS, whose extraordinary
and Truly marvellous exercises have been hitherto exhibited only in the
salons of the nobility and gentry. These performances, which are perfectly unique, comprise, amongst numerous feats equally curious and
interesting, the solution of Problems in Mental Arithmetic, Divination
and Secret Thoughts, Fortune Telling, Exercises in Orthography, Tours
d'Escamotage, &c&c To commence each day at Three o'Clock –
Reserved seats may be obtained at Mr MITCHELL'S Royal Library,
33 Old Bond Street; and at the doors, on the days of performance.

Illustrated London News, 10 JULY 1852

'In the course of the present summer' says a writer in the Gentleman's Magazine for 1772, the Sieur Roman from Paris exhibited his academy of birds in the city of Canterbury, &c. To me their performances seemed wonderful. One appeared as dead, and was held up by the tail or claw without shewing any signs of life; a second stood on its head, with its claws in the air; a third mimicked a Dutch milk-maid going to market with pails on its shoulders; a fourth mimicked a Venetian girl looking out of a window ... and the last bird stood in the midst of some fire-works, which were discharged all around it, without discovering any signs of fear   The birds were linnets, goldfinches and Canary birds.

NATHANIEL WANLEY (1634–80),
The Wonders of the Little World, 1678

In 1772 the celebrated Daniel Wildman exhibited here (Winchester Place, now Pentonville Road) his bees every evening (wet evenings excepted). He made several new and amazing experiments, he rode standing upright, one foot in the neck, with a curious mask of bees on his head and face. He also rode standing upright on the saddle with the bridle in his mouth and by firing a pistol, made one part of the bees march over a table and the other part swarm in the air and return to their proper hive again. Wildman's performances of the 'Bees on Horseback' were also thus described:–

> He with uncommon art and matchless skill
> Commands those insects, who obey his will;
> With bees others cruel means employ,
> They take their honey, and the bees destroy;
> Wildman humanely, with ingenious ease
> He takes the honey, but preserves the bees.

JOHN TIMBS, *English Eccentrics
and Eccentricities*, 1866

Now to conclude, that you may make a horse to pisse when you woulde have him (or at least to straine and move himselfe thereunto) or to leave pissing when you please, you shall for two dayes at the least, watch him, and keepe such distance of times, that hee may never pisse, but when you urge him, and to that end you shall once in two or three houres leade your horse uppe and downe upon straw, so softly, that hee may as it were put out one foot and stand stil, then another and stand stil, yourselfe continually saying unto him Pysse, Pysse, and thus you shal do if it be a whole day togither, till he do pisse or straine himselfe to pisse, and then you shall reward him with bread, and til he do pisse or straine himself to pisse you shal neither moove him in any other lesson nor let him tast foode though it were for a weeke together, and by this meanes after he understands your meaning, you shal no oftner say pisse, but he either will pisse or straine himself to pisse, and then at your pleasure aquainting him with a contrarie worde as No more or such like, which being spoken in threatning sort and accompanied with correction, you shall make him staie his pissing. So you may use him in his eating, drinking, or anie other naturall action, and though these appears verie unnatural, yet they are as easie to be effected, as anie toye whatsoever.

GERVASE MARKHAM (1568–1637),
Cavelarice, or The English Horseman, 1607

It was to King's Cross fair that I moved from Stepney with my conjuring show, and very well I did with it, in spite of counter-attractions – and they were many – especially in the way of performing animals, such as fortune-telling ponies and 'learned pigs.'. . .

Well, now, the making of a learned pig is upon this wise. You get your pig, fat and comfortable-looking and not too old, a fairly long stout stick, a leather strap that will buckle neatly round the pig's neck and has also a small plate and screw rivet that will attach it to the stick. Then you are ready to commence the lessons.

In the end of the stick, not the end to which the strap is attached, you bore a hole, and through this drive a long nail into the floor of your academy so that the stick can move freely round on it in a circle, but in no other way. When the pig's neck is buckled into the strap at the other end of the stick the animal is bound to move in a circle, of which the nailed end of the stick is the centre.

Then with a little cane to direct his movements you induce the pig to walk. Of course he goes round and round and round, for he can move in no other direction, and when he wants to stop, which is often, you just keep him going by gentle taps with the cane. When you have kept him walking round some time you begin to let him stop in his course now and again, but always just before the stop giving a slight click with the fingers. The slightest sound will do, merely the snap of the thumb-nail against the finger-nail is sufficient. The pig will hear it, and in a very short time will stop anywhere in his monotonous walk directly he hears the slight signal.

You then vary the lesson by arranging a pack of cards face upwards just outside the circle, fixed, of course, by the length of the stick, which the pig traverses, and commence to patter as if to an audience somewhat in this style: 'Well, Toby, you see the cards before you. Which is the ace of spades?' Any card you like you can, of course, name. Round goes the pig in his circle, and as he comes opposite the card 'click' go your nails, and he at once stops.

TOBY

THE
SAPIENT PIG,

From the Royal Rooms, Spring Gardens,

The only Scholar of his Race in the World.

THIS MOST EXTRAORDINARY CREATURE

Will Spell and Read, Cast Accounts,

PLAY AT CARDS;

Tell any Person what o'Clock it is to a Minute

BY THEIR OWN WATCH;

ALSO TELL THE AGE OF ANY ONE IN COMPANY,

And what is more Astonishing he will

Discover a Person's Thoughts

A Performance beyond all others the most Incredible.

Mr. HOARE having spent a number of Years in accomplishing this great un-
dertaking, leaves it to a discerning Public, to judge of the laborious task he has
had in bringing the above Animal before them, as of all others in Nature, none
are so obstinate as his species, and it is only by unremitted assiduity and attention,
that he has finally brought to such great perfection what Man never did before.

He is in Colour the most beautiful of his Race, in Symmetry the most Perfect,
in Temper the most Docile; his Nature is so far from being offensive, that he
is pleasing to all who honor him with their presence.

> The silken rob'd peer, and the delicate *belle*,
> Are unsulled by filth, unoffended by smell:
> Toby turns all disdainful from deeds of offence,
> For what would so blast his pretensions to sense.

He EXHIBITS every day at the

Temple Rooms, Fleet-street,

Near TEMPLE BAR, opposite CHANCERY LANE,

At the Hours of **1** and **3**, precisely,

And again in the Evening at **7** and **9** o'Clock.

ADMITTANCE ONE SHILLING.

Just Published, The Life & Adventures of TOBY the SAPIENT PIG,

With his Opinion on Men and Manners,

May be had at the Exhibition Rooms, Price One Shilling.

Printed by H. LYON, John Street, Edgware Road.

'You see, ladies and gentlemen,' you proceed, 'Toby knows the cards. Will someone kindly name a card they would like him to pick out.' Round goes piggy as you patter, and 'click,' you stop him where you like. In two or three days the pig, without the stick or the strap, will commence to move round at a tap from your switch whenever a circle of cards or persons is formed. He will also stop dead at the finger-click until the touch of the switch lets him know he must move on again. Then his education is complete.

You can send him round a circle of people, asking him to pick out the man that likes kissing the girls. In fact, vary your entertainment as you will, the pig will be listening for the 'click,' not to your patter, and will stop directly he hears it, while the audience will not notice the slight sound. With every performance the pig will improve, especially if you accustom him to receive after each show an apple, potato, or some such little luxury.

I have seen first-class learned pigs trained in a week by this simple method. Their intelligence consists almost entirely in having a sharp ear for the 'click' that brings them a welcome stop in their walk. The rest of the performance that so amazes the onlookers is due to the showman's arrangement of his cards, his audience, and his patter.

In my long life I have found many wonderful things, besides the performance of the learned pig, whose entire art and mystery consists of a quick ear and a nimble tongue, and, incidentally, as you will discover if you follow my story, have profited thereby.

'LORD' GEORGE SANGER (1825–1911),
Seventy Years a Showman, 1910

Mrs L., of Bonchurch, brought home to England with her, after a visit to Spain, a little white poodle which she purchased from some professional dog-trainers in Barcelona. It proved an affectionate, intelligent little creature, but displayed no special talent, and for the first two years it was in Mrs L.'s possession she did not know that it had been taught any tricks beyond that of begging for food. At the end of that time, however, two little Italian boys came to spend the afternoon with her. When they entered the drawing-room the poodle was resting unconcernedly on the hearth-rug, but the moment they began to talk to each other in Italian, as they did loudly and volubly, it jumped up, manifested excitement and alarm, and forthwith went through a little performance, throwing a somersault and attempting to stand on its head and walk on its forepaws, after which it ran and cowered in a corner. It was soon calmed and reassured; but whenever during the afternoon one of the boys spoke to it directly in Italian in a loud and commanding voice it made a feeble attempt to perform. It was obvious that it had been trained as an artiste, and that the sound of the Italian language, resembling that of the Spanish tongue employed by its master, set the old and somewhat rusty machinery in motion again.

HUGH MASSINGHAM,
Dogs, Birds and Others, 1921

The people of Sybaris, a city of Calabria, are proverbial on account of their effeminacy; and it said that they taught their horses to dance to the music of the pipe; for which reason, their enemies the Crotonians, at a time when they were at war with them, brought a great number of pipers into the field, and at the commencement of the battle, they played upon their pipes, the Sybarian horses, hearing the sound of the music began to dance; and their riders, unable to manage them as they ought to have done, were thrown into confusion, and defeated with prodigious slaughter. This circumstance is mentioned by Aristotle, and if not strictly true, proves, at least, that the teaching of animals to exceed the bounds of action prescribed by nature was not unknown to the ancients.

JOSEPH STRUTT, *Sports and Pastimes
of the People of England*, 1805

An aged gentleman has mentioned to us that, about fifty years ago, a Frenchman brought to London from eighty to a hundred dogs, chiefly poodles, the remainder spaniels, but all nearly of the same size, and of the smaller kind. On the education of these animals their proprietor had bestowed an immense deal of pains. From puppyhood upwards, they had been taught to walk on their hind-legs, and maintain their footing with surprising ease in that unnatural position. They had likewise been drilled into the best possible behaviour towards each other; no snarling, barking, or indecorous conduct took place when they were assembled in company. But what was most surprising of all, they were able to perform in various theatrical pieces of the character of pantomimes, representing various transactions in heroic and familiar life with wonderful fidelity. The object of their proprietor was, of course, to make money by their performances, which the public were accordingly invited to witness in one of the minor theatres.

Amongst their histrionic performances was the representation of a siege. On the rising of the curtain, there appeared three ranges of ramparts, one above the other, having salient angles and a moat, like a regularly-constructed fortification. In the centre of the fortress arose a tower, on which a flag was flying; while in the distance behind appeared the buildings and steeples of a town. The ramparts were guarded by soldiers in uniform, each armed with a musket or sword, of an appropriate size. All these were dogs, and their duty was to defend the walls from an attacking party, consisting also of dogs, whose movements now commenced the operations of the siege. In the foreground of the stage were some rude buildings and irregular surfaces, from among which there issued a reconnoitring party; the chief, habited as an officer of rank, with great circumspection surveyed the fortification; and his sedate movements, and his consultations with the troops that accompanied him, implied that an attack was determined upon. But these consultations did not pass unobserved by the defenders of the garrison. The party was noticed by a sentinel, and fired upon; and this seemed to be the signal to call every man to his post at the embrasures.

Shortly after, the troops advanced to the escalade; but to cross the moat, and get at the bottom of the walls, it was necessary to bring up some species of pontoon, and accordingly several soldiers were seen engaged in pushing before them wicker-work scaffoldings, which moved on castors towards the fortifications. The drums beat to arms, and the fearful bustle of warfare opened in earnest. Smoke was poured out in volleys from shot-holes; the besieging forces pushed forward in masses, regardless of the fire; the moat was filled with the crowd; and, amid much confusion and scrambling, scaling-ladders were raised against the walls. Then was the grand tug of war. The leaders of the forlorn-hope who first ascended, were opposed with great gallantry by the defenders; and this was perhaps the most interesting part of the exhibition. The chief of the assailants did wonders; he was seen now here, now there, animating his men, and was twice hurled, with ladder and followers, from the second gradation of ramparts; but he was invulnerable, and seemed to receive an accession of courage on every fresh repulse. The scene became of an exciting nature. The rattle of the miniature cannon, the roll of the drums, the sound of trumpets, and the heroism of the actors on both sides, imparted an idea of reality that for the moment made the spectator forget that he was looking on a performance of dogs. Not a bark was heard in the struggle.

After numerous hairbreadth escapes, the chief surmounted the third line of fortifications, followed by his troops; the enemy's standard was hurled down, and the British flag hoisted in its place; the ramparts were manned by the conquerors; and the smoke cleared away – to the tune of 'God Save the King.'

It is impossible to convey a just idea of this performance, which altogether reflected great credit on its contriver, as also on the abilities of each individual dog. We must conclude, that the firing from the embrasures, and some other parts of the mecanique, were effected by human agency; but the actions of the dogs were clearly their own, and showed what could be effected with animals by dint of patient culture. . . .

Another specimen of these canine theatricals was quite a contrast to the bustle of the siege. The scene was an assembly-room, on the sides and the farther end of which seats were placed; while a music-gallery, and a profusion of chandeliers, gave a richness and truth to the general effect. Livery-servants were in attendance on a few of the company, who entered and took their seats. Frequent knockings now occurred at the door, followed by the entrance of parties attired in the fashion of the period. These were, of course, the same individuals who had recently been in the deadly breach; but now all was tranquillity, elegance, and ease. Parties were formally introduced to each other with an appearance of the greatest decorum, though sometimes a young dog would show a slight disposition to break through restraint, but only to the increased amusement of the beholders. Some of the dogs that represented ladies were dressed in silks, gauzes, laces, and gay tasteful ribbons. Some wore artificial flowers, with the flowing ringlets of youth; others wore the powdered and pomatumed head-dress of riper years, with caps and lappets, in ludicrous contrast to the features of the animals. Doubtless the whole had been the result of judicious study and correct arrangement, for the most animated were habited as the most youthful. The animals which represented gentlemen were judiciously equipped; some as youthful, and others as aged beaux, regulated by their degrees of proficiency, since those most youthfully dressed were most attentive to the ladies. The frequent bow, and return of curtsey, produced great mirth in the audience; but when the noses of the animals neared each other, it produced a shriek of delight from the youthful spectators. On a sudden the master of the ceremonies appeared. No doubt he was the chief in the battle fray. He was now an elegant fellow, full of animation; he wore a superb court-dress, and his manners were in agreement with his costume. He approached many of the visitors: to some of the gentlemen he gave merely a look of recognition; to the ladies he was generally attentive; to some he projected his paw familiarly, to others he bowed with respect; and introduced one to another with an air of elegance that surprised and delighted the spectators. There was a general feeling of astonishment at some of the nicer features of the scene, as at the various degrees of intimacy which individuals expressed by their nods and bows of recognition.

As the performance advanced, the interest increased. A little music was heard as from the gallery, but it was soon interrupted by a loud knocking, which announced the arrival of some important visitor, and expectation was raised. Several livery servants entered, and then a sedan-chair was borne in by appropriately dressed dogs; they removed the poles, raised the head, and opened the door of the sedan; forth came a lady, splendidly attired in spangled satin and jewels, and her head decorated with a plume of ostrich feathers! She made a great impression, and appeared as if conscious of her superior attraction; meanwhile the chair was removed, the master of the ceremonies, in his court-dress, was in readiness to receive the *elegante*, the bow and curtsey were admirably interchanged, and an air of elegance pervaded the deportment of both. The band now

'The bow and the curtsey were admirably interchanged.'

struck up an air of the kind to which ball-room companies are accustomed to promenade, and the company immediately quitted their seats and began to walk ceremoniously in pairs round the room. Three of the ladies placed their arms under those of their attendant gentlemen. On seats being resumed, the master of the ceremonies and the lady who came in the sedan-chair arose; he led her to the centre of the room; Foote's minuet struck up; the pair commenced the movements with an attention to time; they performed the crossings and turnings, the advancings, retreatings, and obeisances, during which there was a perfect silence, and they concluded amid thunders of applause. What ultimately became of the ingenious manager with his company, our informant never heard. . . .

Mr S——, having acquired a competency by commercial industry, retired from business, and devoted himself, heart and soul, to the cultivation and enjoyment of music. Every member of his little household was by degrees involved more or less in the same occupation, and even the housemaid could in time bear a part in a chorus, or decipher a melody of Schubert. One individual alone in the family seemed to resist this musical entrancement; this was a small spaniel, the sole specimen of the canine race in the mansion. Mr S—— felt the impossibility of instilling the theory of sounds into the head of Poodle, but he firmly resolved to make the animal bear *some* part or other in the general domestic concert; and by perseverance, and the adoption of ingenious means, he attained his object. Every time that a *false note* escaped either from instrument or voice – as often as any blunder, of whatever kind, was committed by the members of the musical family (and such blunders were sometimes committed intentionally) – down came its master's cane on the back of the unfortunate Poodle, till she howled and growled again. Poodle perceived the meaning of these unkind chastisements, and instead of becoming sulky, showed every disposition to howl on the instant a false note was uttered, without waiting for the formality of a blow. By and by, a mere glance of Mr S——'s eye was sufficient to make the animal howl to admiration. In the end, Poodle became so thoroughly acquainted with, and attentive to, false notes and other musical barbarisms, that the slightest mistake of the kind was infallibly signalised by a yell from her, forming the most expressive commentary upon the misperformance.

When extended trials were made of the animal's acquirements, they were never found to fail, and Poodle became, what she still is, the most famous, impartial, and conscientious connoisseur in the duchy of Hesse. But, as may be imagined, her musical appreciation is entirely negative; if you sing with expression, and play with ability, she will remain cold and impassable. But let your execution exhibit the slightest defect, and you will have her instantly showing her teeth, whisking her tail, yelping, barking, and growling. At the present time, there is not a concert or an opera at Darmstadt to which Mr S—— and his wonderful dog are not invited, or, at least, *the dog*. The voice of the prima donna, the instruments of the band – whether violin, clarionet, hautbois, or bugle – all of them must execute their parts in perfect harmony, otherwise Poodle looks at its master, erects its ears, shows its grinders, and howls outright. Old or new pieces, known or unknown to the dog, produce on it the same effect.

It must not be supposed that the discrimination of the creature is confined to the mere *execution* of musical compositions. Whatever may have been the case at the outset of its training, its present and perfected intelligence extends even to the secrets of composition. Thus, if a vicious modulation, or a false relation of parts, occurs in a piece of music, the animal shows symptoms of uneasy hesitation; and if the error be continued, will infallibly give the grand condemnatory howl. In short, Poodle is the terror of all the middling composers of Darmstadt, and a perfect nightmare to the imagination of all poor singers and players. Sometimes Mr S—— and his friends take a pleasure in annoying the canine critic, by emitting all sorts of discordant sounds from instrument and voice. On such occasions the creature loses all self-command, its eyes shoot forth fiery flashes, and long and frightful howls respond to the immelodious concert of the mischievous bipeds. But the latter must be careful not to go too far; for when the dog's patience is tried to excess, it becomes altogether wild, and flies fiercely at the tormentors and their instruments.

This dog's case is a very curious one, and the attendant phenomena not very easy of explanation. From the animal's power of discerning the correctness of musical composition, as well as of execution, one would be inclined to imagine that Mr S——, in training his dog, had only called into play faculties existing (but latent) before, and that dogs have in them the natural germs of a fine musical ear. This seems more likely to be the case, than that the animal's perfect musical taste was wholly an acquirement, resulting from the training. However this may be, the Darmstadt dog is certainly a marvellous creature, and we are surprised that, in these exhibiting times, its powers have not been displayed on a wider stage. The operatic establishments of London and Paris might be greatly the better, perhaps, of a visit from the critical Poodle.

Chambers's Miscellany of Useful and Entertaining Tracts, 1845

12

Improbabilities

Her words seemed to have a magical effect upon the camel. For it gave a great bound forward that nearly unseated her, and then proceeded in a rapid trot in the direction of the oncoming hounds.

LORD BERNERS (1883–1950),
The Camel, 1936

I give you notice beforehand that I must remain true to the same theme, and record a breakfast at the Duke of Devonshire's at Chiswick . . . the great ornament of the fête was the beautiful Lady Ellenborough. She came in a small carriage drawn by poneys not larger than Kamtschatkad-ale dogs, which she drove herself. From henceforward the doves may be unyoked from the chariot of Venus, and poneys harnessed to it instead.

PRINCE M. L. H. VON PÜCKLER-MUSKAU,
Tour in Germany, Holland and England
in the Years 1826, 1827 and 1828, Vol. II, 1832

(43). *Item,* the skin, head, and legs of a cameleon, perfumed and stuffed. The creature was given me alive in Africa, and it liveth (not by the air, as the report goeth, but) by flies chiefly, as the Moores taught me how to feed it in this manner, by laying in the cage, or sometimes out of the cage in which I kept it, upon a paper some sugar and sweetmeats, which allureth the flies to come to it. The creature hath in its gorge or *gola* a toung that lieth 4 dobled, with a small fibulus button at the end of it, which hath on it a viscous matter. So soon as it seeth the flies at the sweetmeats it darteth forth that toung at a great distance, and with the viscous matter pulleth in the fly to her mouth, and eateth it; and so it will do many, one after the other, so that while we sailed homewards all along the Africa shore, and came out of the Mediterranean Sea by the Streights of Gibralter into the Atlantick Ocean, and then turning north-ward by Spain and Portugall – all that time (I say) that we were in those hot and southerly climates, although it was in January 1662, there were store of flies, and the creature fed on them heartily, and lived well. But as we sailed homeward into the more cold and northern climates, as the flies failed us, so that decayed, and at lenght for want of flies it died; and I had the chirurgeon of the shipp embalm it, and put the skin as you see it.

It seemeth to be a kind of lizard, but is as slow in pace as a tortes, winding its tail about the sticks of the cage, to help and secure its gra-dations. The ribs and the back are boned and scaled like fish. Although the story of its living by the air be fabulous, yet the other story of its changing itself into all colours is very true, as I have seen this of all manner of colours, like silk, and sometimes changeable colours, as the sun happened to shine upon it; and sometimes I have seen it coal-black. But the story is false that it hath a pellucid body, like cristal, and so it will be the colour of scarlet or any other cloth that you lay it upon. No, no such thing; but one way to make it change its colours was to anger it, and put it into a passion, by touching of it with a stick or a bodkin, or the like. Then it would fetch great breaths, many one after another, by which it made itself swell very much, and in its swellings out came the colours of all sorts, which changed as it was more or less provoked to anger. And when the passion was over, it would look as pale as a clout.

It hath no eyelids, and therefore never winketh; but when it sleepeth, the ball of the eye being as round as round can be, it turneth that ball quite round, the inside outward, and so sleepeth. Matthiolus on Dioscorides sayth that it layeth eggs as a tortes doth, and is bred of those eggs.

JOHN BARGRAVE, *Pope Alexander the Seventh and the College of Cardinals, with a Catalogue of Dr Bargrave's Museum*, 1676

Dried in a most frolicsome position, the chameleon still lies in the drawer of the earliest extant 'Cabinet of Curiosities', at Canterbury

I may as well confess that at the mere sight of Mrs Knox's purple bonnet my heart had turned to water. In that moment I knew what it would be like to tell her how I, having eaten her salmon, and capped her quotations, and drunk her best port, had gone forth and helped to steal her horse. I abandoned my dignity, my sense of honour; I took the furze prickles to my breast and wallowed in them.

Mrs Knox had advanced with vengeful speed; already she was in high altercation with Flurry at no great distance from where I lay; varying sounds of battle reached me, and I gathered that Flurry was not – to put it mildly – shrinking from that economy of truth that the situation required.

'Is it that curby, long-backed brute? You promised him to me long ago, but I wouldn't be bothered with him!'

The old lady uttered a laugh of shrill derision. 'Is it likely I'd promise you my best colt? And still more, is it likely that you'd refuse him if I did?'

'Very well, ma'am.' Flurry's voice was admirably indignant. 'Then I suppose I'm a liar and a thief.'

'I'd be more obliged to you for the information if I hadn't known it before,' responded his grandmother with lightning speed; 'if you swore to me on a stack of Bibles you knew nothing about my colt I wouldn't believe you! I shall go straight to Major Yeates and ask his advice. I believe *him* to be a gentleman, in spite of the company he keeps!'

I writhed deeper into the furze bushes, and thereby discovered a sandy rabbit run, along which I crawled, with my cap well over my eyes, and the furze needles stabbing me through my stockings. The ground shelved a little, promising profounder concealment, but the bushes were very thick, and I laid hold of the bare stem of one to help my progress. It lifted out of the ground in my hand, revealing a freshly-cut stump. Something snorted, not a yard away; I glared through the opening, and was confronted by the long, horrified face of Mrs Knox's colt, mysteriously on a level with my own.

Even without the white diamond on his forehead I should have divined the truth; but how in the name of wonder had Flurry persuaded him to couch like a woodcock in the heart of a furze brake? For a full minute I lay as still as death for fear of frightening him, while the voices of Flurry and his grandmother raged on alarmingly close to me. The colt snorted, and blew long breaths through his wide nostrils, but he did not move. I crawled an inch or two nearer, and after a few seconds of cautious peering I grasped the position. They had buried him.

A small sandpit among the furze had been utilised as a grave; they had filled him in up to his withers with sand, and a few furze bushes, artistically disposed round the pit, had done the rest. As the depth of Flurry's guile was revealed, laughter came upon me like a flood; I gurgled and shook apoplectically, and the colt gazed at me with serious surprise, until a sudden outburst of barking close to my elbow administered a fresh shock to my tottering nerves.

Mrs Knox's woolly dog had tracked me into the furze, and was now baying the colt and me with mingled terror and indignation. I addressed him in a whisper, with perfidious endearments, advancing a crafty hand towards him the while, made a snatch for the back of his neck, missed it badly, and got him by the ragged fleece of his hind-quarters as he tried to flee. If I had flayed him alive he could hardly have uttered a more deafening series of yells, but, like a fool, instead of letting him go, I dragged him towards me, and tried to stifle the noise by holding his muzzle. The tussle lasted engrossingly for a few seconds, and then the climax of the nightmare arrived.

Mrs Knox's voice, close behind me, said, 'Let go my dog this instant, sir! Who are you——'

Her voice faded away, and I knew that she also had seen the colt's head.

I positively felt sorry for her. At her age there was no knowing what effect the shock might have on her. I scrambled to my feet and confronted her.

'Major Yeates!' she said. There was a deathly pause. 'Will you kindly tell me,' said Mrs Knox slowly, 'am I in Bedlam, or are you? And *what is that?*'

She pointed to the colt, and that unfortunate animal, recognising the voice of his mistress, uttered a hoarse and lamentable whinny. Mrs Knox felt around her for support, found only furze prickles, gazed speechlessly at me, and then, to her eternal honour, fell into wild cackles of laughter.

So, I may say, did Flurry and I. I embarked on my explanation and broke down; Flurry followed suit and broke down too. Overwhelming laughter held us all three, disintegrating our very souls. Mrs Knox pulled herself together first.

'I acquit you, Major Yeates, I acquit you, though appearances are against you. It's clear enough to me you've fallen among thieves.' She stopped and glowered at Flurry. Her purple bonnet was over one eye. 'I'll thank you, sir,' she said, 'to dig out that horse before I leave this place. And when you've dug him out you may keep him. I'll be no receiver of stolen goods!'

She broke off and shook her fist at him. 'Upon my conscience, Tony, I'd give a guinea to have thought of it myself!'

SOMERVILLE AND ROSS (1858–1949 and 1862–1915), *Some Experiences of an Irish RM*, 1899

In South America there are some remarkable animals which are popularly known under the name of 'Coaiti-mondi.' They are allied to the racoons, and, like those creatures, are excellent tree-climbers. They have very long and mobile noses, and in consequence have been called by the generic name of Nasua, *i.e.* Nosey. They are easily tamed, and in their own country are often kept in a domesticated state.

I have now the pleasure of presenting to the reader an account of a tame coaiti-mondi which lived in this country. . . .

'Kiko was the most affectionate and amusing pet we ever had. He very soon found his way into the garden, and used to grub up the earth with his flexible snout for the purpose of digging out the great lob-worms, which he sucked up with great glee. He also was clever at finding and crunching snails.

'But "Tim," our gardener, thought that he did more harm than good, and, whenever he saw Kiko digging, used to pelt him. Then Kiko would run up a poplar tree, break off twigs, and throw them at Tim, as he worked, uttering the while little derisive squeaks of "Kiko-kiko."

'I regret to say that there was nothing which Kiko loved so much in life as a good *stink*. I can use no other word. A rotten egg was a joy to him – but not for ever, as we soon washed him – but it really was a joy to him as *long* as it lasted.

'He would not eat it – for it was much too precious to be disposed of in that manner; but he would dip his long nose in it, and spread it carefully over his hairy tail. When he had used up every drop, he would turn round at intervals, bring his nose to his perfumed tail, and with grunts of pleasure would take long, long sniffs.

'Once, among the men-folk of the household, there was a grand pipe cleaning, they washing out their meerschaums with gin. You cannot imagine what a walking nosegay Kiko made of himself. If anyone who was smoking a pipe stooped down to pat him, Kiko would fork out the contents with one claw, and do it so swiftly that there was no time for interference. Of course, my father and brother, having once been served so, were always on their guard; but, even then, Kiko was often one too many for them.

'When his nose became soiled from digging in the garden, he would

clean it on his tail, rubbing it up and down just as a knife is cleaned on a board. . . .

'One day, my mother detected him in stealing and beat him for it; whereupon he disappeared whimpering sadly. In an hour or so, she felt sorry that she had whipped him, and set off in search of Kiko, hoping to comfort him. She might have spared herself the trouble, for when we found him he was sitting up in a waste-paper basket, hugging a rolled-up pancake, which he had stolen from the pantry, and nibbling the top. He looked exactly like Punch with his staff.

'He once stole an egg while cooking was going on. My mother saw him, and chivied him all over the house until at last she cornered him. He still kept the egg in his mouth, and when he saw that he must be caught, he deliberately raised his head as high as he could, dashed the egg down on the carpet, and "kikoed" frantically with glee.

'The little animal would always find out a scented handkerchief, roll it up into a tight ball, sit upright, hold the ball to his nose with his fore-paws, and sniff at it with eyes closed in ecstasy. After enjoying it for some time, he would turn round and rub it gently up and down his tail.

'The little animal would always find out a scented handkerchief and sniff it with eyes closed in ecstacy.'

'This tail was a funny piece of furniture, for it never moved to the right or left when he turned round, but stuck out stiffly in an arch behind. He was very fond of lying inside the fender, and whenever we perceived a smell of burning, we always knew that it was only Kiko's tail raking out the lower bar as he turned. He never seemed to feel a burn, and once quite roasted the tip, which he ate, and enjoyed immensely. It soon got well again, and there did not seem to be any sore.

'Most of our friends had a horror of Kiko, which he knew, and used to take advantage of it. He would gallop furiously at them, making extraordinary noises, with his mouth wide open (he had tremendous teeth) and his nose straight up in the air. Of course they used to rush away, or get on chairs and call for help, and then Kiko used to double up with fun and chuckle most humanly.

'A favourite trick of his was to lie in wait on the top of the garden wall, leap on the back of anyone who was passing underneath, and cling round their neck. It was done for pure mischief, as he never treated any of his friends in that fashion. . . .

'If any of us touched Kiko, he had the power of distinguishing the owner of the hand, even though he did not see it.

'Sometimes my mother would take him on her lap and hide his head so that he could neither see nor hear. Then we children, six of us, would all stand round with our hands spread out, and one would poke or pinch him. He always darted round and seized the hand that had touched him, squeaking loudly at the same time.

'He never bit one of us, but always understood that we were playing a game, and enjoyed it thoroughly. We always kept our hands well mixed up, so I suppose that he must have been guided by scent.

'He would insist on following us to church, and then had to be popped into a high truck while we made our way out of his sight. When at last he managed to climb out, he generally made a few morning calls on the neighbouring cottages. They had a wholesome horror of him, and the women would shriek in their own tongue, "Mae diawl yn dod" (pronounced "My jowl in dode") – *i.e.* "The devil is coming." Then they would skedaddle upstairs, and let him work his wicked will on their Sunday's dinner.

'Though he went by the name of "Capt. F——'s devil," he was a most affectionate little creature, and used to smuggle himself under our coats and aprons. A favourite position of his was to sit in a lady's arms, with his paws around her neck, and his nose safely stowed away in the coils of her hair.

'Of his nose he took the greatest care, and, whenever he was frightened, always grasped it with both his hands. If he were scolded he would sit up, take his nose between his paws, whimper, and shed great tears. He reminded me of Hans Christian Andersen's tale of "Punchinello," because he always looked most comical when he was most miserable.

'I am sorry to say that he was given away because he destroyed the top of the garden wall by picking out the mortar and throwing the stones down. He soon changed hands again, because whenever he saw a strange woman he would frighten her by getting under her dress and scrambling up her petticoats.'

REV. J. G. WOOD (1827–89),
Petland Revisited, 1890

A friend of Edward James's, Arthur Ryall, was in hospital in
Nice with double pneumonia and a telegram had been sent
to his mother in New York

. . . she came straight over on the *Ile de France* which arrived at Villefranche only five days after I had sent her the telegram. When she arrived at the hospital, I was amusing Arthur, who by now was convalescent, with a pet bear I had been given by Mrs Benjamin Guinness. It was a Russian bear, which for a while I had kept at Lady Hadfield's villa, La Sungarelli, because my mother wouldn't have it in hers. On one occasion it got loose in the garden and then during a dinner-party for twenty people it pulled the entire table-cloth off, and all the glass and the wine and all the cutlery crashed onto the floor, terrifying everybody. Even Lady Hadfield, who was sweet and generous and sorry for me because she knew how miserable I was with my mother, thought this was a bit too much and, smiling sweetly, said in her Virginian drawl, 'Edward, dear, I think perhaps the bear is *de trop*.' Anyway, theré I was, cheering up Arthur, and to my horror Mrs Ryall arrived in the room. I immediately dropped the bear, which was very frisky, and it got under Arthur's bed and started chewing his slippers. Mrs Ryall noticed something going on and asked, 'What kind of an animal have you got under the bed? Is it a dog?'

'Well,' I replied, 'a sort of dog,' and put my arm under the bed to pick up the bear which so far had been friendly with me; but he didn't know whose arm it was and bit me to the bone. I gave a little cry of astonishment, blood pouring down my arm, and Mrs Ryall exclaimed, 'But what is that? The dog's savage!'

'Actually it's not exactly a dog, it's a bear.'

Mrs Ryall immediately leaped up on one of the radiators against the wall and stood there shrieking. . . . Then, as I was binding up my arm in a towel, the bear came out from under the bed, ran along the corridor of the second floor of the hospital and bit the matron in the calf. I nearly went to gaol, and was not allowed to see Arthur again until he left hospital. The bear ended up in a zoo in Cannes.

EDWARD JAMES, *Swans Reflecting Elephants*,
1982, EDITED BY GEORGE MELLY

When I paid my visit to Lady W—— this morning, she had just received a great cargo of curiosities from one of her sons, who is travelling in South America. Among them was a lion-monkey, with a tail and mane like those of the king of beasts, on a body not larger than that of a rat. Instead of the disagreeable smell of most of his tribe, this little fellow exhales musk and cinnamon; and . . . perfumed the room like a pastille.

PRINCE M. L. H. VON PÜCKLER-MUSKAU,
Tour in Germany, Holland and England
in the Years 1826, 1827 and 1828, Vol. II, 1832

Tiglath Pileser, the bear, was about six months old when he entered Christ Church, where he lived in a corner of a court beside Fell's Buildings. He was provided with cap and gown, and in this costume was taken to wine parties, or went boating with his master, to the wonderment of the children in Christ Church Meadow, who would follow them down the walk leading to the boats, regardless of expostulations and threats, until sometimes the bear was turned loose and shambled after them, whereupon they fled.

Tig, as he was familiarly called, took part in the proceedings of the British Association at Oxford in 1847, attending in cap and gown the garden party at the Botanic Gardens, and receiving a visit from Lord Houghton, then Mr Monckton Milnes, who attempted to mesmerise him in his corner. This made the bear furious, but he gradually yielded to the influence, and at last fell senseless on the ground.

Of this meeting Sir Charles Lyell wrote: 'In the evening we had an immense party at the Botanic Gardens. Young Buckland had a young bear dressed up as a student of Christ Church, with cap and gown, whom he formally introduced to me and successively to the Prince Canino (Charles Buonaparte), Milne Edwards, member of the French Institute, and Sir T. Acland. The bear sucked all our hands and was very caressing. Amid our shouts of laughter in the garden by moonlight, it was diverting to see two or three of the dons, who were very shy, not knowing how far their dignity was compromised.'

Tig at last fell under the censure of the Dean of Christ Church. 'Mr Buckland,' the Dean is reported to have said, 'I hear you keep a bear in college; well, either you or your bear must go.'

So Tig was sent to Islip, a living held by Dean Buckland, seven miles from Oxford, and lived there for some months with the eagle and Jacko, who had also been rusticated. Tig sometimes rode out on horseback with his master, at other times he took walking exercise, following any who would allow him to suck their fingers. This unsubstantial solace was so much to his taste that it was needful to provide a lump of sugar to induce him to let go.

From Islip, Tig once visited Canon and Mrs Rawlinson at the neighbouring village of Merton, and was first put in the stable; but the horses became so excited that he was removed to an outhouse. Soon after a maid, running across the fields breathless and capless, called Mrs Rawlinson back from her walk, crying out that the bear had got loose; they were all afraid to catch him; he was in the kitchen, and the leg of mutton roasting for luncheon would be eaten up. Returning instantly, there appeared a doubt whether the bear was in the kitchen or the scullery; the cook had shut herself up in one, and the bear in the other. Both were, however, soon released, and the bear again consigned to his outhouse. He soon escaped again, and walked up into the nursery, where the first-born baby was sleeping. Father and mother ran upstairs and met Tig walking out of the room, leaving the child quietly sleeping.

After this Tig was sent back to his master. Led by a chain, he trotted quietly along until in one of the fields a flock of sheep came in view.

This temptation he could not withstand, he was off in an instant after them. The poor old shepherd, his dog and the sheep, all fled, with the bear after them, enjoying the fun. At last Tig was caught by his conductor, who picked him up and carried him the rest of the way to Islip.

At Islip, news came one day that the bear had broken his chain and was in the grocer's shop, devouring the sugar and sweetstuff, and terrifying the shop-woman out of her wits. She had retired to the back parlour, leaving him in possession of the shop, and had sent a messenger out at the back door for relief. After this the bear developed such a proclivity for the sweetstuff shop, to the damage of the woman's nerves and his master's pocket, that in November 1847 he was sent to the Zoological Gardens, where he died some time after in an effort to cut his teeth.

GEORGE C. BOMPAS,
Life of Frank Buckland, 1885

Perhaps the event by which the British public at home have most cause to remember Murray's Consul-Generalship is the important part he took in securing a hippopotamus for the Zoological Society. . . . By the friendly cooperation of the Peninsular and Oriental Company a chamber was constructed on their steamship *Ripon*, in which the hippopotamus was conveyed to England, and deposited in the Zoological Society's Gardens on 25th May 1850. Obaysch, for he was so named after the place of his capture, was supplied with a mate, Adhela, in 1853, and lived twenty-eight years in Regent's Park, dying in 1878. Mr Phillip Lutlet Sclater, the present secretary to the Society, remembers Murray frequently visiting his old pet, shouting to him in Arabic, when the enormous creature would come towards him, grunting loudly in recognition of one of his earliest friends.

SIR HERBERT MAXWELL, BART, MP, FRS,
Hon. Sir Charles Murray KCB: A Memoir, 1898

Now there had arisen a similar difficulty about this camel. Antonia had evidently set her heart on keeping it. The Vicar, as he paced up and down the room, was once more assailed by a sense of guilt with respect to their childlessness. The horrid suspicion that perhaps he was to blame for it often weighed upon his mind. Of course Antonia had never reproached him; she was so loyal and devoted a wife that, even had she possessed positive scientific proofs, she would never have dreamt of doing so. The Vicar felt that he owed Antonia every form of consolation that it lay within his power to offer. She had always been an excellent wife and the comfortable income she had inherited from her parents had certainly been of great assistance to him. Perhaps it was rather unworthy of him to object to her keeping a camel if she really wished to. He began to relent. A further paroxysm of sobbing broke down the last barrier of his opposition.

'Well, well,' he said, patting his wife's heaving shoulders, 'if you really want to keep it, we will see how we get on. Perhaps, after all, it won't be a source of so much trouble as I had imagined. I will postpone my letter to the Zoo.'

And so the camel remained for the time being at Slumbermere Vicarage.

A few days passed and Antonia felt a growing desire to go for another ride on the camel, for her own pleasure as well as for practical purposes. Toby, the pony, was no better, and the Vet. had said that he feared it would never again be fit for service.

When Antonia informed her husband, with a certain apprehension, that she thought of going for a ride on the camel again, she found to her intense relief that he made no objection. He even offered to help her in saddling and harnessing the beast. He was in an excellent humour that day. He had thought of a very good riddle for the Parish Magazine; it had come to him that morning, after breakfast, in the closet.

Q. Why is Sunday unlike any other day?

A. Because it is the Lord's day.

He was delighted with it. These little inspirations always afforded him intense pleasure, and, after all, it was not so easy as one might imagine to think of something bright and original for the Parish Magazine.

So the Vicar was in the best of tempers. He made several jokes during the harnessing of the camel, and when finally he saw Antonia perched on its back he said to her in his most jocular vein, 'Really my dear, I can hardly claim to be an authority on the subject, but you seem to me to have an excellent seat on a camel. All the same, dear,' he added more seriously, 'if I were you, I don't think I should ride through the village.'

Antonia thought as he did, and she decided to take the same direction she had taken before, through the little wood beyond the churchyard. She had hardly emerged from it into the open fields on the other side when she was startled by the sudden blast of a horn, accompanied by the distant sound of galloping horses. She saw, through the trees, the flash of a scarlet coat and a pack of hounds moving swiftly over the brow of the hill. She realised that a fox hunt was in progress.

Now Antonia, on account of her great love of animals, was very much averse to blood sports, but she objected far less to foxhunting than to the other forms of harrying wild beasts. Firstly because it gave pleasure to a large number of ladies and gentlemen, some of whom were her personal friends and who were regular in their attendance at Divine Service. And secondly because foxes very frequently made incursions into her hen coops. Only quite recently one of them had carried off a very fine cock with which she was hoping to win a prize at the local poultry show. But when, a few moments later, the hunted fox passed within a couple of yards of her, in an obviously exhausted condition, her heart was smitten with compassion.

'Poor creature,' she cried aloud, 'I hope it may escape!'

Her words seemed to have a magical effect upon the camel. For it gave a great bound forward that nearly unseated her, and then proceeded in a rapid trot in the direction of the oncoming hounds. It appeared as though it were trying to place itself between the hounds and the fox. The hounds, however, were so intent on their pursuit that they paid little attention to the monster that was attempting to bar their passage and they surged past it, dividing on both sides like a rushing stream meeting a barge. The camel lashed out at them with an unexpected ferocity, severely injuring two of the hounds who lay writhing and whimpering on the ground.

At that moment the Master, Captain Jollyboy, came galloping up; he seemed to be in a fine rage. His usually rubicund countenance was deepened by wrath to a dark purple. He was using the most frightful language. A stream of the foulest expletives flowed from his lips. Antonia had never heard such expressions. He shook his fist and raised his hunting crop with a menacing gesture. The camel turned suddenly and faced its aggressor, whereupon Captain Jollyboy's horse was seized with panic, wheeled round and galloped off in the opposite direction. Captain Jollyboy was powerless to hold it and he was carried for nearly a quarter of a mile before he was able to pull up. The apparition of the camel had a no less disastrous effect on the rest of the hunting field. In every direction horses reared up into the air, unseating their riders; others bucked, kicked up their heels and bolted. The ground was strewn with fallen ladies and gentlemen struggling in the mud. One or two of them were quite badly injured. The hunting field assumed the aspect of a battle field after an unsuccessful cavalry charge.

Poor Antonia was completely dazed by what had happened. It seemed to her like an appalling nightmare. She was saved however from any responsible decision by the camel itself, who appeared to be quite unperturbed by the violence and magnitude of the catastrophe it had caused, and set off gently but firmly in the direction of home, bearing the unresisting Antonia away from the scene of carnage. She was almost in a fainting condition when she reached the Vicarage.

LORD BERNERS (1883–1950),
The Camel, 1936

The novelty of a man ascending to the clouds by the assistance of a quantity of inflammable air, contained in a balloon of thirty three feet diameter, was a curiosity which this country had never beheld, and of course both the credulous and the infidel attended. . . .

Mr Lunardi was accompanied in his aerial passage by a couple of pigeons, a cat, and a favourite lap-dog. . . . When the grand machine appeared superbly floating in the newly subdued element, and the cradle containing the bold Aerial Navigator was seen depending from it, astonishment filled the multitude, and *awful silence filled the air*, which the next instant was in tremulation with the most impassioned bursts of applause. . . .

Mr Lunardi appeared perfectly composed, and as the balloon went up, bowed most gracefully, and calmly waved his flag to the admiring and wonder-struck spectators. . . . Being evidently too much encumbered, he threw it out. Soon after one of his oars broke from the pivot, and he threw that down also; but . . . made use of the other occasionally to direct his course. . . .

When Mr Lunardi had gained the utmost altitude of his ascension, he felt so strong a propensity for sleeping, that it was with the utmost difficulty he could keep himself awake; the cold at this time became so intensely piercing, as to render Mr Lunardi's situation in it almost insupportable. . . . The cat was . . . benumbed . . . and had not Mr Lunardi's regard for his dog led him to afford him the warmth of his bosom, the animal would inevitably have perished.

The prospects were grand and awful beyond the power of imagination. . . .

Le Cheval Aréonaute.

After Mr Lunardi had been up . . . about an hour and a half, the thermometer stood at 35 degrees, when the atmosphere was so cold, that icicles were upon his clothes, and he was fearful his balloon would burst; at this time he drank several glasses of wine . . . on throwing out some air the thermometer rose to 50, when the atmosphere was delightful. . . . At Northaw . . . he threw out his cat . . . which was taken up alive. . . .

Lunardi's Grand Aerostatic Voyage through the Air,
containing a complete and circumstantial Account of the
Grand Aerial Flight made by that enterprising Foreigner
in his Air Balloon on Sept. 15th 1784

The origin of the Jack Russell

It was on a glorious afternoon towards the end of May, when strolling round Magdalen meadow with Horace in hand, but Beckford in his head, he emerged from the classic shade of Addison's Walk, crossed the Cherwell in a punt, and passed over in the direction of Marston, hoping to devote an hour or two to study in the quiet meads of that hamlet, near the charming slopes of Elsfield, or in the deeper and more secluded haunts of Shotover Wood. But before he had reached Marston a milkman met him with a terrier – such an animal as Russell had as yet only seen in his dreams; he halted, as Actæon might have done when he caught sight of Diana disporting in her bath; but, unlike that ill-fated hunter, he never budged from the spot till he had won the prize and secured it for his own. She was called Trump, and became the progenitress of that famous race of terriers which, from that day to the present, have been associated with Russell's name at home and abroad.

EDWARD W. L. DAVIES,
Memoir of the Rev. John Russell, 1883

As company for the monkeys and myself for many years past I have had a 'Jemmy.' All my suricates I call 'Jemmys.' The Latin name is *Suricata Zenick*. Jemmy is a very pretty little beast, somewhat like a small mongoose, or very large rat. His head is as like the head of a hedgehog as can be imagined. His colour is light brown with a darker stripe down the sides. He is an African animal, and lives in burrows on the plains, whence he is sometimes called the 'prairie dog,' or meercatze.

I should like now to say something of the habits of this pretty little fellow. Jemmy the Third (for I have previously had two Jemmys) was allowed the free range of the whole house. He was full of curiosity and restlessness, running continually from one room to another. He very seldom walked; his pace, on the contrary, was a short gallop, or rather canter. When on the move he always gave tongue like a hound on a

'A pretty little fellow.'

scent. It was impossible to describe his melodious cry in words. When handled and petted he would utter a sharp bark, not unlike that of a dog, and if he was in a very good humour I could by imitating him make him bark alternately with myself. His great peculiarity was his wonderfully intelligent and observant look. He had the habit also of sitting up on his tail like a kangaroo; his fore-paws on this occasion were like a dog's when begging. He was very fond of warmth, and would sit up inside the fender and warm himself, occasionally leaning back against the fender and looking round with the satisfied air of an old gentleman reposing after dinner.

When the morning sun came into the room 'Jemmy' would go and sit in the sunbeams and look out of the window at the passing cabs and omnibuses. While doing this he had a way of turning round very sharply, and looking with his little pig's eyes at me and back, as much as to say, 'What do you think of that?' When breakfast came up he would dance round me on his hind legs watching for something. I often put him on the breakfast-table, for if I did not put him up he would climb up uninvited. It was very amusing to see him go and smell the egg, and, in his own language, swear at it for being hot. He could not understand its being hot enough to burn his nose; raw eggs were his special favourites. His great delight was to scratch about among the lumps of sugar. He was also very fond of cream, and it was most amusing to see him try to get the little drops of cream I had left for him out of the cream-can, as left by the milk-woman. I am obliged to have my cream in this little can, otherwise the cats, Judy the marmoset, or somebody else would be sure to have it before I come down. I placed the cream-can on the floor, and it was fun to see Jemmy try to force it open with his teeth to get the cream out; he used quite to lose his patience with this metal cream-can.

After breakfast Jemmy generally had a stand-up fight with the monkeys. He would inspect, from the outside, the bottom of the monkey-cage. If he discovered any portion of the monkeys' breakfast which he thought might suit him, he would immediately try to steal it by thrusting his arms through the bars. The monkeys invariably resented this indignity. The old, crippled monkey, Carroty Jane, could only make eyes and faces at him. The wicked, impudent Little Jack would jump up and down like an india-rubber ball, all the time well inside the cage, where Jemmy could not get at him. When Jemmy was fighting the monkeys he would stand on his hind legs and show his lovely white carnivorous teeth at them, turning up his sharp mole-like nose in a most contemptuous manner, all the time keeping up a continuous bark, into which fun the parrot generally entered, and barked like Jemmy also. . . .

Jemmy was always fond of getting under anything or into any kind of hole; his great delight was to get into a boot, and when he got to the end he scratched as though he wanted to get farther into the burrow. Frequently I found my boot going round the room, propelled apparently by some internal machinery. This machinery was Master Jemmy.

Jemmy was a greedy little fellow. John could not bring up any kind of food into my room without Jemmy. He would watch the cook broiling the chop down stairs, and when John brought it up would follow close to his heels; and what between Jemmy's pretty begging manner, the monkey's plaintive cries, and the parrot's demand, it often happened that I myself got very little of the chop.

I had hoped to have written a fuller biography of our poor little Jemmy the Third, but, alas! Jemmy was taken with a fit. I did everything I could to relieve the poor little fellow, but the fits were too much for him, and I have since been busily occupied in making his skin into a mat and his bones into a skeleton. The last Jemmy died of eating cotton wool; this Jemmy died, I think, of eating too much, for he was as fat as a little bacon pig, and weighed 2 lbs., a great weight for such a little animal. It is curious how fond I become of dear little animals such as Jemmy, and how much I miss his pretty little ways as I sit in the Monkey room writing this memoir of my little pet.

FRANK BUCKLAND (1826–80),
Notes and Jottings from Animal Life, 1882

The war was even worse than I expected. Mud and lice. Hunger and cold. Pain and waste. Whenever I think of it I remember especially the beautiful horses lifting their heads and feet high with life one minute, lying torn and bloody the next. Dying for what – for a quarrel they didn't make – for glory they didn't want – for prizes they wouldn't share?

When their tormented eyes asked us this question the only answer we could give them was a straight quick bullet.

For us men it wasn't much better.

After twenty months I had a furlough, two weeks. First was a bath; second, clean clothes; third, get Kola and go home.

I rode in a carriage to Vanno's house. Before I got out I could see it was empty, the door padlocked, the windows boarded up. I knocked and shook at the latch. No answer.

Once again someone at the fountain helped me, this time an old man.

Vanno. Yes, he knew him. At last Vanno, too, had been called for a soldier. Gone about two months. Where or what regiment, he couldn't say. The old lion died. Some Syrians, he thought, had come for the monkeys. But a bear, a smart bear, he smiled to think of that animal it was so smart, Vanno himself had taken it away. Where he could not say. . . .

I went back to Tiflis and spent the days following the trail of some travelling shows into the country but no Kola was with them.

My furlough finished and I met the other soldiers who were going back to the front with me the next day. They had a carriage and they wanted to drive about five miles outside town to a little inn in the woods for a farewell celebration.

'Come on,' they urged me. 'There's music and a green lawn for dancing and all we can eat for the last time maybe until the war is over.'

So I went along, and time did pass better in somebody's company than my own. When we came to the inn it was crowded but the host led us to a table under a little green arbor and laid a clean cloth and gave us fresh trout and cucumbers and salt cheese.

'And wine!' I called out across the garden to the waiter. 'Bring us wine to drink with it.'

The answer to me was a great bellowing roar that sent everybody to their feet. Then we heard wood splinter and bricks crash. From the kitchen, the cook came screaming. 'Run, run for your lives.'

Something plunged after him. Men crowded back. A woman fainted. I stood on a chair and looked over the arbor. Long before my head understood my heart knew. I jumped across the table and pushed my way through the crowd.

'Kola,' I cried, and he ran into my arms.

When I could make him stop hugging me I looked around. The proprietor stood ready with a gun aimed at us, the cook beside him with a cleaver. The women were still screaming, hands over their eyes so they wouldn't see me killed.

'Please everybody,' I said, 'be at ease. This is my bear and he doesn't hurt anybody. I lost him and he found me. He recognized my voice when I spoke and he's excited. But how,' I said to the inn-keeper, 'did *you* get him? I left him with Vanno, The Bear Man.'

'The young soldier tells the truth,' the man spoke to the whole garden. 'As I stand here, it *is* his bear, and no ordinary animal either. So my brother Vanno told me, and so I found out for myself. When Vanno was called to the army he came to me, "I have arranged for the other animals but you must keep the bear and keep him well until I return or you will answer to me and to the bear's brother who is also a soldier." Then to prove the rareness of this beast he showed me a bankbook. Ladies and gentlemen, it is true, may I lose heaven if I lie, to the account of Kola, a bear, is deposited one hundred and fifty rubles in the Tiflis Government Bank. Now, young sir, let me chain him again.'

'No,' I said, 'he will stay quietly beside me.'

And he did. We made a place for him at our table. We ate and we drank and we danced and Kola crying with joy did all his tricks over and over and hugged me until I hardly had breath left in me.

I stayed until at last he fell asleep from so much food and wine and we could put him on the chain. Then I had to tell him good-by again.

'Keep him,' I said to the innkeeper. 'Send me news of him until I come for him.' I gave him my address.

The next morning I was supposed to go direct to the troop train but I couldn't. I had to have one more glimpse of Kola first. I took a carriage back to the inn. When I drove in, the place was in an uproar. The proprietor ran here and there. The cook chattered. The waiters looked behind the doors. Finally the story came out. When they carried Kola's food to him that morning his post was splintered, his chain broken. Kola had run away.

'He's gone to find me,' I said.

I looked until I found his tracks and we followed them half a mile up the mountain to where the deep forest began. There I told the others to wait. I walked in alone calling as I went. Far back in among the rocks I thought I heard a rustle – yes – and a low growl. I stood in the half-darkness and spoke.

'Kola? Kola, if that's you, don't come out. Go back into the woods and be your own bear. For you it's a better chance than tied on a chain. Wait for me there and the day wars are over I'll come for you again. Stay hidden and wait.'

Then I turned and went back down the road to town alone. And in the forest my Kola still waits.

GEORGE AND HELEN PAPASHVILY,
Anything Can Happen, 1985

In 1857, the late Mr Ainley, Surgeon, of Bingley, informed me that there was a very singular hatch of the domestic duck in his neighbourhood, every one of which had some congenital defective formation. I desired my friend to procure the whole hatch for me, in order that I might have them 'mounted' as specimens of 'sports of nature.' Mr Ainley, intrinsically shrewd, and well skilled in ornithology, and whose heart was immersed in natural history pursuits, returned home immediately, anxiously hoping to lay hold of the whole hatch, but, to his great disappointment and regret, only just in time to secure a solitary male . . .

On becoming possessed of this bird, which fortunately arrived at my house when the Squire was spending the day with me, it was forthwith carefully examined by Mr Waterton and myself, and found to be devoid of a particle of web between its toes. My old friend, who had a keen eye for the discovery of any peculiarity, was in an ecstacy of delight, and instantly, on this discovery, proposed that it should go on to the lake at Walton Hall, until matured and dressed in its most captivating plumage, when he would then show me a specimen of his handiwork in taxidermy, if we could then harden our hearts to return a verdict of execution by chloroform.

Mr Waterton became much attached to this bird, which was, on its arrival at Walton Hall, named by him, and, consequently, by all in the establishment, 'Doctor Hobson,' a name which it continued to retain so long as it floated as a living mallard on the lake, and, indeed, singularly enough, for a short period after its death.

In 1860, a somewhat ludicrous circumstance, connected with this water-fowl, occurred. A farm servant, hat in hand, and with elongated visage, approached the Squire, saying, 'If you ple-ase Squire, Doctor Hobson is de-ad.' My good old friend, who was always much more careful of my life than of his own, instantaneously pictured *my* death and not that of the bird. He urgently interrogated the hand as to the source of his information, and to the trustworthiness of his authority for the assertion he had made; when the reply was, 'I seed him mysen liggin de-ad, all on a lump at dam he-ad.' The mystery, hitherto inexplicable to my well-wisher, at once vanished, and Mr Waterton, combining the grave with the gay, immediately wrote an exceedingly clever and most entertaining letter of condolence and congratulation to me, in which, in melancholy strain he regretted the death of this favourite bird, but rejoiced that his friend was not 'liggin all on a lump at dam he-ad,' and added, 'I shall, in due time, "set up" your namesake, so that no taxidermist in existence shall have just cause to display a fault-finding physiognomy, and that is saying a good deal, as we bird-stuffers are admitted to be a snarling and pugnacious tribe.'

In July, 1860, the Squire, on presenting me with this beautifully-preserved mallard, declared that he felt as much delighted on the occasion as if he were treading on enchanted ground. Its attitude is perfect, whilst the various elevations and depressions of feathers give the appearance of actual life. . . .

I may add, that one duck in this extraordinary brood was hatched with its head reversed, having its bill, as regards its horizontal position, appearing and indeed actually situated immediately above its tail, so that when food was placed on the ground, behind its tail, this duck always had to seize it by turning a somersault.

RICHARD HOBSON, *Charles Waterton:*
His Home, Habits and Handiwork, 1866

The manatee belongs to the class *Sirenia*, but it is very puzzling to know what she is, whether a pachyderm or a cetacean. I think she may be said to be a little of both. She is purely an aquatic animal, and when seen in the water her head reminds one of something between a mole and a pig. Her body is terminated by a large tail, the shape of a lady's fan. She swims with it moving it up and down, with the same action as a porpoise, and not sideways like a fish. Take a pig, tie his hind legs and curly tail, and flatten them into a broad, flat appendage, like a beaver's tail; turn his fore-feet into paddles like a turtle's flippers; cut off his ears, give him valvular nostrils like a seal, reduce his eyes to one-fourth, and then you will have a manatee. . . .

Mr Bartlett was much pleased to find the manatee feed so well; she would eat lettuces and vegetable marrows all day. She got quite tame like a sheep, and would follow Mr Bartlett round the pond and eat from his hand. It was very interesting to remark the extreme quiet with which this animal, one can hardly say swims, but rather gently glides through the water. Its skin is covered with two kinds of hair, soft and bristly. The appearance of the back reminds one of a prickly pear. The nostrils are most peculiar; they are situated at the extreme end of the nose, and the two valves seem to rise from the inside, with exactly the same quiet motion as does the hydraulic lift when it rises level with the platform at the Great Western station.

FRANK BUCKLAND (1826–80),
Notes and Jottings from Animal Life, 1882

'Something between a mole and a pig.'

13

Of Beasts and Men

Adolphus Cooke ... believed in reincarnation. This
made him unfailingly courteous to his father, whom he
believed had returned to earth in the guise of a turkey
cock. He instructed his men servants to take off their
hats to this bird every time they passed it, while women
had to genuflect and bend one knee.

PETER SOMERVILLE-LARGE,
Irish Eccentrics, 1975

On one of the first days of the year 1880, in the early afternoon, husband and wife went for a walk in the copse on the little hill above Rylands. They were still at this time like lovers in their behaviour and were always together. . . .

Hearing the hunt, Mr Tebrick quickened his pace so as to reach the edge of the copse, where they might get a good view of the hounds if they came that way. His wife hung back, and he, holding her hand, began almost to drag her. Before they gained the edge of the copse she suddenly snatched her hand away from his very violently and cried out, so that he instantly turned his head.

Where his wife had been the moment before was a small fox, of a very bright red. It looked at him very beseechingly, advanced towards him a pace or two, and he saw at once that his wife was looking at him from the animal's eyes. You may well think if he were aghast: and so maybe was his lady at finding herself in that shape, so they did nothing for nearly half-an-hour but stare at each other, he bewildered, she asking him with her eyes as if indeed she spoke to him: 'What am I now become? Have pity on me, husband, have pity on me for I am your wife.'

So that with his gazing on her and knowing her well, even in such a shape, yet asking himself at every moment: 'Can it be she? Am I not dreaming?' and her beseeching and lastly fawning on him and seeming to tell him that it was she indeed, they came at last together and he took her in his arms. She lay very close to him, nestling under his coat and fell to licking his face, but never taking her eyes from his. . . .

One fancy that came to him, because he was so much more like a lover than a husband, was that it was his fault, and this because if anything dreadful happened he could never blame her but himself for it.

So they passed a good while, till at last the tears welled up in the poor fox's eyes and she began weeping (but quite in silence), and she trembled too as if she were in a fever. At this he could not contain his own tears, but sat down on the ground and sobbed for a great while, but between his sobs kissing her quite as if she had been a woman, and not caring in his grief that he was kissing a fox on the muzzle. . . .

. . . after that he went about getting his experiment ready, which was this. In the garden he gathered together a nosegay of snowdrops, those being all the flowers he could find, and then going into the village of Stokoe bought a Dutch rabbit (that is a black and white one) from a man there who kept them.

When he got back he took her flowers and at the same time set down the basket with the rabbit in it, with the lid open. Then he called to her: 'Silvia, I have brought some flowers for you. Look, the first snowdrops.'

At this she ran up very prettily, and never giving as much as one glance at the rabbit which had hopped out of its basket, she began to thank him for the flowers. Indeed she seemed indefatigable in shewing her gratitude, smelt them, stood a little way off looking at them, then thanked him again. Mr Tebrick (and this was all part of his plan) then took a vase and went to find some water for them, but left the flowers beside her. He stopped away five minutes, timing it by his watch and listening very intently, but never heard the rabbit squeak. Yet when he went in what a horrid shambles was spread before his eyes. Blood on the carpet, blood on the armchairs and antimacassars, even a little blood spurtled on to the wall, and what was worse, Mrs Tebrick tearing and growling over a piece of the skin and the legs, for she had eaten up all the rest of it. The poor gentleman was so heartbroken over this that he was like to have done himself an injury, and at one moment thought of getting his gun, to have shot himself and his vixen too. Indeed the extremity of his grief was such that it served him a very good turn, for he was so entirely unmanned by it that for some time he could do nothing but weep, and fell into a chair with his head in his hands, and so kept weeping and groaning.

After he had been some little while employed in this dismal way, his vixen, who had by this time bolted down the rabbit skin, head, ears and all, came to him and putting her paws on his knees, thrust her long muzzle into his face and began licking him. But he, looking at her now with different eyes, and seeing her jaws still sprinkled with fresh blood and her claws full of the rabbit's fleck, would have none of it. . . .

'Oh Silvia, Silvia, would you had never done this! Would I had never tempted you in a fatal hour! Does not this butchery and eating of raw meat and rabbit's fur disgust you? Are you a monster in your soul as well as in your body? Have you forgotten what it is to be a woman?'

Meanwhile, with every word of his, she crawled a step nearer on her belly and at last climbed sorrowfully into his arms. His words then seemed to take effect on her and her eyes filled with tears and she wept most penitently in his arms, and her body shook with her sobs as if her heart were breaking. This sorrow of hers gave him the strangest mixture of pain and joy that he had ever known, for his love for her returning with a rush, he could not bear to witness her pain and yet must take pleasure in it as it fed his hopes of her one day returning to be a woman. So the more anguish of shame his vixen underwent, the greater his hopes rose, till his love and pity for her increasing equally, he was almost wishing her to be nothing more than a mere fox than to suffer so much by being half-human.

At last he looked about him somewhat dazed with so much weeping, then set his vixen down on the ottoman, and began to clean up the room with a heavy heart. He fetched a pail of water and washed out all the stains of blood, gathered up the two antimacassars and fetched clean ones from the other rooms. While he went about this work his vixen sat and watched him very contritely with her nose between her two front paws, and when he had done he brought in some luncheon for himself, though it was already late, but none for her, she having lately so infamously feasted. But water he gave her and a bunch of grapes. . . .

The summer was over and Mr Tebrick noticed this now for the first time and was astonished. . . .

That night he slept indoors, but in the morning early he was awoken by the sound of trotting horses, and running to the window saw a farmer riding by very sprucely dressed. Could they be hunting so soon, he wondered, but presently reassured himself that it could not be a hunt already.

He heard no other sound till eleven o'clock in the morning when suddenly there was the clamour of hounds giving tongue and not so far off neither. At this Mr Tebrick ran out of his house distracted and set open the gates of his garden, but with iron bars and wire at the top so the huntsmen could not follow. There was silence again; it seems the fox must have turned away, for there was no other sound of the hunt. Mr Tebrick was now like one helpless with fear, he dared not go out, yet could not stay still at home. There was nothing that he could do, yet he would not admit this, so he busied himself in making holes in the hedges, so that Silvia (or her cubs) could enter from whatever side she came.

At last he forced himself to go indoors and sit down and drink some tea. While he was there he fancied he heard the hounds again; it was but a faint ghostly echo of their music, yet when he ran out of the house it was already close at hand in the copse above.

Now it was that poor Mr Tebrick made his great mistake, for hearing the hounds almost outside the gate he ran to meet them, whereas rightly he should have run back to the house. As soon as he reached the gate he saw his wife Silvia coming towards him but very tired with running and just upon her the hounds. The horror of that sight pierced him, for ever afterwards he was haunted by those hounds – their eagerness, their desperate efforts to gain on her, and their blind lust for her came at odd moments to frighten him all his life. Now he should have run back, though it was already late, but instead he cried out to her, and she ran straight through the open gate to him. What followed was all over in a flash, but it was seen by many witnesses.

The side of Mr Tebrick's garden there is bounded by a wall, about six feet high and curving round, so that the huntsmen could see over this wall inside. One of them indeed put his horse at it very boldly, which was risking his neck, and although he got over safe was too late to be of much assistance.

His vixen had at once sprung into Mr Tebrick's arms, and before he could turn back the hounds were upon them and had pulled them down. Then at that moment there was a scream of despair heard by all the field that had come up, which they declared afterwards was more like a woman's voice than a man's. But yet there was no clear proof whether it was Mr Tebrick or his wife who had suddenly regained her voice. When the huntsman who had leapt the wall got to them and had whipped off the hounds Mr Tebrick had been terribly mauled and was bleeding from twenty wounds. As for his vixen she was dead, though he was still clasping her dead body in his arms.

Mr Tebrick was carried into the house at once and assistance sent for, but there was no doubt now about his neighbours being in the right when they called him mad.

For a long while his life was despaired of, but at last he rallied, and in the end he recovered his reason and lived to be a great age, for that matter he is still alive.

DAVID GARNETT (1892–1981),
Lady into Fox, 1923

. . . the pranks of Mr Waterton were as unexpected as they were endless. Dr Hobson tells us that 'To show the playful levity of my octogenarian friend, I may mention a circumstance which occurred without a moment's warning [*sic*] of a very unexpected nature. In the north-east corner of the first entrance hall of the mansion stood a table, on which to place the hats and greatcoats, gloves, etc., of arriving visitors, which was covered by a large cloth, hanging down to the floor. On seeing me drive up the bridge in front of the house, the Squire has more than once secretly crept on all fours like a dog under the table, waiting in order that I might place my greatcoat upon this table; and whilst I was thus unsuspiciously engaged, he has, in his private retreat, commenced to growl like a savage dog behind the cloth, and has seized my legs in such a practically canine manner that I really had no idea at the time but that some fierce dog was attacking my lower extremities.'

At last, after his legs had experienced the sharpness of the aged Squire Waterton's teeth on several occasions, Dr Hobson thought it better to 'drop a hint of the fortuitous hazard incurred, stating that many instances were recorded, on undoubted authority, where even permanent aberration of the mind had been the result of such a sudden and unexpected shock to the nervous system'. Overcome with remorse at this revelation of the risk that had been run, Mr Waterton promised not to bite his friend's legs again, and Dr Hobson's reason remained unimpaired. In future, the Squire welcomed Dr Hobson by 'dancing down the whole length of the broad flagged walk, and from time to time (even when the snow lay deep upon the ground) throwing one of his loose slippers from his foot high up in the air above his head and expertly catching it in his hand in its descent. . . . The wetness of the flags underfoot,' we are told, 'or a shower overhead, never constituted any impediment to an exploit of this character.'

EDITH SITWELL (1887–1964),
English Eccentrics, 1933

Caligula . . . loved *Prasinus* the Cochman so wel, that for good wil to the master, he bid his horse to supper, gave him wine to drink in cups of estate, set barly graines of golde before him to eate, and swore by no bugs, that hee would make him a Consul: which thing (saith *Dion*) had bin performed, had hee not bin prevented by suddain death.

STEPHEN GOSSON,
The Schoole of Abuse, 1579

'I have more relatives in the South than any Blue Ridge Rabbit.'

ABRAHAM LINCOLN IN MICHAEL CURTIS'S
FILM *'Virginia City'*, 1940

When Dr Buckland was at Palermo on his wedding tour in 1826, he, as all strangers did, visited the shrine, and with his keen eyes saw in a moment that the bones never belonged to Rosalia. 'Those are the bones of a goat,' he said, 'not of a woman!' Of course the priests were greatly scandalised, and declared that the saint would not permit him to see what only the faithful could discern. From that time, however, the bones were enclosed in a casket, and neither faithful nor heretics were any longer permitted to scan the sacred relics too closely.

MRS E. O. GORDON, *Life of Dean Buckland,* 1894

Under the influence of a rare Austrian wine Dean Spanley could be persuaded to talk about his life as a dog:-

'Winds, you know, blew down paths and between trees, and we noted what was at the other end of the wind. And we hadn't gone far when there was a strange scent indeed, quite close, and there was a big hole in the chalk and the scent coming pouring out of it. "You have big game in your woods," said my friend.

'"Very big game indeed," I said.

'And we went up to the hole. There was a badger at home down there, and he seemed to be asleep. So we decided to wake him up. We just put our mouths right into the hole and barked at him. He was a long way down, but he must have heard us. We barked at him for ten minutes. It was the greatest fun. Then we went on through the wood, and presently what should we see but a very showy young rabbit who was out on a long walk, no doubt for social reasons. So we chased the young fellow all through the wood and back to his own house. It was a very popular neighbourhood, and we didn't stop and dig: too many passages, you know, running in all directions. So I said: "Let's come and hunt a large bad animal."'

'He bid his horse to supper.'

'It was a pig that I meant, but that's how one puts it. You know the sniff beside the other fellow's face, and the beckoning of the head, that means that?'

I merely nodded: I was not going to interrupt the Dean just now with a request for explanations. And I looked at Wrather, in case he was going to do so; but Wrather sat silent and interested.

'He said, "Let's." And I said, "I know where there is one." And I ran on in front.

'So we came to the pig's house and looked in through his door at him and shouted, "Pig." He didn't like that. He looked just like a pig; he was a pig; and he knew it. He came towards his door saying silly surly things in a deep voice. You know the kind of talk. And we just shouted, "Pig. Pig. Pig." Both of us, for nearly half an hour. It was perfectly splendid and we enjoyed it immensely.'

'What did he say?' asked Wrather.

'What could he say?' said the dean. 'He knew he was a pig. But he didn't like being told about it. I've seldom enjoyed myself more. It made up for that fool of a rabbit that was so silly in the hedge on the hill.

'Then we went to the back of the stables and rolled in something nice, till our coats were smooth and we both had a beautiful scent. Then we killed a hen for the fun of it. It was a lovely evening.'

LORD DUNSANY, My Talks With Dean Spanley,
1936

Adolphus [Cooke] left three wills. . . . His second will left the property to a cousin, Dr Wellington Purdon, who lived nearby. Dr Purdon was a follower of hounds, and one day the Westmeath pack came into Cookesborough, the Doctor behind them, and killed a fox in front of Adolphus. He was most upset. He did not always see himself spending the after-life sitting and reading improving books. Sometimes he imagined that he would return to earth as a bird, but increasingly he felt that he might become a fox. He hated hunting, and changed his will again to exclude Dr Purdon, at the same time ordering his men to dig a number of exceptionally deep foxholes and trenches lined with stone which he could use in case of need. He was worried about being hunted in the next life, but consoled himself with the reflection that he had a good knowledge of the topography of the district. It may have been to no avail. A fox was killed by the Westmeaths in the kitchen of Cookesborough very soon after he died. Dease's *History of the Westmeath Hunt* remarks that 'the kitchen was a fit and proper place to find a Cooke'. (Adolphus may have known that thirty miles away at Larch Hill, Kilcock, a Mr Watson had also built himself an earth for use after his death. This earth is highly ornamental; the great mound which covers the dome-shaped chamber is decorated with a pepperpot tower, probably erected by Mr Watson's widow, who had an obsession for building follies. After his death food was regularly laid out for Mr Watson beside his earth, and hounds were directed never to draw his coverts.)

The third will left the Cookesborough estate to a younger son of the Earl of Longford. The Honourable Mr Edward Pakenham had very few prospects and was perfectly willing to hyphenate his name and become Pakenham-Cooke in order to obtain the inheritance. Dr Purdon contested this will on the grounds that Adolphus had been of unsound mind. The case was first heard at the Common Pleas Summer Assizes at Mullingar. Among the witnesses called was a Dr William Williams, who disliked Dr Purdon and was prepared to swear that Adolphus was sane. He had known him for thirty-six years, and although he admitted that he was a difficult patient – after he cured him of a serious illness, he was never invited to the house again – he contended that the old gentleman enjoyed making himself out queerer than he was. Dr Williams was cross-examined about the occasion when Cooke informed him that he was becoming a screech owl.

Doctor: I told him that I admired screech owls very much.

Counsel: Do you admire screech owls?

Doctor: Well, I said I liked places that had birds and crows and rooks . . . that they generally accompanied old demesnes and old families.

Counsel: Can you give me the exact words he used when he said his voice was becoming like that of a screech owl?

Doctor: He said, 'This is the first day I perceived my voice becoming like that of a screech owl.' He was very hoarse at the time.

His Lordship (intervening): Did you ever hear a man saying he was as hoarse as a raven?

Doctor: I did.

His Lordship: Now, when Mr Cooke said his voice was becoming like that of a screech owl, do you think he supposed he was a screech owl?

Doctor: I do not.

The verdict was in favour of Pakenham-Cooke, concluding that Adolphus was not mad. The judge summed up: 'I believe that his belief as to what might happen to him after his death is no proof of want of capacity if there is any other proof of capacity. If a man believes he will turn into a successful screech owl after his death, that is no proof that he is incapable.'

PETER SOMERVILLE-LARGE,
Irish Eccentrics, 1975

'Miss Reynard, la belle, *has a perfect seat.'*

FOX'S FROLIC

Miss Reynard, *la belle*, has a perfect seat,
Well-fitting foxgloves and habit neat,
'I may be,' she says, 'the last of my race,
But first in the field is to-day my place.'
She's fit for an Emperor, Duke, or Count,
Looking 'A1' on her ca-nine mount.
Princess, or Queen, she is worthy to be,
If either can ride as well as she.

SIR FRANCIS BURNAND (1836–1917)

Soon afterwards I made the acquaintance . . . of Doctor Campbell. . . .
A more savage-looking old gentleman I never saw, wild blood-shot eyes
and cruel lips, the flushed face of a drunkard, all covered with hair like
a monkey, and a long, unkempt beard. He was said to be over eighty,

the retired old English chemist told me he looked exactly the same thirty years ago when first he arrived in Rome. Nobody knew from where he came, it was rumoured he had been a surgeon in the Southern army in the American war.... One day I found him standing by my carriage patting Tappio.

'I envy you that dog,' he said abruptly in a rough voice. 'Do you like monkeys?'

I said I loved monkeys.

He said I was his man, he begged me to come and have a look at his monkey who had been scalded almost to death by upsetting a kettle of boiling water.

We climbed up to his flat at the top of the corner house of Piazza Mignanelli. He begged me to wait in his salon and appeared a minute later with a monkey in his arms, a huge baboon all wrapped up in bandages.

'I am afraid he is very bad,' said the old doctor in quite a different voice, tenderly caressing the emaciated face of his monkey. 'I do not know what I shall do if he dies, he is my only friend. I have brought him up on the bottle since he was a baby, his dear mother died when she gave birth to him. She was almost as big as a gorilla, you never saw such a darling, she was quite human. I do not mind in the least cutting my fellow creatures to pieces, I rather like it, but I have no more courage left in me to dress his scalded little body, he suffers so horribly when I try to disinfect his wounds that I cannot stand it any longer. I am sure you like animals, will you take him in hand?'

We unwrapped the bandages soaked with blood and pus, it was a pitiful sight, his whole body was one terrible wound.

'He knows you are a friend or he would not sit as still as he does, he never allows anybody but me to touch him. He knows everything, he has more brains than all the foreign doctors in Rome put together. He has eaten nothing for four days,' he went on, with a tender expression in his blood-shot eyes. 'Billy, my son, won't you oblige your papa by trying this fig?'

I said I wished we had a banana, there was nothing monkeys liked better.

He said he would telegraph at once to London for a bunch of bananas, never mind the cost.

I said it was a question of keeping up his strength. We poured a little warm milk into his mouth, but he spat it out at once.

'He cannot swallow any more,' groaned his master, 'I know what it means, he is dying.'

We improvised ... a sort of feeding tube and this time he kept the milk to the delight of the old doctor.

Billy got slowly better. I saw him every day for a fortnight, and I ended by becoming quite fond both of him and his master. Soon I found him sitting in his specially constructed rocking-chair on their sunny terrace by the side of his master, a bottle of whisky on the table between them. The old doctor was a great believer in whisky to steady one's hand before an operation. To judge from the number of empty whisky bottles in the corner of the terrace his surgical practice must have been considerable. Alas! they were both addicted to drink, I had often caught Billy helping himself to a little whisky and soda out of his master's glass. The doctor had told me whisky was the best possible tonic for monkeys, it had saved the life of Billy's beloved mother after her pneumonia. One evening I came upon them on their terrace, both blind drunk. Billy was executing a sort of negro dance on the table round the whisky bottle, the old doctor sat leaning back in his chair clapping his hands to mark the time, singing in a hoarse voice:

'Billy, my son, Billy, my sonn, soooooooonny!' They neither heard nor saw me coming. I stared in consternation at the happy family. The face of the intoxicated monkey had become quite human, the face of the old drunkard looked exactly like the face of a gigantic gorilla. The family likeness was unmistakable.

'Billy, my son, Billy, my sonn, soooooooonny!'

Was it possible? No, of course it was not possible but it made me feel quite creepy. . . .

AXEL MUNTHE (1857–1949),
The Story of San Michele, 1929

In a place in *Holland,* called *Wirkham,* being a neuter Towne; as lying betwixt *Holland* and those parts belonging to the Empire, on the River Rhine, lived one *Ioachim Skinker,* whose wife name was *Parnel* . . . it hapned, that in the yeere 1618, she was safely delivered of a Daughter, all the limbes and lineaments of her body, well featur'd and proportioned, only her face, which is the ornament and beauty of all the rest, had the Nose of a Hog, or Swine . . .

. . . much confluence of people came to see the progedy, which wearied the Father, and cast a blush upon the cheekes of the good woman the mother: some desirous to heare her speake, whose language was onely the Dutch Hoggish Houghs, and the Piggs French Owee, Owee, for other words she was not able to utter; which bred in some pitty, in others laughter, according to their severall dispositions.

Others were importunate to see her feede, then milke and the like was brought unto her in a silver Trough; to which she stooped and eate, just as a Swine doth in his swilling Tub; which the more mirth it bred in the Spectators, increased in the Parents the more melancholy: insomuch that he bethought himselfe to finde out some meanes (if it were possible) either to mend or end his sorrowes. . . .

Fairburn's portrait of the pig-faced lady.

After much reasoning *Pro* and *Con*, they concluded to put her into very rich and costly habit (but her face still vaild and covered) and to give out that what gentleman of fashion or quality soever, would take her to his bed after loyall Matrimony, (for she was at this time betwixt sixteene and seventeene years of age, and therefore marriageable) should receive for a Dowry with her, forty thousand pound, payed downe in Starling and Currant money. *This was a baite sufficient to make every Fish to bite at, for no sooner was this publickely divulged, but there came Suitors of all sorts; insomuch that his Gates were thronged as at an Outcry, or rather as a Lottery, every one in hope to carry away the great Prize of forty thousand pound; for it was not the person, but the prize at which they aimed.*

One thinkes with him selfe, so the body bee handsome, though her countenance bee never so course and ugly, all are alike in the night; and in the day time, put her head but in a blacke bagge, and what difference betwixt her and another woman. Another comforteth him selfe thus: *That if shee cannot speake, shee cannot chide; and theefore hee shall be sure not to have a scold to his wife.* Another apprehends, *That if shee feede but on wash and the like, shee will not be very chargeable to him for her Dyet; and therefore hee shall have a good bargaine by the match.*

These and the like inducements were so farre prevalent with them, that they came from divers places, and Countries, to solicite her for marriage: some from *Italy*, some from *France*, some from *Scotland*, some from *England*, and every one of them howsoever debosht in their means, and more studdying upon her meanes then the maide, put all their Fortunes upon the hazard of an handsome suite, to appeare like gentlemen; because otherwise they could have no admittance into her company.

But to give you better information, then by these suppositians, or rather stupid conceits for her order of Diet; *She doth eate all those meates which commonly we doe feede on, and doth digest it very naturally; onely shee doth feede in a Silver Trough, or Bole; which is always carried with her wheresoever shee doth remove: and if shee doth want any thing that shee hath a mind unto, bee it Apparrell, or Dyet, she doth write her mind; and by that means, (as wee are given to understand, by those which have seene her) she hath all things to her desire.*

Amongst some Sutors came a *Scotch man* being a Captaine, who having hazarded the greatest part of a months pay upon one Suite of Cloths, was desirous to see this Gentlewoman, and was received by the Parents; who thinking him to be some great Leard in his Country, gave him generous entertainment: she was brought unto him with her face covered, and in an habit which might well have suited the greatest Lady in the Land; who admiring her feature and proportion, was much inamoured of her person, but desirous to see her face discovered, when hee beheld it; hee would stay no other conference but ran away without further answer, saying: they must pardon him, for hee could *indure no Porke*.

Next came a Sowce-man, borne in *England*, having accomodated himselfe for the same adventure, and presuming that loving Sowce so well, no Hogs-face could affright him, he presently at the sight of her could endure her company no longer, and at his farewell, said, so long as I have known *Rumford*, I never saw such a Hogsnout, but whensoever my stomach shall serve for any such dish, I will never enter upon any raw, but I will be sure it shall be either well boyled or rosted.

A Taylor came, having borrowed (whether with leave, or without I know not) a costly suite of one of his best customers: and he vowed notwithstanding, all impediments whatsoever interdicting him, hee would enjoy her, and marry her, and sleep with her close as Hogs in pease-straw, but though hee liked her feature when her face was discovered, he gave back, and vowed he would not go through stitch with any such businesse. I should but lose my self in writing, and tyre the Reader in turning over many Voluminous leaves of paper, to shew you here many severall men and of sundry conditions, came in a kinde of jealousie one of another, to purchase this masse or magazine of money: every one ambitious after the portion, but noe one amongst them amorous of the person, whose countenance was so farre from seeming lovely to them, that it appeared altogether lothsome.

A certaine Relation of the Hog-faced Gentlewoman
called Mistris Tannakin Skinker, who was borne at
Wirkham a Neuter Towne betweene the Emperour and
the Hollander, scituate on the river Rhyne, 1640

Now I must press on with my narrative, and so come to another event that looms big in my life – the great fair in Hyde Park that was held when Queen Victoria came to the throne. . . .

At this fair there was exhibited almost for the last time a freak that had puzzled and amazed the public for a considerable period. This was Madame Stevens, 'the Pig-faced Lady,' concerning whom I have one of my promised exposures to make.

Madame Stevens was really a fine brown bear, the paws and face of which were kept closely shaved, the white skin under the fur having a close resemblance to that of a human being. Over the paws were fitted white gloves, with well-stuffed fingers, so that the pig-faced lady seemed to have nice plump white arms above them.

The bear was strapped in a chair at the back of the caravan, clothed in female dress, shawl, cap, the poke bonnet of the time, etc. In front was a table at which the seeming lady sat, her paws being laid upon it, and all the rest of the body from the arms of the chair downwards hidden by drapery. Under the table was concealed a boy with a short stick to make the pig-faced lady talk.

When all was ready, and the booth full of spectators, the showman would commence his patter thus, as he pulled aside the curtains:

'I call your attention, ladies and gentlemen, to the greatest wonder of the world! Behold and marvel! Madame Stevens, the pig-faced lady, who is now in her eighteenth year. I believe that is correct, miss?' (Here the hidden boy would prod the bear, who gave a grunt.) 'As you see, ladies and gentlemen, the young lady understands what is said perfectly, though the peculiar formation of her jaws has deprived her of the power of uttering human speech in return.

'You were born at Preston in Lancashire?' (Another prod and another grunt.) 'Quite so. And you enjoy good health and are very happy?' (Another prod and grunt.) 'You are inclined, I suppose, as other ladies, to be led by some gentleman into the holy bonds of matrimony?' (Here the boy would give an extra prod, causing the bear to grunt angrily.) 'What, no! Well, well, don't be cross because I asked you!'

This would be sure to raise a laugh and expressions of wonder. Then a plate would be passed round to receive contributions 'to buy the lady small comforts and luxuries,' as the showman said. After this he would conclude as follows:

'Now, Miss Stevens, you will return thanks to the ladies and gentlemen for coming to see you!' The boy would use his stick and the bear would growl loudly. The doors of the caravan were then thrown open, and as the sightseers poured out the showman would rush to the front, shouting: 'Hear what they say! Hear what they all say about Madame Stevens, the wonderful pig-faced lady!'

This show and some others of its class were stopped by the authorities at the following Camberwell fair, and the pig-faced lady became only a memory, lots of people, to their dying day, believing that such a person really existed.

'LORD' GEORGE SANGER (1825–1911),
Seventy Years a Showman, 1910

Adolphus Cooke . . . believed in reincarnation. This made him unfailingly courteous to his father, whom he believed had returned to earth in the guise of a turkey cock. He instructed his men servants to take off their hats to this bird every time they passed it, while women had to genuflect and bend one knee. . . .

The belief in reincarnation led to confusion about the identity of certain creatures. The turkey cock received due reverence because he was considered to be the late Robert Cooke. But there were doubts, which led to the reprieve of the dog Gusty.

Gusty, a large red setter, was one of Adolphus's favourite dogs, an affectionate animal, well liked by the staff and friendly with strangers. But he was apt to stray. On each occasion Adolphus would send a dozen men from Cookesborough out to scour the countryside to find him. When he was brought home his master would severely reprimand him for his wandering habits and love of low company, and command that he should be kept in solitary confinement for three days and placed on short rations. After several such imprisonments Adolphus decided to give Gusty one last chance. In the presence of most of his workmen, who were summoned as witnesses, Gusty was warned that if he strayed once again he would be hanged like any common criminal, and to impress the point on him he was shown the rope and the tree.

Gusty took no notice, and soon afterwards was found near Mullingar late in the evening with some common dogs. Adolphus ordered his trial to take place the following morning in the great hall at ten o'clock with all his staff present. Billy Dunne and Tom Cruise acted as special advisers and a jury was chosen. Two labourers gave evidence of how Gusty had been found and how he had resisted arrest. The jury went into another room, and after an absence of two hours they returned with a verdict of 'Guilty of Misbehaviour', which they thought would best please the judge. On passing sentence Adolphus emphasized the heinousness of the crime for which Gusty was convicted – ingratitude to a good master who had fed and nurtured him with tender care since he was a puppy. Gusty was told that after he was dead a tombstone would be erected over his grave with the following inscription:

Executed for high crimes and misdemeanours
GUSTY

Once the favourite setter dog of
Adolphus Cooke, Esq.,
Cookesborough,

And it is earnestly hoped that his sad fate will be a warning to other dogs against so offending.
Tuesday, 8th May, 1860.

But who was to hang him? None of the staff was eager to do the job, since they were fairly sure that the hangman would be dismissed after the execution. In due course a man called 'The Bug Mee' consented to dispose of him. 'To plaze your honour I'll hang him; and I'd hang the missus and childer too, if it came to that.' He was instructed to carry out the execution next morning at one of the seven lime trees a quarter of a mile from the house. He disappeared towards them with Gusty and the rope, but returned a little while later with Gusty alive. Adolphus met him and thundered: 'How was it you did not carry out my instructions?' The Bug Mee replied: 'Your honour, I was knotting the rope on his neck when he put the heart across me. He began speaking to me in some kind of foreign language. So, I said to myself, I'd bring him back to you because there is something in him.' 'So, Mee,' Adolphus replied, 'you do believe as I do.' 'Your honour, wouldn't anyone believe if he saw what I saw? Who knows, but it's the ould gentleman himself that is living with?' 'You can be right, Mee.' Go Gusty was brought back into the house, and like the turkey cock he lived to a ripe old age.

PETER SOMERVILLE-LARGE,
Irish Eccentrics, 1975

When Richard was in India he at one time got rather tired of the daily Mess, and living with men, and he thought he should like to learn the manners, customs, and habits of monkeys, so he collected forty monkeys of all kinds of ages, races, species, and he lived with them, and he used to call them by different offices. He had his doctor, his chaplain, his secretary, his aide-de-camp, his agent, and one tiny one, a very pretty, small, silky-looking monkey, he used to call his wife, and put pearls in her ears. His great amusement was to keep a kind of refectory for them, where they all sat down on chairs at meals, and the servants waited on them, and each had its bowl and plate, with the food and drinks proper for them. He sat at the head of the table, and the pretty little monkey sat

by him in a high baby's chair, with a little bar before it. He had a little whip on the table, with which he used to keep them in order when they had bad manners, which did sometimes occur, as they frequently used to get jealous of the little monkey, and try to claw her. He did this for the sake of doing what Mr Garner is now doing, that of ascertaining and studying the language of monkeys, so that he used regularly to talk to them, and pronounce their sounds afterwards, till he and the monkeys at last got quite to understand each other. He obtained as many as sixty words, I think twenty more than Mr Garner – that is, leading words – and he wrote them down and formed a vocabulary, meaning to pursue his studies at some future time. Mr Garner has now the advantage of phonographs, and all sorts of appliances. Had Richard been alive, he could have helped him greatly. Unfortunately his monkey vocabulary was burnt in Grindlay's fire.

ISABEL BURTON, *The Life of Captain Sir Richard*
Burton by His Wife Isabel, 1893

Whenever you observe an animal closely you feel as if a human being sitting inside were making fun of you.

ELIAS CANETTI (b. 1905),
Journals of Canetti, 1942

14

Sentiments of Beauty, Wonder and Passion

Dauphin: When I bestride him I soar, I am a hawk; he trots the air; the earth sings when he touches it; the basest horn of his hoof is more musical than the pipe of Hermes.

WILLIAM SHAKESPEARE (1564–1616),
Henry V, ACT III, SCENE VI

On the threshold of this year (1884), I had an extraordinary adventure. On the morning of New Year's Day, I had been to wish Anna, who lived, as I have said, in the Rue de Vaugirard, a happy New Year. I was coming back in high spirits, pleased with myself, with Heaven and with all mankind, interested by everything, amused by anything and blessed with all the unbounded wealth of the future. That day, why I cannot tell, instead of going along the Rue Placide, as I usually did, I went down a little street on the left, parallel to it – just for fun, for the mere pleasure of changing. It was near noon; the air was clear and the sun, which was almost hot, shone straight across the street, so that one pavement was bright and the other dark.

Half-way down, I left the sunny side, thinking I would try the shady. I was feeling so joyful that I went along singing and skipping, my eyes turned up to Heaven. And then, all at once, I saw coming down to me, as though in answer to my happiness, a little, fluttering, golden thing, falling through the shade like a bit of sunshine. Nearer and nearer it came, hovered for a moment, and then settled on my cap, Holy Ghostwise. I put up my hand and a charming little canary-bird nestled into it; the little creature was palpitating like my own heart, which was as light in my bosom and as winged as any bird. Surely the excess of my joy had become visible, though perhaps not to the dull senses of man; surely, for eyes of any discernment, my whole being must be shining with the brightness of a decoy mirror, and it was my radiance that had drawn this creature down from Heaven.

I ran home in delight to my mother, carrying my canary with me; but what chiefly excited and uplifted me was the thrilling assurance that the bird had been sent by Heaven especially in order to mark me out. I was already more than inclined to think I had a vocation of a mystic nature; henceforth I felt I was bound by a kind of secret pact, and when I heard my mother making plans for my future, wishing, for instance, that I might enter the Forestry Department, which she thought particularly suited to my tastes, I would acquiesce half-heartedly, for politeness' sake, as one lends oneself to a game, but knowing all the while that the vital interest lies elsewhere. I was on the verge of saying to my mother: 'How can I possibly dispose of my future? Don't you know I haven't the right to? Don't you realise I am one of the elect?' I think, indeed, that one day when my mother spoke to me about the choice of a profession, I did say something of the kind.

The canary, which was a hen, was put into a large cage along with a family of finches which I had brought from La Roque, and lived on very good terms with them. But the most surprising part of the story remains to be told. A few days later, as I was on my way to Batignolles, where Monsieur Richard now lived, just as I was going to cross the Boulevard Saint-Germain, what should I suddenly see drop slantwise down into the middle of the roadway – had I gone crazy? – but another canary! I darted after it; but this bird (it had no doubt escaped from the same cage) was a little wilder than the first and flew away to avoid me, not in a single steady flight indeed, but with little pauses, skimming along the ground like a bird accustomed to a cage and bewildered by its liberty. I ran after it for some time along the tram-lines and it escaped me three times, but at last I managed to clap my cap over it, just as a tram was on the point of running over us both.

This chase made me late for my lesson; wild with joy, in a state of delirium, I ran all the way to my professor's, holding my canary fast in my hands. I had no difficulty with Monsieur Richard, who was easily distracted; the lesson hour was delightfully spent in looking for a tiny cage in which to carry my bird back to the Rue de Commaille. As for me, I had been longing for a mate for my canary, and to see one drop again from Heaven was nothing short of a miracle. That such exquisite adventures should befall me of all people in the world filled me with more frantic pride than if I had accomplished some wonderful feat myself. There was no doubt about it, I was predestined. I walked about now with my head in the air, expecting, like Elijah, my pleasure and my food to fall from Heaven.

My canaries bred, and a few weeks later, the cage, big as it was, was crowded with my pensioners. On Sundays, when my cousin Edouard was allowed out of school, we used to let them fly about my room together; there they disported themselves, letting their droppings fall wherever they felt inclined, and perching on our heads, on the furniture, on the branches we had brought home for them from the Bois de Boulogne or the forest of Meudon, and which we fastened in the drawers and stuck horizontally into key-holes or vertically into pots.

ANDRÉ GIDE (1869–1951),
Si le grain ne meurt (If I Die), 1926,
TRANS. DOROTHY BUSSY

The aviary [at Denbigh Castle], which elsewhere is filled with gold pheasants and other foreign birds, was here more usefully tenanted; and was exclusively devoted to cocks and hens, geese, ducks, peacocks, and pigeons. It was however, from its extraordinary cleanliness and nice adaptation, a very pretty and agreeable sight. German housewives, listen and wonder! Twice-a-day are the yards, which are provided with the most beautiful receptacles of water, – the separate houses, pigeon-holes, &c., – twice a-day are they cleaned: the straw nests of the hens were so pretty; the perches on which the fowls roost, so smooth and clean; the water in the stone basins which served as duck-ponds, so clear; the barley and the boiled rice (equal to Parisian 'riz au lait') so tempting, that one thought one's self in the Paradise of fowls. They enjoyed, too, the freedom of Paradise: here were no clipped wings; and a little grove of high trees, close by their house, formed their pleasure-ground. Most of them were still poised in air, waving to and fro on the topmost boughs, when we arrived: but scarcely did they espy the rosy little Fanny tripping towards them, with dainties in her apron, like a beneficent fairy, than they flew down in a tumultuous cloud, and ran to her feet, pecking and fluttering. I felt a sort of pastoral sensibility come over me, and turned homewards, to get rid of my fit of romance before breakfast.

PRINCE M. L. H. VON PÜCKLER-MUSKAU,
Tour in Germany, Holland and England
in the Years 1826, 1827 and 1828, Vol. I, 1832

FÉNG HOU

She is absurdly small – a homoeopathic dose of a dog. Nothing prevents her being carried in the sleeve, as Nature and Art intended her to be. But she is small only in figure: in all else she is as large as a Newfoundland – in fidelity and courage and spirit and protectiveness and appetite (proportionately) and love of ease – while in brain-power she is larger. . . .

When she runs from room to room she beats the floor with her forepaws with a gallant little rocking-horse action. When she runs over grass she makes a russet streak like a hare, with the undulating ripple of a sea-serpent, and her soft pads reverberate like muffled hoofs. When she is not running she is asleep. When she sleeps the most comfortable place in the room is hopelessly engaged until she wakes. However deeply she may be sleeping, she wakes directly her particular friend leaves the room, her religion being sociability. Left alone, she screams. Put out of the house alone, she encircles it with the speed of thought, seeking an open door or window. The sunlight through her tongue is more than rubies. . . . She has brought into a house hitherto unconscious of it the delectable piquancy of Pekin.

Having done all that was possible to make Féng Hou at home, to give her an environment free from discontent, and butcher's bills rising like a Handley Page: having done all this, it was something more than a shock to be favoured with a translation of the rhapsodical pearls of wisdom dropped from the lips of Her Imperial Majesty Tzu Hsi, the late Dowager Empress of Western China, for the guidance of the master of her kennel. One saw at once how much was still to do if Féng Hou was to be worthy of her race. I quote this most delightful document, the very flower of Chinese solicitude and fancy:–

'PEARLS DROPPED FROM THE LIPS OF HER IMPERIAL MAJESTY,
TZU HSI,
DOWAGER EMPRESS OF THE FLOWERY LAND.

'Let the Lion Dog be small: let it wear the swelling cape of dignity around its neck: let it display the billowing standard of pomp above its back.

'Let its face be black: let its fore-legs be shaggy: let its forehead be straight and low, like unto the brow of an Imperial righteous harmony boxer.

'Let its eyes be large and luminous: let its ears be set like the sails of a war-junk: let its nose be like that of the monkey god of the Hindus.

'Let its fore-legs be bent, so that it shall not desire to wander far, or leave the Imperial precincts.

'Let its body be shaped like that of a hunting lion spying for its prey.

'Let its feet be tufted with plenty of hair, that its footfall may be soundless: and for its standard of pomp let it rival the whisk of the Tibetan's yak, which is flourished to protect the Imperial litter from the attacks of flying insects.

'Let it be lively, that it may afford entertainment by its gambols; let it be timid, that it may not involve itself in danger; let it be domestic in its habits, that it may live in amity with the other beasts, fishes, or birds that find protection in the Imperial Palace. And for its colour, let it be that of the lion, a golden sable, to be carried in the sleeve of a yellow robe, or the colour of a red bear, or a black or a white bear, or striped like a dragon, so that there may be dogs appropriate to every costume in the Imperial wardrobe.

'Let it venerate its ancestors and deposit offerings in the canine cemetery of the Forbidden City on each new moon.

'Let it comport itself with dignity; let it learn to bite the foreign devils instantly.

'Let it be dainty in its food, that it shall be known for an Imperial dog by its fastidiousness.

'Sharks' fins and curlews' livers and the breasts of quails, on these it may be fed; and for drink give it the tea that is brewed from the spring buds of the shrub that groweth in the province of Hankow, or the milk of the antelopes that pasture in the Imperial parks. Thus shall it preserve its integrity and self-respect; and for the day of sickness let it be anointed with the clarified fat of the leg of a sacred leopard, and give it to drink a throstle's egg-shell full of the juice of the custard-apple in which have been dissolved three pinches of shredded rhinoceros horn, and apply it to piebald leeches.

'So shall it remain; but if it die, remember thou too art mortal.'

E. V. LUCAS (1868–1938),
'The More I See of Men . . .', 1927

Mrs Septimus Small, known in the Forsyte family as Aunt Juley, . . . took by force of habit the path which led her into the then somewhat undeveloped gardens of Kensington. . . .

Dear, dear! That little white dog was running about a great deal. Was it lost? Backwards and forwards, round and round! What they called –

she believed – a Pomeranian, quite a new kind of dog. And, seeing a bench, Mrs Septimus Small bent, with a little backward heave to save her 'bustle', and sat down to watch what was the matter with the white dog. . . . Of late – owing to poor Tommy's (their cat's) disappearance, very mysterious – she suspected the sweep – there had been nothing but 'Polly' at Timothy's to lavish her affection on. This dog was draggled and dirty, as if it had been out all night, but it had a dear little pointed nose. She thought, too, that it seemed to be noticing her, and at once had a swelling-up sensation underneath her corsets. Almost as if aware of this, the dog came sidling, and sat down on its haunches in the grass, as though trying to make up its mind about her. Aunt Juley pursed her lips in the endeavour to emit a whistle. The veil prevented this, but she held out her gloved hand. 'Come, little dog – nice little dog!' It seemed to her dear heart that the little dog sighed as it sat there, as if relieved that at last someone had taken notice of it. But it did not approach. The tip of its bushy tail quivered, however, and Aunt Juley redoubled the suavity of her voice: 'Nice little fellow – come then!'

The little dog slithered forward, humbly wagging its entire body, just out of reach. Aunt Juley saw that it had no collar. Really, its nose and eyes were sweet!

'Pom!' she said. 'Dear little Pom!'

The dog looked as if it would let her love it, and sensation increased beneath her corsets. . . .

And then, without either of them knowing how, her fingers and the nose were in contact. The dog's tail was now perfectly still; its body trembled. Aunt Juley had a sudden feeling of shame at being so formidable; and with instinct inherited rather than acquired, for she had no knowledge of dogs, she slid one finger round an ear and scratched. It *was* to be hoped he hadn't fleas! And then! The little dog leaped on her lap. It crouched there just as it had sprung, with its bright eyes upturned to her face. A strange dog – her dress – her Sunday best! It *was* an event! . . .

The dog, who had no say whatever in the matter, put out a pink strip of tongue and licked her nose. Aunt Juley had the exquisite sensation of being loved; and, hastily, to conceal her feelings, bore it lolling over her arm away. She bore it upstairs, instead of down, to her room which was at the back of dear Ann's, and stood, surrounded by mahogany, with the dog still in her arms. . . . It was panting, and every now and then with its slip of a tongue it licked her cheek, as if to assure itself of reality. Since the departure of Septimus Small ten years ago, she had never been properly loved, and now that something was ready to love her, they wanted to take it away. . . . And, again, all that was maternal in Aunt Juley swelled, beneath the dark violet of her bosom sprinkled with white hairs.

JOHN GALSWORTHY (1867–1933),
On Forsyte 'Change, 1930

Sympathy with the lower animals is one of the noblest virtues with which man is endowed.

CHARLES DARWIN (1809–82)

... If he hears armour clang in the distance,
He can't keep still, the ears prick up, the limbs quiver,
He drinks the air, he jets it in hot steam out of his nostrils.
The mane is thick, and tumbles on the right shoulder when tossed:
The spine runs over the loins, sunk between two ridges;
The solid hoof makes a deep clatter and hurls up divots.
Such a horse was Cyllarus, that Pollux broke in; and such were
The two-yoke team of the War-god
And the horses of great Achilles, mentioned by Greek poets.
Such the guise that Saturn assumed to escape his wife,
A horse-mane streaming over his neck as he streaked away
And made the peaks of Pelion resound with a stallion's neighing. ...

... Look how a horse shudders in his whole frame, if the familiar
Scent is but borne downwind!
Nothing will hold him now – neither bridle nor blows of the whip,

'He drinks the air, he jets it in hot steam out of his nostrils.'

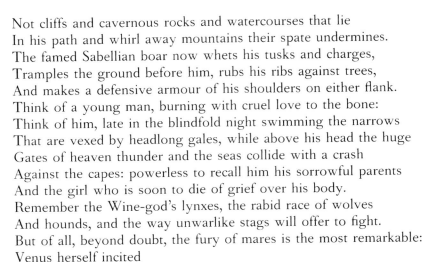

Not cliffs and cavernous rocks and watercourses that lie
In his path and whirl away mountains their spate undermines.
The famed Sabellian boar now whets his tusks and charges,
Tramples the ground before him, rubs his ribs against trees,
And makes a defensive armour of his shoulders on either flank.
Think of a young man, burning with cruel love to the bone:
Think of him, late in the blindfold night swimming the narrows
That are vexed by headlong gales, while above his head the huge
Gates of heaven thunder and the seas collide with a crash
Against the capes: powerless to recall him his sorrowful parents
And the girl who is soon to die of grief over his body.
Remember the Wine-god's lynxes, the rabid race of wolves
And hounds, and the way unwarlike stags will offer to fight.
But of all, beyond doubt, the fury of mares is the most remarkable:
Venus herself incited
The chariot-team that day they champed the limbs of Glaucus.
In heat, they'll range over Gargarus and across the roaring
Ascanius, they'll climb mountains and swim rivers.
The moment that flame is kindled within their passionate flesh
(In spring above all, when warmth returns to their bones) the whole
    herd
Wheels to face the west wind high up there on the rocks;
They snuff the light airs and often without being mated
Conceive, for the wind – astounding to tell – impregnates them. . . .

VIRGIL (70–19 BC), *Georgics*, BOOK III,
TRANS. CECIL DAY LEWIS

This tortoise was the result of a fancy which had occurred to him shortly
before leaving Paris. Looking one day at an Oriental carpet aglow with
iridescent colours, and following with his eyes the silvery glints running
across the weft of the wool, which was a combination of yellow and
plum, he had thought what a good idea it would be to place on this
carpet something that would move about and be dark enough to set off
these gleaming tints.

Possessed by this idea, he had wandered at random through the streets
as far as the Palais-Royal, where he glanced at Chevet's display and
suddenly struck his forehead – for there in the window was a huge
tortoise in a tank. He had bought the creature; and once it had been left
to itself on the carpet, he had sat down and subjected it to a long scrutiny,
screwing up his eyes in concentration.

Alas, there could be no doubt about it: the nigger-brown tint, the raw
Sienna hue of the shell, dimmed the sheen of the carpet instead of bringing out its colours; the predominating gleams of silver had now lost
nearly all their sparkle and matched the cold tones of scraped zinc along

the edges of this hard, lustreless carapace.

He bit his nails, trying to discover a way of resolving the marital discord between these tints and preventing an absolute divorce. At last he came to the conclusion that his original idea of using a dark object moving to and fro to stir up the fires within the woollen pile was mistaken. The fact of the matter was that the carpet was still too bright, too garish, too new looking; its colours had not yet been sufficiently toned down and subdued. The thing was to reverse his first plan and to deaden those colours, to dim them by the contrast of a brilliant object that would kill everything around it, drowning the gleams of silver in a golden radiance. Stated in these terms, the problem was easier to solve; and Des Esseintes accordingly decided to have his tortoise's buckler glazed with gold.

Back from the workshop where the gilder had given it board and lodging, the reptile blazed as brightly as any sun, throwing out its rays over the carpet, whose tints turned pale and weak, and looking like a Visigothic shield tegulated with shining scales by a barbaric artist.

At first, Des Esseintes was delighted with the effect he had achieved; but soon it struck him that this gigantic jewel was only half-finished and that it would not be really complete until it had been encrusted with precious stones.

From a collection of Japanese art he selected a drawing representing a huge bunch of flowers springing from a single slender stalk, took it to a jeweller's, sketched out a border to enclose this bouquet in an oval frame, and informed the astonished lapidary that the leaves and petals of each and every flower were to be executed in precious stones and mounted on the actual shell of the tortoise. . . .

He made up his bouquet in this way: the leaves were set with gems of a strong and definite green – asparagus-green chrysoberyls, leek-green peridots, olive-green olivines – and these sprang from twigs of almandine and uvarovite of a purplish red, which threw out flashes of harsh, brilliant light like the scales of tartar that glitter on the insides of wine-casks.

For the flowers which stood out from the stem a long way from the foot of the spray, he decided on a phosphate blue; but he absolutely refused to consider the Oriental turquoise which is used for brooches and rings, and which, together with the banal pearl and the odious coral, forms the delight of the common herd.

He chose only turquoises from the West – stones which, strictly speaking, are simply a fossil ivory impregnated with coppery substances and whose celadon blue looks thick, opaque, and sulphurous, as if jaundiced with bile.

This done, he could now go on to encrust the petals of such flowers as were in full bloom in the middle of his spray, those closest to the stem, with translucent minerals that gleamed with a glassy, sickly light and glinted with fierce, sharp bursts of fire. . . .

Now Des Esseintes sat gazing at the tortoise where it lay huddled in a corner of the dining-room, glittering brightly in the half-light. . . . He touched it; it was dead. Accustomed no doubt to a sedentary life, a modest existence spent in the shelter of its humble carapace, it had not been able to bear the dazzling luxury imposed upon it, the glittering cape in which it had been clad, the precious stones which had been used to decorate its shell like a jewelled ciborium.

> J.-K. HUYSMANS (1848–1907), *A Rebours*,
> 1884, TRANS. ROBERT BALDICK, 1959

Theseus: My hounds are bred out of the Spartan kind,
So flew'd, so sanded; and their heads are hung
With ears that sweep away the morning dew;
Crook-knee'd, and dew-lapp'd like Thessalian bulls;
Slow in pursuit, but match'd in mouth like bells,
Each under each. A cry more tuneable
Was never holla'd to, nor cheer'd with horn,
In Crete, in Sparta, nor in Thessaly:
Judge when you hear.

> WILLIAM SHAKESPEARE (1564–1616),
> *A Midsummer Night's Dream*, ACT IV, SCENE I

Cats, no less liquid than their shadows,
Offer no angles to the wind.
They slip, diminished, neat, through loopholes
Less than themselves; will not be pinned

To rules or routes for journeys; counter
Attack with non-resistance; twist
Enticing through the curving fingers
And leave an angered, empty fist.

They wait, obsequious as darkness
Quick to retire, quick to return;
Admit no aim or ethics; flatter
With reservations; will not learn

To answer to their names; are seldom
Truly owned till shot and skinned.
Cats, no less liquid than their shadows,
Offer no angles to the wind.

> A. S. J. TESSIMOND (1902–62)

When the dray arrived at the Zoological Gardens from the railway station, we were all of course very anxious to see the new arrival. When the tarpauling was taken off, there was discovered a huge box, made of strong deal boards, like a diminutive railway horse-box, and in this Hippo had travelled all the way (with an occasional bucket of water thrown over him) from Alexandria, thereby proving an important fact, that he can dispense with the bath without other prejudice than a rough skin. How to get him into his sleeping-apartment was the question. Salama assured his master that Hippo would follow him anywhere. One side of the box, therefore, was taken off, and out the poor frightened beast walked. Salama gave him his hand to smell, and he trotted after his kind protector with a long, steady, calf-like trot, swinging from side to side, while he kept his head close to his master, staring about him like a frightened deer.

He was about the size of a very large bacon hog, only higher on the legs. From not having been able to have a bath for six weeks or more, his skin had assumed a curious appearance; the back, instead of being soft, slimy, and india-rubber-like, was quite hard and dry, and the skin was peeling off from it as from the bark of a tree; it was, in fact, much more like a bit of an old forest oak than of a water-loving animal. It was of course expected that the moment Hippo smelt and saw the water he would rush into it; but no – he merely went up to it and smelt it with a look of curiosity, as though he had never seen water before; and it was not till the Arab advanced himself partially into the water that Hippo would follow. He soon came out again, and was only persuaded to go right in to the deep part of the water by the Arab walking round the edge of the tank. Hippo then began to find out where he was, and how comfortable the warm clean water was. Down he went to the bottom, like a bit of lead; then up he came with a tremendous rush and a vehement snorting; then a duck under, then up again, prancing and splashing in the water after the manner of Neptune's sea-horses that are harnessed to his chariot in the old pictures of the worthy marine deity. I never recollect to have seen any creature, either man or beast, so supremely happy for a short time as was poor travel-worn Hippo after his long voyage of so many thousand miles.

Coming out of the water, Hippo smelt about for food; mangold-wurzel was given him, and mightily did he enjoy it.

FRANK BUCKLAND (1826–80),
Curiosities of Natural History, 1866

Early next morning when Farah brought me in my tea, Juma came in with him and carried the fawn in his arms. It was a female, and we named her Lulu, which I was told was the Swahili word for a pearl.

Lulu by that time was only as big as a cat, with large quiet purple eyes. She had such delicate legs that you feared they would not bear being folded up and unfolded again, as she lay down and rose up. Her ears were smooth as silk and exceedingly expressive. Her nose was as black as a truffle. Her diminutive hoofs gave her all the air of a young Chinese lady of the old school, with laced feet. It was a rare experience to hold such a perfect thing in your hands.

Lulu soon adapted herself to the house and its inhabitants and behaved as if she were at home. During the first weeks the polished floors in the rooms were a problem in her life, and when she got outside the carpets her legs went away from her to all four sides; it looked catastrophic but she did not let it worry her much and in the end she learnt to walk on the bare floors with a sound like a succession of little angry finger-taps. She was extraordinarily neat in all her habits. She was headstrong already as a child, but when I stopped her from doing the things she wanted to do, she behaved as if she said: Anything rather than a scene.

Kamante brought her up on a sucking-bottle, and he also shut her up at night, for we had to be careful of her as the leopards were up round the house after nightfall. So she held to him and followed him about. From time to time when he did not do what she wanted, she gave his thin legs a hard butt with her young head, and she was so pretty that you could not help, when you looked upon the two together, seeing them as a new paradoxical illustration to the tale of the Beauty and the Beast. On the strength of this great beauty and gracefulness, Lulu obtained for herself a commanding position in the house, and was treated with respect by all.

. . . Now my dogs understood Lulu's power and position in the house. The arrogance of the great hunters was like water with her. She pushed them away from the milk-bowl and from their favourite places in front of the fire. I had tied a small bell on a rein round Lulu's neck, and there came a time when the dogs, when they heard the jingle of it approaching through the rooms, would get up resignedly from their warm beds by the fireplace, and go and lie down in some other part of the room. Still nobody could be of a gentler demeanour than Lulu was when she came and lay down, in the manner of a perfect lady who demurely gathers her skirts about her and will be in no one's way. She drank the milk with a polite, pernickety mien, as if she had been pressed by an overkind hostess. She insisted on being scratched behind the ears, in a pretty forbearing way, like a young wife who pertly permits her husband a caress.

When Lulu grew up and stood in the flower of her young loveliness she was a slim delicately rounded doe, from her nose to her toes unbelievably beautiful. She looked like a minutely painted illustration to Heine's song of the wise and gentle gazelles by the flow of the river Ganges.

KAREN BLIXEN (1883–1962),
Out of Africa, 1937

They were at play, she and her cat,
And it was marvellous to mark
The white paw and the white hand pat
Each other in the deepening dark.

PAUL VERLAINE (1844–96),
TRANS. ARTHUR SYMONS

A further pleasure-ground [at Woburn Abbey], with the finest trees and many beautiful surprises, – among others pretty children's gardens, and a grass garden in which all sorts of gramineous plants were cultivated in little beds, forming a sort of chequer-work, – led to the Aviary. This consists of a large place fenced in, and a cottage, with a small pond in the centre, all dedicated to the feathered race. Here the fourth or fifth attendant awaited us, (each of whom expects a fee, so that you cannot see such an establishment under some pounds sterling,) and showed us first several gay-plumed parrots and other rare birds, each of whom had his own dwelling and little garden. These birds' houses were made of twigs interwoven with wire, the roof also of wire, the shrubs around evergreen, as were almost all the other plants in this inclosure. As we walked out upon the open space which occupies the centre, our Papageno whistled, and in an instant the air was literally darkened around us by flights of pigeons, chickens, and heaven knows what birds. Out of every bush started gold and silver, pied and common, pheasants; and from the little lake a black swan galloped heavily forward, expressing his strong desire for food in tones like those of a fretful child. This beautiful bird, raven black with red feet and bill, was exceedingly tame, ate his food 'chemin faisant' out of the keeper's pocket, and did not leave us for a moment while we were sauntering about the birds' paradise, only now and then pushing away an intrusive duck or other of the vulgar herd, or giving a noble gold pheasant a dig in the side. A second interesting but imprisoned inhabitant of this place was Hero, an African crane, a creature that looks as if it were made of porcelain, and frequently reminded me in his movements of our departed dancing Ballerino. The incident of his history which had gained him his lofty name was unknown to the keeper.

PRINCE M. L. H. VON PÜCKLER-MUSKAU,
Tour in Germany, Holland and England,
in the Years 1826, 1827 and 1828, Vol. III, 1832

Don Pierrot never went to bed until I came in. He waited for me inside the door, and as I entered the hall he would rub himself against my legs and arch his back, purring joyfully all the time. Then he proceeded to walk in front of me like a page, and if I had asked him, he would certainly have carried the candle for me. In this fashion he escorted me to my room and waited while I undressed; then he would jump on the bed, put

The fishing cat.

his paws round my neck, rub noses with me, and lick me with his rasping little pink tongue, while giving vent to soft inarticulate cries, which clearly expressed how pleased he was to see me again. Then when his transports of affection had subsided, and the hour for repose had come, he would balance himself on the rail of the bedstead and sleep there like a bird perched on a bough. . . .

Don Pierrot had a companion of the same race as himself, and no less white. All the imaginable snowy comparisons it were possible to pile up would not suffice to give an idea of that immaculate fur, which would have made ermine look yellow.

I called her Seraphita, in memory of Balzac's Swedenborgian romance. The heroine of that wonderful story, when she climbed the snow peaks of the Falberg with Minna, never shone with a more pure white radiance. Seraphita had a dreamy and pensive character. She would lie motionless on a cushion for hours, not asleep, but with eyes fixed in rapt attention on scenes invisible to ordinary mortals.

Caresses were agreeable to her, but she responded to them with great reserve, and only to those of people whom she favoured with her esteem, which it was not easy to gain. She liked luxury, and it was always in the newest armchair or on the piece of furniture best calculated to show off her swan-like beauty, that she was to be found. Her toilette took an immense time. She would carefully smooth her entire coat every morn-ing, and wash her face with her paw, and every hair on her body shone like new silver when brushed by her pink tongue. If anyone touched her she would immediately efface all traces of the contact, for she could not endure being ruffled. Her elegance and distinction gave one an idea of aristocratic birth, and among her own kind she must have been at least a duchess. She had a passion for scents. She would plunge her nose into bouquets, and nibble a perfumed handkerchief with little paroxysms of delight. She would walk about on the dressing-table sniffing the stoppers of the scent-bottles, and she would have loved to use the violet powder if she had been allowed.

Such was Seraphita, and never was a cat more worthy of a poetic name.

THÉOPHILE GAUTIER (1811–72),
The White and Black Dynasties,
TRANS. LADY CHANCE

The somewhat imperfect shape of his favorite horse was instantly manifest to Vronsky's eyes. Frou Frou was of medium size, with slender bones; her breast was narrow, though the breast-bone was prominent; the crupper was rather tapering; and the legs, particularly the hind-legs, considerably bowed. The muscles of the legs were not large, but the flanks were very enormous on account of the training she had had, and the smallness of her belly. The bones of the legs below the knee seemed not thicker than a finger, seen from the front: they were extraordinarily large when seen sidewise. The whole steed seemed squeezed in and lengthened out. But she had one merit that outweighed all her faults: she had good blood, – was a thoroughbred, as the English say. Her muscles stood out under a network of veins, covered with a skin as smooth and soft as satin: her slender head, with prominent eyes, bright and animated; her delicate, mobile nostrils, which seemed suffused with blood, – all the points of this noble animal had something energetic, decided, and keen. She was one of those creatures such as never fail to fulfil their promise owing to defect in mechanical construction. Vronsky felt that she understood him while he was looking at her. When he came in, she was taking long breaths, turning her head round, and showing the whites of her bloodshot eyes, and trying to shake off her muzzle, and dancing on her feet as though moved by springs.

'You see how excited she is,' said the Englishman.

'Whoa, my loveliest, whoa!' said Vronsky, approaching to calm her; but the nearer he came, the more nervous she grew; and only when he had caressed her head, did she become tranquil. He could feel her muscles strain and tremble under her delicate, smooth skin. Vronsky patted her beautiful neck, and put into place a bit of her mane that she had tossed on the other side; and then he put his face close to her nostrils, which swelled and dilated like the wings of a bat. She snorted, pricked up her ears, and stretched out her long black lips to seize his sleeve; but when she found herself prevented by her muzzle, she began to caper again.

'Quiet, my beauty, quiet,' said Vronsky, calming her; and he left the stable with the re-assuring conviction that his horse was in perfect condition.

But the nervousness of the steed had taken possession of her master. Vronsky felt the blood rush to his heart, and, like the horse, he wanted violent action. It was a sensation at once strange and joyful.

LEO TOLSTOY (1828–1910),
Anna Karenina, 1875–6

A little beyond Abingdon I came out upon the line of my walk *from* London; thus virtually completing the tour. I therefore proceeded directly to Windsor....

'The Queen's Dairy'! How saxon and homelike sounds that term! The Queen's cows 'with crumpled horns;' brindled cows, spotted, red-faced, white-faced, mottled, brown and dun, coming in from pasture at eve with whisking tails, and eyes soft, gentle, round and honest. The Queen's milk-maids, with rosy cheeks, patting the meditating *mullies* with white, soft hands and voices of kindly accent. The Queen's milk-pails, with her crown mark upon them all, so pure and sweet in their polished hoops. The Queen's milk-pans, shelved in long rows, with the cream-lily's golden leaf, like another *Victoria Regina*, overlaying the luscious deep an inch or two below the brim. The Queen's churns, so surpassing all that Dutch housewifery ever dreamed of in purification and polish of wood and brass. The Queen herself, in straw bonnet and thick-soled shoes, walking up and down the dairy-room, dropping happy and smiling looks into pails and pans of milk and cream; perhaps anon stamping a roll of new-made butter with her wife's seal manual for the royal table; thinking the while of dairies and pastures far and near, of Alderneys, Devonshires, Herefords and Shorthorns, and their comparative graces and merits. *The Queen's Dairy!* The very name seems to link her queenhood to the happiest and homeliest experiences of rural life; to attach her, by a sensible lien of industrial sympathy, to all the farmers' wives in the British Empire; to introduce her into the daily fellowship of their feelings and interests; to morning and evening walks on their rustic levels of care, learning what milk, butter and cheese mean, and all the minute details of their production....

To say that it is a little marble temple polished after the similitude of a palace, would convey a sense of its cool whiteness and purity, but not its aspect of softness. The walls, the long marble tables, the fountains, the statuary of rustic life, and all the finely-sculptured allegories look as if wrought from new milk petrified just as the cream began to rise to the surface. Or as if, looking into the basined pools of the soft white fluid circling around the interior, like great fluent pearls strung for a bracelet, they had gradually assimilated themselves to the medium that reflected their faces, and had taken up both its softness of look and sweetness of savor. It was truly a beautiful sight, that would dazzle and delight the eyes of our Orange County dairy-women....

The cows were nearly all in the pasture when I visited their stabling; but a good number of calves were in the stalls or boxes, of different breeds and ages, all looking as bright and sleek as possible. I was struck with the eclectic character of the names they bore. The floral and fairy kingdoms of nature, heroes and heroines of ancient mythology, history and poetry supplied most of this interesting nomenclature; and this made it all the more interesting to me to see that *Uncle Tom's Cabin* had furnished two or three names, and that 'Eva' and 'Topsy' had their place in the rank of chosen celebrities.

ᴇʟɪʜᴜ ʙᴜʀʀɪᴛᴛ, *A Walk from London to Land's End and Back*, 1865

One day, on paying Mr Waterton a casual visit, and immediately observing that his temper was somewhat ruffled, – which irritated condition his countenance always speedily betrayed, – I enquired if he was poorly, or if any thing had distressed him, when he excitedly replied, 'Yes, I am grieved to the back bone, Mr ——, whom you would just now meet in the carriage-road, and who professes to be enchanted and in raptures with the works of God's creation, has just left the house; and, what do you think? he coolly turned up his nose at my Bahia toad, calling it "an ugly brute."

'That a gentleman, avowing himself a lover of natural history, and pretending an anxiety to work in the same vineyard with me, should profanely designate one of God's creation "an ugly brute," was enough "to put me out" for a week, – so I left him in the staircase to his own cogitations.'

ʀɪᴄʜᴀʀᴅ ʜᴏʙꜱᴏɴ, *Charles Waterton: His Home, Habits and Handiwork*, 1866

Dauphin: What a long night this is! I will not change my horse with any that treads but on four pasterns. *Ça, ha!* He bounds from the earth as if his entrails were hairs: *le cheval volant*, the Pegasus, *qui a les narines de feu!* When I bestride him, I soar, I am a hawk: he trots the air; the earth sings when he touches it; the basest horn of his hoof is more musical than the pipe of Hermes.
Orleans: He's of the colour of the nutmeg.
Dauphin: And of the heat of the ginger. It is a beast for Perseus: he is pure air and fire; and the dull elements of earth and water never appear in him but only in patient stillness while his rider mounts him: he is indeed a horse; and all other jades you may call beasts.
Constable: Indeed, my lord, it is a most absolute and excellent horse.
Dauphin: It is the prince of palfreys; his neigh is like the bidding of a monarch and his countenance enforces homage.

Orleans: No more, cousin.

Dauphin: Nay, the man hath no wit that cannot, from the rising of the lark to the lodging of the lamb, vary deserved praise on my palfrey: it is a theme as fluent as the sea; turn the sands into eloquent tongues, and my horse is argument of them all. 'Tis a subject for a sovereign to reason on, and for a sovereign's sovereign to ride on; and for the world – familiar to us, and unknown – to lay apart their particular functions and wonder at him. I once writ a sonnet in his praise and began thus: 'Wonder of nature!' –

Orleans: I have heard a sonnet begin so to one's mistress.

Dauphin: Then did they imitate that which I composed to my courser; for my horse is my mistress.

WILLIAM SHAKESPEARE (1564–1616),
Henry V, ACT III, SCENE VI

My own dog is an Alsatian bitch. Her name is Tulip. Alsatians have a bad reputation; they are said to bite the hand that feeds them. Indeed Tulip bit my hand once, but accidentally; she mistook it for a rotten apple we were both trying to grab simultaneously. One of her canines sank into my thumb-joint to the bone: when I held it under the tap afterwards I could see the sinews exposed. We all make mistakes and she was dreadfully sorry. She rolled over on the grass with all her legs in the air; and later on, when she saw the bandage on my hand, she put herself in the corner, the darkest corner of the bedroom, and stayed there for the rest of the afternoon. One can't do more than that. . . .

It is necessary to add that she is beautiful. People are always wanting to touch her, a thing she cannot bear. Her ears are tall and pointed, like the ears of Anubis. How she manages to hold them constantly erect, as though starched, I do not know, for with their fine covering of mouse-grey fur they are soft and flimsy; when she stands with her back to the sun it shines through the delicate tissue, so that they glow shell-pink as though incandescent. Her face also is long and pointed, basically stone-grey but the snout and lower jaw are jet black. Jet, too, are the rims of her amber eyes, as though heavily mascara'd, and the tiny mobile eye-brow tufts that are set like accents above them. And in the midst of her forehead is a kind of Indian caste-mark, a black diamond suspended there, like the jewel on the brow of Pegasus in Mantegna's 'Parnassus', by a fine dark thread, no more than a pencilled line, which is drawn from it right over her poll midway between the tall ears. A shadow extends across her forehead from either side of this caste-mark, so that, in certain lights, the diamond looks like the body of a bird with its wings spread, a bird in flight.

These dark markings symmetrically divide up her face into zones of pale pastel colours, like a mosaic, or a stained-glass window; her skull, bisected by the thread, is two primrose pools, the centre of her face light grey, the bridge of her nose above the long, black lips fawn, her cheeks white, and upon each a *patte de mouche* has been tastefully set. A delicate white ruff, frilling out from the lobes of her ears, frames this strange, clownish face, with its heavily leaded features, and covers the whole of her throat and chest with a snowy shirt front.

For the rest, her official description is sable-grey: she is a grey dog wearing a sable tunic. Her grey is the grey of birch bark; her sable tunic is of the texture of satin and clasps her long body like a saddle-cloth. No tailor could have shaped it more elegantly; it is cut round the joints of her shoulders and thighs and in a straight line along the points of her ribs, lying open at the chest and stomach. Over her rump it fits like a cap, and then extends on in a thin strip over the top of her long tail down to the tip. Viewed from above, therefore, she is a black dog; but when she rolls over on her back she is a grey one. Two dark ribbons of fur, descending from her tunic over her shoulders, fasten it at her sternum, which seems to clip the ribbons together as with an ivory brooch.

J. R. ACKERLEY (1896–1967),
My Dog Tulip, 1965

We hastened accordingly to secure a sight of the giraffe, which was led out before us by two Moors who had accompanied her from Africa. A wonderful creature indeed! You know her form; but nothing can give an idea of the beauty of her eyes. Imagine something midway between the eye of the finest Arab horse, and the loveliest Southern girl, with long and coal-black lashes, and the most exquisite beaming expression of tenderness and softness, united to volcanic fire. The giraffe is attached to man, and is extremely 'gentle' and good-natured. Her appetite is good, for she daily sucks the milk of three cows who were lying near her. She uses her long bright-blue tongue like a trunk, in which way she took from me my umbrella, which she liked so much that she would not give it up again.

PRINCE M. L. H. VON PÜCKLER-MUSKAU,
Tour in Germany, Holland and England, in the Years
1826, 1827, and 1828, Vol. IV, 1832 `

What keen eyes they have! these busy little workers, flying hither and thither, over hill and valley, in the early spring days. House-hunting, that is what they are doing. In at your window, under the eaves of the barn, getting in the most inconceivable and, sometimes, unwelcome places. Nothing is beneath their notice; no, not even an old, discarded curtain tassel, as a friend tells me who has seen the tassel.

Perhaps it was once one of the much-prized treasures of some small girl,

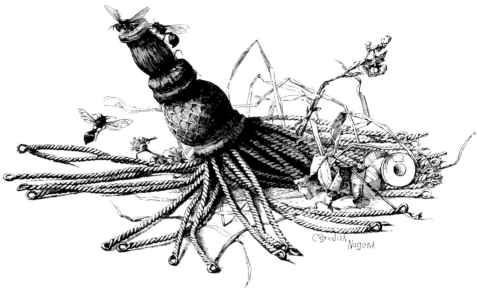

The wasps in their aesthetic home.

rambling through the loose hay, with her arms so full of toys that the treasure dropped, and was lost forever to the fond eyes of its owner. There it lay, unseen and useless, until, one day, a busy wasp came buzzing around the barn-yard, and, being a wasp of high æsthetic taste, this odd-looking, pretty-colored object in the long grass attracted its attention and gave it a most brilliant idea.

First taking a peep in at the top, it disappeared from view, only to reappear at the other end; then, the inspection revealing all that its cultivated taste demanded, flying off, with a satisfied buzz, to return with a whole colony of its fellow-workers, ready to begin on the new home.

So the wasp and its family worked day after day, from early morn until dusk, flying back and forth to their tasseled home, first making the cells for their eggs and food, then, all being snug and tight, hurrying off again to have the store-rooms well filled with provisions for the few who would live until another spring. . . .

Long since the little occupants deserted their aesthetic home, while the tassel, with the house still complete, reposes in the South Kensington Museum of Natural History, a lasting relic of the industry of those æsthetic wasps.

Jack-in-the-Pulpit, St Nicholas Magazine, June 1889

The air of heaven is that which blows between a horse's ears.

ARAB PROVERB

15

and Fancies, Myths and Magic

A blue cloud of incense was wafted up into Félicité's room. She opened her nostrils wide and breathed it in with a mystical, sensuous fervour. Then she closed her eyes. Her lips smiled. Her heart-beats grew slower and slower, each a little fainter and gentler, like a fountain running dry, an echo fading away. And as she breathed her last, she thought she could see, in the opening heavens, a gigantic parrot hovering above her head.

GUSTAVE FLAUBERT (1821–80),
Un Coeur Simple FROM *Trois Contes,* TRANS.
ROBERT BALDICK

An old Danish shipowner sat and thought of his young days and of how he had, when he was sixteen years old, spent a night in a brothel in Singapore. He had come in there with the sailors of his father's ship, and had sat and talked with an old Chinese woman. When she heard that he was a native of a distant country she brought out an old parrot, that belonged to her. Long, long ago, she told him, the parrot had been given her by a highborn English lover of her youth. The boy thought that the bird must then be a hundred years old. It could say various sentences in the languages of all the world, picked up in the cosmopolitan atmosphere of the house. But one phrase the old China-woman's lover had taught it before he sent it to her, and that she did not understand, neither had any visitor ever been able to tell her what it meant. So now for many years she had given up asking. But if the boy came from far away perhaps it was his language, and he could interpret the phrase to her.

The boy had been deeply, strangely moved at the suggestion. When he looked at the parrot, and thought that he might hear Danish from that terrible beak, he very nearly ran out of the house. He stayed on only to do the old Chinese woman a service. But when she made the parrot speak its sentence, it turned out to be classic Greek. The bird spoke its words very slowly, and the boy knew enough Greek to recognise it; it was a verse from Sappho:

> 'The moon has sunk and the Pleiads,
> And midnight is gone,
> And the hours are passing, passing,
> And I lie alone.'

The old woman, when he translated the lines to her, smacked her lips and rolled her small slanting eyes. She asked him to say it again, and nodded her head.

KAREN BLIXEN (1883–1962),
Out of Africa, 1937

Conradin was ten years old, and the doctor had pronounced his professional opinion that the boy would not live another five years. The doctor was silky and effete, and counted for little, but his opinion was endorsed by Mrs De Ropp, who counted for nearly everything. Mrs De Ropp was Conradin's cousin and guardian, and in his eyes she represented those three-fifths of the world that are necessary and disagreeable and real; the other two-fifths, in perpetual antagonism to the foregoing, were summed up in himself and his imagination. One of these days Conradin supposed he would succumb to the mastering pressure of wearisome necessary things – such as illnesses and coddling restrictions and drawn-out dulness. Without his imagination, which was rampant under the spur of loneliness, he would have succumbed long ago.

Mrs De Ropp would never, in her honestest moments, have confessed to herself that she disliked Conradin, though she might have been dimly aware that thwarting him 'for his good' was a duty which she did not find particularly irksome. Conradin hated her with a desperate sincerity which he was perfectly able to mask. Such few pleasures as he could contrive for himself gained an added relish from the likelihood that they would be displeasing to his guardian, and from the realm of his imagination she was locked out – an unclean thing, which should find no entrance.

In the dull, cheerless garden, overlooked by so many windows that were ready to open with a message not to do this or that, or a reminder that medicines were due, he found little attraction. The few fruit-trees that it contained were set jealously apart from his plucking, as though they were rare specimens of their kind blooming in an arid waste; it would probably have been difficult to find a market-gardener who would have offered ten shillings for their entire yearly produce. In a forgotten corner, however, almost hidden behind a dismal shrubbery, was a disused tool-shed of respectable proportions, and within its walls Conradin found a haven, something that took on the varying aspects of a playroom and a cathedral. He had peopled it with a legion of familiar phantoms, evoked partly from fragments of history and partly from his own brain, but it also boasted two inmates of flesh and blood. In one corner lived a ragged-plumaged Houdan hen, on which the boy lavished an affection that had scarcely another outlet. Further back in the gloom stood a large hutch, divided into two compartments, one of which was fronted with close iron bars. This was the abode of a large polecat-ferret, which a friendly butcher-boy had once smuggled, cage and all, into its present quarters, in exchange for a long-secreted hoard of small silver. Conradin was dreadfully afraid of the lithe, sharp-fanged beast, but it was his most treasured possession. Its very presence in the tool-shed was a secret and fearful joy, to be kept scrupulously from the knowledge of the Woman, as he privately dubbed his cousin. And one day, out of Heaven knows what material, he spun the beast a wonderful name, and from that moment it grew into a god and a religion. The Woman indulged in religion once a week at a church near by, and took Conradin with her, but to him the church service was an alien rite in the House of Rimmon. Every Thursday, in the dim and musty silence of the tool-shed, he worshipped with mystic and elaborate ceremonial before the wooden hutch where dwelt Sredni Vashtar, the great ferret. Red flowers in their season and scarlet berries in the winter-time were offered at his shrine, for he was a god who laid some special stress on the fierce impatient side of things, as opposed to the Woman's religion, which, as far as Conradin could observe, went to great lengths in the contrary direction. And on great festivals powdered nutmeg was strewn in front of his hutch, an important feature of the offering being that the nutmeg had to be stolen.

These festivals were of irregular occurrence, and were chiefly appointed to celebrate some passing event. On one occasion, when Mrs De Ropp suffered from acute toothache for three days, Conradin kept up the festival during the entire three days, and almost succeeded in persuading himself that Sredni Vashtar was personally responsible for the toothache. If the malady had lasted for another day the supply of nutmeg would have given out.

The Houdan hen was never drawn into the cult of Sredni Vashtar. Conradin had long ago settled that she was an Anabaptist. He did not pretend to have the remotest knowledge as to what an Anabaptist was, but he privately hoped that it was dashing and not very respectable. Mrs De Ropp was the ground plan on which he based and detested all respectability.

After a while Conradin's absorption in the tool-shed began to attract the notice of his guardian. 'It is not good for him to be pottering down there in all weathers,' she promptly decided, and at breakfast one morning she announced that the Houdan hen had been sold and taken away overnight. With her short-sighted eyes she peered at Conradin, waiting for an outbreak of rage and sorrow, which she was ready to rebuke with a flow of excellent precepts and reasoning. But Conradin said nothing: there was nothing to be said. Something perhaps in his white set face gave her a momentary qualm, for at tea that afternoon there was toast on the table, a delicacy which she usually banned on the ground that it was bad for him; also because the making of it 'gave trouble,' a deadly offence in the middle-class feminine eye.

'I thought you liked toast,' she exclaimed, with an injured air, observing that he did not touch it.

'Sometimes,' said Conradin.

In the shed that evening there was an innovation in the worship of the hutch-god. Conradin had been wont to chant his praises, tonight he asked a boon.

'Do one thing for me, Sredni Vashtar.'

The thing was not specified. As Sredni Vashtar was a god he must be supposed to know. And choking back a sob as he looked at that other empty corner, Conradin went back to the world he so hated.

And every night, in the welcome darkness of his bedroom, and every evening in the dusk of the tool-shed, Conradin's bitter litany went up: 'Do one thing for me, Sredni Vashtar.'

Mrs De Ropp noticed that the visits to the shed did not cease, and one day she made a further journey of inspection.

'What are you keeping in that locked hutch?' she asked. 'I believe it's guinea-pigs. I'll have them all cleared away.'

Conradin shut his lips tight, but the Woman ransacked his bedroom till she found the carefully hidden key, and forthwith marched down to the shed to complete her discovery. It was a cold afternoon, and Conradin had been bidden to keep to the house. From the furthest window of the dining-room the door of the shed could just be seen beyond the corner of the shrubbery, and there Conradin stationed himself. He saw the Woman enter, and then he imagined her opening the door of the sacred hutch and peering down with her short-sighted eyes into the thick straw bed where his god lay hidden. Perhaps she would prod at the straw in her clumsy impatience. And Conradin fervently breathed his prayer for the last time. But he knew as he prayed that he did not believe. He knew that the Woman would come out presently with that pursed smile he loathed so well on her face, and that in an hour or two the gardener would carry away his wonderful god, a god no longer, but a simple brown ferret in a hutch. And he knew that the Woman would triumph always as she triumphed now, and that he would grow ever more sickly under her pestering and domineering and superior wisdom, till one day nothing would matter much more with him, and the doctor would be proved right. And in the sting and misery of his defeat, he began to chant loudly and defiantly the hymn of his threatened idol:

Sredni Vashtar went forth,
His thoughts were red thoughts and his teeth were white.
His enemies called for peace, but he brought them death.
Sredni Vashtar the Beautiful.

And then of a sudden he stopped his chanting and drew closer to the window-pane. The door of the shed still stood ajar as it had been left, and the minutes were slipping by. They were long minutes, but they slipped by nevertheless. He watched the starlings running and flying in little parties across the lawn; he counted them over and over again, with one eye always on that swinging door. A sour-faced maid came in to lay the table for tea, and still Conradin stood and waited and watched. Hope had crept by inches into his heart, and now a look of triumph began to blaze in his eyes that had only known the wistful patience of defeat. Under his breath, with a furtive exultation, he began once again the pæan of victory and devastation. And presently his eyes were rewarded: out through that doorway came a long, low, yellow-and-brown beast, with eyes a-blink at the waning daylight, and dark wet stains around the fur of jaws and throat. Conradin dropped on his knees. The great polecat-ferret made its way down to a small brook at the foot of the garden, drank for a moment, then crossed a little plank bridge and was lost to sight in the bushes. Such was the passing of Sredni Vashtar.

'Tea is ready,' said the sour-faced maid; 'where is the mistress?'

'She went down to the shed some time ago,' said Conradin.

And while the maid went to summon her mistress to tea, Conradin fished a toasting-fork out of the sideboard drawer and proceeded to toast himself a piece of bread. And during the toasting of it and the buttering of it with much butter and the slow enjoyment of eating it, Conradin listened to the noises and silences which fell in quick spasms beyond the dining-room door. The loud foolish screaming of the maid, the answering chorus of wondering ejaculations from the kitchen region, the scuttering footsteps and hurried embassies for outside help, and then, after a lull, the scared sobbings and the shuffling tread of those who bore a heavy burden into the house.

'Whoever will break it to the poor child? I couldn't for the life of me!' exclaimed a shrill voice. And while they debated the matter among themselves, Conradin made himself another piece of toast.

'SHREDNI VASHTAR', SAKI (1870–1916)

Mary Tofts, who claimed to have given birth to thirteen rabbits in the 1720s.

It is thought that ass's milk removed wrinkles from the face, making the skin white and soft, and it is well known that some women every day bathe their cheeks in it seven times, keeping carefully to that number. Poppaea, wife of the Emperor Nero, began this custom, even preparing her bath-tubs with the milk, and for this purpose she was always attended by troops of she-asses.

PLINY THE ELDER (AD 23–79),
Natural History, TRANS. H. RACKHAM

His name was Loulou. His body was green, the tips of his wings were pink, his poll blue, and his breast golden.

Unfortunately he had a tiresome mania for biting his perch, and also used to pull his feathers out, scatter his droppings everywhere, and upset his bath water. He annoyed Mme Aubain, and so she gave him to Félicité for good.

Félicité started training him, and soon he could say: 'Nice boy! Your servant, sir! Hail, Mary!' . . .

As the result of a chill [Félicité] had an attack of quinsy, and soon after that her ears were affected. Three years later she was deaf, and she spoke at the top of her voice, even in church. . . .

Imaginary buzzings in the head added to her troubles. Often her mistress would say: 'Heavens, how stupid you are!' and she would reply: 'Yes, Madame', at the same time looking all around her for something.

The little circle of her ideas grew narrower and narrower, and the pealing of bells and the lowing of cattle went out of her life. Every living thing moved about in a ghostly silence. Only one sound reached her ears now, and that was the voice of the parrot.

As if to amuse her, he would reproduce the click-clack of the turn-spit, the shrill call of a man selling fish, and the noise of the saw at the joiner's across the way; and when the bell rang he would imitate Mme Aubain's 'Félicité! The door, the door!'

They held conversations with each other, he repeating *ad nauseam* the three phrases in his repertory, she replying with words which were just as disconnected but which came from the heart. In her isolation, Loulou was almost a son or a lover to her. He used to climb up her fingers, peck at her lips, and hang on to her shawl; and as she bent over him, wagging her head from side to side as nurses do, the great wings of her bonnet and the wings of the bird quivered in unison. . . .

One morning in the terrible winter of 1837, when she had put him in front of the fire because of the cold, she found him dead in the middle of his cage, hanging head down with his claws caught in the bars. He had probably died of a stroke, but she thought he had been poisoned with parsley, and despite the absence of any proof, her suspicions fell on Fabu.

She wept so much that her mistress said to her: 'Why don't you have him stuffed?' . . .

At last he arrived – looking quite magnificent, perched on a branch screwed into a mahogany base, one foot in the air, his head cocked to one side, and biting a nut which the taxidermist, out of a love of the grandiose, had gilded.

Félicité shut him up in her room. . . .

With the aid of a wall-bracket, Loulou was installed on a chimney-breast that jutted out into the room. Every morning when she awoke, she saw him in the light of the dawn, and then she remembered the old days, and the smallest details of insignificant actions, not in sorrow but in absolute tranquillity. . . .

In church she was forever gazing at the Holy Ghost, and one day she noticed that it had something of the parrot about it. This resemblance struck her as even more obvious in a colour-print depicting the baptism of Our Lord. With its red wings and its emerald-green body, it was the very image of Loulou.

She bought the print and hung it in the place of the Comte d'Artois, so that she could include them both in a single glance. They were linked together in her mind, the parrot being sanctified by this connexion with the Holy Ghost, which itself acquired new life and meaning in her eyes. God the Father could not have chosen a dove as a means of expressing Himself, since doves cannot talk, but rather one of Loulou's ancestors. And although Félicité used to say her prayers with her eyes on the picture, from time to time she would turn slightly towards the bird. . . .

Her eyes grew weaker. The shutters were not opened any more. Years went by. . . .

For fear of being evicted, Félicité never asked for any repairs to be done. The laths in the roof rotted, and all through one winter her bolster was wet. After Easter she began spitting blood. . . .

The time to set up the altars of repose was drawing near.

The first altar was always at the foot of the hill, the second in front of the post office, the third about half-way up the street. There was some argument as to the siting of this one, and finally the women of the parish picked on Mme Aubain's courtyard. . . .

From Tuesday to Saturday, the eve of Corpus Christi, she coughed more and more frequently. In the evening her face looked pinched and drawn, her lips stuck to her gums, and she started vomiting. At dawn the next day, feeling very low, she sent for a priest.

Three good women stood by her while she was given extreme unction. . . .

The clergy appeared in the courtyard. Mère Simon climbed on to a chair to reach the little round window, from which she had a full view of the altar below.

It was hung with green garlands and adorned with a flounce in English needle-point lace. In the middle was a little frame containing some relics, there were two orange-trees at the corners, and all the way along stood silver candlesticks and china vases holding sunflowers, lilies, peonies, foxgloves, and bunches of hydrangea. This pyramid of bright colours stretched from the first floor right down to the carpet which was spread out over the pavement. Some rare objects caught the eye: a silver-gilt sugar-basin wreathed in violets, some pendants of Alençon gems gleaming on a bed of moss, and two Chinese screens with landscape decorations. Loulou, hidden under roses, showed nothing but his blue poll, which looked like a plaque of lapis lazuli.

The churchwardens, the choristers, and the children lined up along the three sides of the courtyard. The priest went slowly up the steps and placed his great shining gold sun on the lace altar-cloth. Everyone knelt down. There was a deep silence. And the censers, swinging at full tilt, slid up and down their chains.

A blue cloud of incense was wafted up into Félicité's room. She opened her nostrils wide and breathed it in with a mystical, sensuous fervour. Then she closed her eyes. Her lips smiled. Her heart-beats grew slower and slower, each a little fainter and gentler, like a fountain running dry, an echo fading away. And as she breathed her last, she thought she could see, in the opening heavens, a gigantic parrot hovering above her head.

GUSTAVE FLAUBERT (1821–80),
'*Un Coeur Simple*', FROM *Trois Contes*,
TRANS. ROBERT BALDICK

'Her heart-beats grew slower and slower . . .'

At the Spring Meeting of 1804, Mr Whalley's KING PIPPIN was brought on the Curragh of Kildare to run. He was a horse of the most extraordinary savage and vicious disposition. His particular propensity was that of *flying at and worrying* any person who came within his reach, and if he had an opportunity, he would get his head round, seize his rider by the leg with his teeth, and drag him down from his back. For this reason he was always ridden with what is called *a sword*; which is a strong flat stick, having one end attached to the cheek of the bridle, and the other to the girth of the saddle, a contrivance to prevent a horse of this kind from getting at his rider.

King Pippin had long been difficult to manage and dangerous to go near to, but on the occasion in question he could not be got out to run at all. *Nobody could put the bridle upon his head.* It being Easter Monday, and consequently a great holiday, there was a large concourse of people assembled at the Curragh, consisting principally of the neighbouring peasantry; and one countryman, more fearless than the rest of the lookers-on, forgetting, or perhaps never dreaming that the better part of courage is discretion, volunteered his services to bridle the horse. No sooner had he committed himself in this operation, than King Pippin seized him somewhere about the shoulders or chest, and, says Mr Watts (Mr Castley's informant) 'I know of nothing I can compare it to, so much as a dog shaking a rat.' Fortunately for the poor fellow, his body was very thickly covered with clothes, for on such occasions an Irishman of this class is fond of displaying his wardrobe, and if *he has three coats at all in the world*, he is sure to put them all on.

This circumstance in all probability saved the individual who had so gallantly volunteered the forlorn hope. His person was so deeply enveloped in extra-teguments, that the horse never got fairly hold of his skin, and I understand that he escaped with but little injury, beside the sadly rent and totally ruined state of his holyday toggery.

The Whisperer was sent for, who, having arrived, was shut up with the horse all night, and in the morning he exhibited this hitherto ferocious animal, following him about the course like a dog – lying down at his command – suffering his mouth to be opened, and any person's hand to be introduced into it – in short, as quiet almost as a sheep.

He came out the same meeting, and won his race, and his docility continued satisfactory for a considerable time; *but at the end of about three years his vice returned*, and then he is said to have *killed a man*, for which he was destroyed.

It may not be uninteresting, in this connexion, to give some account of this tamer of quadruped vice. However strange and magical his power may seem to be, there is no doubt of the truth of the account that is given of him. . . . [We] give the following extract from Croker's Fairy Legends and Traditions of Ireland, Part II, p. 200, for his performances seem the work of some elfin sprite, rather than of a rude and ignorant horse-breaker.

'He was an awkward, ignorant rustic of the lowest class, of the name of Sullivan, but better known by the appellation of the Whisperer. His occupation was horse-breaking. The nickname he acquired from the vulgar notion of his being able to communicate to the animal what he wished by means of a whisper . . . As far as the sphere of his control extended, the boast of *veni, vidi, vici*, was more justly claimed by Sullivan than even by Cæsar himself.

'How his art was acquired, and in what it consisted, is likely to be for ever unknown, as he has lately (about 1810) left the world without divulging it. . . . The wonder of his skill consisted in the celerity of the operation, which was performed in privacy, without any apparent means of coercion. Every description of horse, or even mule, whether previously broken or unhandled, whatever their peculiar habits or vices might have been, submitted without show of resistance to the magical influence of his art, and in the short space of half an hour became gentle and tractable. . . .

'When sent for to tame a vicious beast, for which he was either paid according to the distance, or generally two or three guineas, he directed the stable, in which he and the object of the experiment were, to be shut, with orders not to open the door until a signal was given. After a *tête-à-tête* of about half an hour, during which little or no bustle was heard, the signal was made, and, upon opening the door, the horse appeared lying down, and the man by his side, playing with him like a child with a puppy dog. From that time he was found perfectly willing to submit to any discipline – however repugnant to his nature before.' 'I once,' continues Mr Croker, 'saw his skill tried on a horse, which could never before be brought to stand for a smith to shoe him. . . . I observed that the animal appeared terrified whenever Sullivan either spoke to or looked at him; how that extraordinary ascendancy could have been obtained, is difficult to conjecture.

'In common cases this mysterious preparation was unnecessary. He seemed to possess an instinctive power of inspiring awe, the result, perhaps, of natural intrepidity, in which, I believe, a great part of his art consisted; though the circumstance of the *tête-à-tête* shows that, on particular occasions, something more must have been added to it. A faculty like this would, in some hands, have made a fortune, and I understand that great offers were made to him, for the exercise of his art abroad. But hunting was his passion. He lived at home in the style most agreeable to his disposition, and nothing could induce him to quit Duhallow and the fox-hounds.'

WILLIAM YOUATT, *The Horse*, 1843

Careful observers may foretell the hour
(By sure prognostics) when to dread a shower;
While rain depends, the pensive cat gives o'er
Her frolics, and pursues her tail no more.

JONATHAN SWIFT (1667–1745)

Rosslyn's own pet scientific experiment was drawing to a conclusion. Since he was expelled from one school for keeping white mice in his pencil box, and another for rats in his boot hole, he had been breeding and interbreeding the different basic colours on the lines of Mendel's sweetpeas, and had produced every colour of the rainbow – except green. True the blue model were only as blue as a kerry blue terrier, the red only as red as a red setter, the lavender only as lavender as lavender waistcoats, but the champagne and strawberry roans were more aptly named. The apricot vole, which appeared at a Rotary dinner, was said to be distinctly striking.

Now he bred a green mouse, as green as, if not greener than, a green monkey. To one whose glasses were always rose-tinted, who saw good – which is what he wanted to see – in all men, the mouse seemed as green as a field of sprouting corn in June. He was interviewed by the Eastbourne Herald, which wrote:

'After breeding 50 generations Dr Rosslyn Bruce has succeeded in producing the first Green Mouse. This mouse was grass green and its first offspring was bottle green. Now he has the first family from the bottle green mouse. Their chief tendency is back to grass green, but two of the family are white – the first sign of the original Japanese white strain for many generations of this family.

"It is obvious to me that I have to get more yellow and blue into my strain to improve the green," said the rector, "so I shall have to introduce further crosses. I am still hopeful of producing something approaching perfection in green mice. Some of my latest young mice may be better than I think. It is early yet to be quite sure of their colour."'

The national press took it up immediately, and even the American press. ENGLISH RECTOR GETS GREEN MOUSE AFTER 50 TRIES was a headline in the Chicago Tribune. . . .

On Christmas Day, his thirty-third Christmas at Herstmonceux, Rosslyn took the services. . . . Quintin brought a bottle of champagne. 'I knew he was dying,' he wrote many years afterwards, 'and wanted him to have a happy Christmas. . . . I offered him a tame mouse, borrowed from one of my children, and he was much more pleased. He put it on the white counterpane of his bed and said, 'Look at the light through its beautiful little ears. Isn't that like the gates of heaven?'

VERILY ANDERSON, *The Last of the Eccentrics -
A life of Rosslyn Bruce*, 1972, TRANS. KATE HARRIS

'I am still hopeful of producing something approaching perfection in green mice.'

I have – or I should say regretfully – had a friend, once well known in many walks of life (only in all of them she ran rather than walked) her name, Fanny Currey. Painter, rider, florist, speaker, fighting-suffragist, dashing and dauntless. In one of her many trades I was accustomed to participate, and thus it chanced that one fine afternoon she and I had gone forth with our paints to the lovely mountain country above Lismore, in the County of Waterford, where Fanny had her home. . . .

On the road below us, motionless, on one of the steepest pinches of the long hill, was a cart laden high with sacks of flour, in the shafts a big hairy-heeled mare, her ears back, her feet firmly planted, while her driver on foot dragged at her head, yelling curses and thrashing her with a long ash plant. Fanny cast aside '*les instruments de sa fatigue*' and scrambled down into the road.

'I can't stand this! I'll try the Charm!' she called back to me, running down the hill and hurrying to the mare's head. I followed her. To save bloodshed. The carter's blood, not Fanny's.

'Leave her to me!' she ordered the carter, imperiously. 'Stop hitting her!'

Taken by surprise, the carter dropped his stick and the reins, staring in entire bewilderment.

Fanny took the mare, wild-eyed and sweating, by the reins, going up close to her head and patting her dripping neck, and whispering close to her ear. What she whispered I know not. I can only affirm that in what felt to me no longer than two or three minutes, the big mare unhesitatingly put down her head in apparently cheerful acceptance, and started away up the hill, making light of her load, filled apparently with an enthusiastic determination to oblige. The carter, snatching up the long rope reins, rushed forward with the cart, shouting blessings and gratitude, and continued so to rush and to shout until he and his cart were out of sight.

Then I laid hold of Fanny and demanded an explanation. 'It's a charm,' she said, shortly. 'My father gave an American, a man named Rarey, thirty sovereigns for it. I can't tell you more than that.'

In response to my protests she expanded a little.

'I tried it once on a jibbing hansom-horse in Oxford Street. The police tried to stop me. They had cleared the road and stopped all traffic, but I dodged under a bobby's arm and got up to the horse. The charm worked at once. The driver shouted thanks at me –'

I could see the scene. Fanny, small and wiry, darting eel-like under the big policeman's outstretched arm, scudding across the empty street, defying Law and Authority with the dash and success learned in the brave days of the Suffrage War.

SOMERVILLE AND ROSS (1858–1949 and 1862–1915), 'Happy Days!', 1946

Beasts were often condemned to be burned alive; and strangely enough, it was in the latter half of the seventeenth century, an age of comparative enlightenment, that this cruel penalty seems to have been most frequently inflicted. Occasionally a merciful judge adhered to the letter of the law and curbed its barbarous spirit by sentencing the culprit to be slightly singed and then to be strangled before being committed to the flames. Sometimes brutes were doomed to be buried alive. Thus we have the receipt of 'Phélippart, sergeant of high justice of the city of Amiens,' for the sum of sixteen soldi, in payment for services rendered in March 1463, in 'having buried in the earth two pigs, which had torn and eaten with their teeth a little child in the faubourg of Amiens, who for this cause passed from life to death (*étoit allé de vie a trépas*).' In 1557, on the 6th of December, a pig in the Commune of Saint-Quentin was condemned to be 'buried all alive' (*enfoui tout vif*), 'for having devoured a little child in l'hostel de la Couronne.' Again, a century earlier, in 1456, two pigs were subjected to this punishment, 'on the vigil of the Holy Virgin,' at Oppenheim on the Rhine, for having killed a child. . . . By the advice of a French veterinary doctor, who was quartered there with the army of General Moreau, the town bull was buried alive at the crossroads in the presence of several hundred persons. We are not informed whether this sacrifice proved to be a sufficiently 'powerful medicine' to stay the epizoötic plague; the noteworthy fact is that the superstitious rite was prescribed and performed, not by an Indian magician or an African sorcerer, but by an official of the French republic. . . .

In 1394, a pig was hanged at Mortaign for having sacrilegiously eaten a consecrated wafer; and in a case of infanticide, it is expressly stated in the plaintiff's declaration that the pig killed the child and ate of its flesh, 'although it was Friday,' and this violation of the *jejunium sextae,* prescribed by the Church, was urged by the prosecuting attorney and accepted by the court as a serious aggravation of the porker's offence.

E. P. EVANS, *The Criminal Prosecution and
Capital Punishment of Animals,* 1907

THE GARDENER AND THE HOG

A gard'ner, of peculiar taste,
On a young Hog his favour plac'd,
Who fed not with the common herd,
His tray was to the hall prefer'd,
He wallow'd underneath the board,
Or in his master's chamber snor'd,
Who fondly stroak'd him ev'ry day,
And taught him all the puppy's play;
Where'er he went, the grunting friend
Ne'er fail'd his pleasure to attend.
As on a time, the loving pair
Walk'd forth to tend the garden's care,
The master thus addrest the swine.
My house, my garden, all is thine:
On turnips feast whene'er you please,
And riot in my beans and pease,
If the potatoe's taste delights,
Or the red carrot's sweet invites,
Indulge thy morn and evening hours,
But let due care regard my flowers;
My tulips are my garden's pride.
What vast expence those beds supply'd!
The Hog by chance one morning roam'd
Where with new ale the vessels foam'd;
He munches now the steaming grains,
Now with full swill the liquor drains;
Intoxicating fumes arise,
He reels, he rolls his winking eyes,
Then stagg'ring through the garden scowers,
And treads down painted ranks of flowers,

With delving snout he turns the soil,
And cools his palate with the spoil.
The Master came, the ruin spy'd.
Villain, suspend thy rage, he cry'd:
Hast thou, thou most ungrateful sot,
My charge, my only charge forgot?
What, all my flowers! No more he said,
But gaz'd, and sigh'd, and hung his head.
The Hog with stutt'ring speech returns.
Explain, Sir, why your anger burns;
See there, untouch'd your tulips strown,
For I devour'd the roots alone!
At this, the Gard'ner's passion grows;
From oaths and threats he fell to blows;
The stubborn brute the blows sustains,
Assaults his leg and tears the veins.
Ah, foolish swain, too late you find
That sties were for such friends design'd!
Homeward he limps with painful pace,
Reflecting thus on past disgrace;
Who cherishes a brutal mate
Shall mourn the folly soon or late.

JOHN GAY (1685–1732), *Fables*, 1727

I find that a heavy cold clears up if the sufferer kisses a mule's muzzle.

PLINY THE ELDER (AD 23–79),
Natural History, TRANS. H. RACKHAM

My nurse declared that I and my brother were cured of the measles by having hair cut from the nape of each of our necks, and then separately placed between two slices of bread and butter. She says she anxiously watched for a strange dog to pass (no other being efficacious). She then gave him the bread and butter, and as he ate it without loathing, she was sure we should be cured. He then went away, and of course never came again, for he died of the measles – miserably, no doubt, poor fellow, having travelled off with the disease of three affected children.

Notes and Queries, 6 OCTOBER 1855

JUBILATE AGNO

For I will consider my Cat Jeoffrey.

For he is the servant of the Living God, duly and daily serving him.

For at the First glance of the glory of God in the East he worships in his way.

For is this done by wreathing his body seven times round with elegant quickness.

For then he leaps up to catch the musk, which is the blessing of God upon his prayer....

CHRISTOPHER SMART (1722–71)

'The handsome horse shown is in the possession of the longest foretop, mane, and tail in the world. He was born about seven years ago in the State of Oregon. His tail is now nine feet long; his foretop is five and one-half feet long; while his mane measures exactly seven feet and ten inches.

'"Linus" is perfectly formed and weighs about fourteen hundred pounds. His hair, which is "done up" when he is not receiving visitors, continues to grow, though now very slowly.' (St Nicholas *Magazine*, March 1891)

At this same season Fayoles, fourth king of Numidia, sent to Grandgousier from the land of Africa the greatest and most enormous mare that ever was seen; and she was the most monstrous too, since you know very well that Africa is always producing some new monstrosity. She was as big as six elephants, she had her hoofs divided into toes, like

Julius Caesar's horse, and pendant ears, like the goats of Languedoc, and a little horn on her rump. For the rest, her coat was of a burnt sorrel with dapple-grey spots and, what is more, she had the most fearsome tail. For it was as large, more or less as the pillar of Saint Mars, near Langès, and as square; and its tufts were as spiky, in every respect, as blades of wheat.

If this description astounds you, you will be even more astounded by what I tell you about the tail of the Scythian ram, which weighed more than thirty pounds, or about that of the Syrian sheep, to whose rump (if Thenaud speaks true) they have to fix a little cart to carry it, it is so long and heavy. You clods from the flat lands possess nothing to compare with that. . . .

So they passed joyfully along the highway, always in high spirits, till they came above Orléans, at which place there was a great forest a hundred and five miles long and fifty-one miles wide, or thereabouts. This forest was horribly abundant and copiously swarming with ox-flies and hornets, so that it was an absolute brigands' lair for the poor mares, asses, and horses. But Gargantua's mare handsomely avenged all the outrages ever perpetrated there on the beasts of her kind, by a trick of which they had not the slightest inkling beforehand. As soon as they had entered this forest and the hornets had opened their attack she threw out her tail, and at her first skirmish swatted them so completely that she swept down the whole wood. Crossways and lengthways, here and there, this way and that, to front and to side, over and under, she swept down the trees as a mower does the grass, so that since that time there has been neither wood nor hornets, and the whole country has been reduced to a plain.

FRANÇOIS RABELAIS (*c.* 1494–1553),
Pantagruel, 1532

16

ave a
Good Laugh

'Run, girls, run! Fetch the dogs! I've a rat in my bustle!'

SOMERVILLE AND ROSS (1858–1949 and
1862–1915), 'Happy Days!', 1946

I heard a very ridiculous story a few days ago: Mr Page, brother to Sir Gregory, going to visit Mr Edward Walpole, a tame goat which was in the street followed him unperceived when he got out of the coach into the house. Mr Walpole's servant, thinking the goat came out of Mr Page's coach, carried it into the room to Mr Walpole, who thought it a little odd Mr Page should bring such a visitor, as Mr Page no less admired his choice of so savoury a companion; but civility, a great disguiser of sentiments, prevented their declaring their opinions, and the goat, no respecter of persons or furniture, began to rub himself against the frame of a chair which was carved and gilt, and the chair, which was fit for a Christian, but unable to bear the shock of a beast, fell almost to pieces. Mr Walpole thought Mr Page very indulgent to his dear crony the goat, and wondering he took no notice of the damage, said he fancied tame goats did a great deal of harm, to which the other said he believed so too: after much free and easy behaviour of the goat, to the great detriment of the furniture, they came to an explanation, and Mr Goat was turned downstairs . . .

LETTER TO THE DUCHESS OF PORTLAND,
17 DECEMBER 1738, FROM ELIZABETH
MONTAGU (1720–1800)

'Returning from the University of Giessen in October 1845, I brought with me,' he wrote, 'about a dozen green tree-frogs, which I had caught in the woods near the town. The Germans call them Laub-Frosch or leaf-frog; they are most difficult things to find, on account of their colour so much resembling the leaves on which they live. I have frequently heard one singing in a small bush, and, though I have searched carefully, have not been able to find him; the only way is to remain quite quiet till he again begins his song. After much ambush work, at length I collected a dozen frogs and put them in a bottle. I started at night on my homeward journey by the diligence, and I put the bottle containing the frogs into the pocket inside the diligence. My fellow-passengers were sleepy old smoke-dried Germans. Very little conversation took place, and, after the first mile, every one settled himself to sleep, and soon all were snoring. I suddenly awoke with a start, and found all the sleepers had been roused at the same moment. On their sleepy faces were depicted fear and anger. What had woke us all up so suddenly? The morning was just breaking, and my frogs, though in the dark pocket of the coach, had found it out, and, with one accord, all twelve of them had begun their morning song. As if at a given signal, they one and all of them began to croak as hard as ever they could. The noise their united concert made, seemed, in the closed compartment of the coach, quite deafening: well might the Germans look angry; they wanted to throw the frogs, bottle and all, out of the window, but I gave the bottle a good shaking, and made the frogs keep quiet. The Germans all went to sleep again, but I was obliged to remain awake to shake the frogs when they began to croak. It was lucky that I did so, for they tried to begin their concert again two or three times.'

GEORGE C. BOMPAS, *Life of Frank Buckland,* 1885

The cook had a saying that went with him down the years: 'I laughed enough to kill a military horse.'

REGINALD POUND,
A. P. Herbert, a Biography, 1976

When the owls first joined our family they were less than six weeks old, but already giving promise that their ultimate size would be impressive. My parents, who had never seen a full-grown horned owl, had no real idea as to just *how* impressive they could be, and I preserved a discreet silence on the subject. . . .

. . . my parents had decided that we should spend a weekend at Emma Lake, a resort area far to the north. We loaded our camping gear aboard Eardlie, our Model A, and the six of us – Mother, Father, myself, Mutt, and the two owls – set out.

Having ridden in the car on several previous occasions, the owls had developed a preference in the matter of seating arrangements. Their chosen roost was the back of the rumble seat, where they were exposed to the full force of the slip stream. They loved it, for it offered them the same exhilarating thrill that all small boys experience when they thrust a hand out of a car window and let the wind act on it as it does upon the wing of an aircraft. My owls exploited this adventure to the limit. As soon as the car was in motion they would extend their great pinions as if in flight. If they then slanted the leading edges downward, the rush of air would force them into a squatting position. But when they tipped the leading edges upward, they would be lifted clean off the seat, and only the grip of their talons would keep them from soaring aloft like kites.

There was not sufficient room to allow both of them to bob up and down together, so they learned to alternate. While one was going down, the other would be coming up, in rhythmic frequency. Intoxicated by the rush of air, they would often break into song, and my father, caught up in the spirit of the thing, would punctuate their excited hootings with blasts on Eardlie's horn. . . .

In the summer of Wol's second year, Saskatoon was enriched by the arrival of a young curate who had just graduated from a divinity school in the east. The curate was of an earnest persuasion and he made it his first duty to pay a call on every one of the members of our parish (to which he was attached). It was a warm and balmy summer afternoon when he reached our house.

He rang our doorbell, and Mother was pleased to welcome him, for he was a well-favored youth. He had the high-domed forehead that is so often the mark of the stage cleric, but the hair above it was black and curly. Mother invited him into the living room for a chat and a cup of tea.

The young man made himself discreetly comfortable on our chesterfield, a massive and antique piece of furniture which was so placed that it faced the fireplace, with its back to the open window some six feet behind. Balancing a cup gracefully in his hand, the young divine engaged Mother in conversation, the burden of which was concerned with my own lamentable absences from Sunday School.

Wol had been spending that afternoon ant bathing. This was a peculiar pastime in which he sometimes engaged, and which consisted of tearing an anthill apart and then fluffing the mixture of dust and angry ants through his feathers. He appeared to find the sensation gratifying, although its purpose seemed inscrutable to us. At any rate, he finished his bath about 4 p.m. and, feeling in an amiable mood, decided to come into the house and tell Mother about it.

To this day my mother swears she did not see him in sufficient time to warn her visitor. I believe that she saw him well enough, but was just too petrified to open her mouth.

Now Wol, in his maturity, had become a sentimental bird, possessed of the habit of leaping lightly to one's shoulder, there to balance himself while he tenderly nibbled the nearest human ear with his great beak, breathing harshly but affectionately into the face of his companion the while. Everyone who was acquainted with Wol knew of this habit, and some deplored it, for Wol was a carnivore and as a result he had the most atrocious breath.

The flight of an owl is noiseless; there is no warning rustle of pinions. Wol's arrival on the windowsill was as silent as the arrival of a puff of thistledown. He paused a moment in the opening and then, spying a pair of tempting shoulders on the chesterfield, launched himself across the narrow intervening space.

The object of his attentions shot straightway into the air and began leaping ecstatically about the room. It was an ill-considered action on his part, for Wol lost his balance and, with a purely involuntary reflex action, tightened up his talons.

The curate now demonstrated that he was both an athletic and a vocal youth. He howled, and his bouncing became wilder. Wol, clinging for dear life now, and deeply disturbed by his reception, tightened up his grip once more and then – and it could have happened only as a result of his surprise and indignation – he forgot for the first and last time in his mature life that he was housebroken.

FARLEY MOWAT (b. 1921), *The Dog Who Wouldn't Be*, 1957

There is a photograph, taken in the eighties of the last century, that represents a match which was played at Hong Kong, between two girls, and two young soldiers. The girls, I suppose, received points from the soldiers; they had undoubtedly handicapped themselves. The rules that apply to ladies' hats in church, applied also to hats on tennis grounds. These girls wore large hats with tall trimmings like the sails of a racing yacht, and equally suited to hold the wind. They had tight-fitting bodices, buttoned from throat to waist, with long sleeves, long voluminous white serge skirts with (no doubt) an under petticoat and these were surmounted by excrescences that the latest Paris fashion had recently inflicted upon the submissive world of women. These were called '*Tournures*,' or (less elegantly and quite senselessly) 'bustles.' A bustle may be described as a small pillow or cushion that varied in size in accordance with the ambition of the wearer to be more or less in the fashion. Parisian fancy had capriciously decided to admire a generous posterior outline, and such a development was characterized as a 'Hottentot beauty.' In order to acquire this attraction a bustle was attached to the person at the back below the waist, and was thus intended to supply that which Nature had denied to European ladies, while bestowing it in rather superfluous abundance on the female Hottentot. It was pretty soon found by its wearers to have the disadvantage of acting very rapidly either as a live coal or, more gradually, as a Spanish-fly blister.

A humane and inventive dressmaker (qualities unusual in dressmakers) was, however, presently inspired to substitute for the cushion a large wire sponge-basket. The idea was adopted, and expanded variants of the sponge-basket were put on the market and found a ready sale. The vagaries of fashion have seldom any practical value, but I have heard of an emergency when a bustle (or *tournure*) (contrary to its usual habit) came to the front and proved itself good at need.

This was in the case of a sporting and courageous young woman who, in the course of a rat-hunt in which she and her sisters were engaged, suddenly became aware of a happening or assault, in rear, and instantly realised its import. Instead of abandoning herself to panic, she stood quite still and shouted to her sisters.

'Run, girls, run! Fetch the dogs! I've a rat in my bustle!'

It is a nice point whether the bustle (or *tournure*) had been seized by the rat as a place of safety, or as a strategic point from which to launch an assault, but its indomitable wearer is said to have regarded it as a rat-trap, and with both hands to have pressed it closely to her person. Her confidence in the staunchness of the sponge-basket was rewarded. The dogs got the rat.

SOMERVILLE AND ROSS (1858–1949 and
1862–1915), '*Happy Days!*', 1946

Exclusively belonging to the cook, although a favourite with the whole crew, my friend (a Cercopithecus monkey from Senegal) had been at first kept by means of a cord . . . but, as he became more and more tame, his liberty was extended, till at last he was allowed the whole range of the ship, with the exception of the captain's and passengers' cabins. . . .

Two days in each week, the pigs, which formed part of our live stock, were allowed to run about the deck for exercise, and then Jack was particularly happy: hiding himself behind a cask, he would suddenly spring on to the back of one of them, his face to the tail, and away scampered his frightened steed. Sometimes an obstacle would impede the gallop, and then Jack, loosening the hold which he had acquired by digging his nails into the skin of the pig, industriously tried to uncurl its tail, and if he were saluted by a laugh from some one near by, he would look up with an assumed air of wonder, as much as to say, What can you find to laugh at? When the pigs were shut up, he thought it his turn to give others a ride, and there were three little monkeys, with red skins and blue faces, whom he particularly favored: I frequently met him with all of them on his back at the same time, squeaking and huddling together, and with difficulty preserving their seat; when he suddenly stopped, and seemed to ask me to praise the good-natured action which he was performing. He was, however, jealous of all those of his brethren who came in contact with me, and freed himself from two of his rivals by throwing them into the sea. One of them was a small Lion monkey, of great beauty and extreme gentleness, and immediately after I had been feeding him, Jack called him with a coaxing, patronizing air; but as soon as he was within reach, the perfidious creature seized him by the nape of his neck, and, as quick as thought, popped him over the side of the ship. We were going at a brisk rate, and although a rope was thrown out to him, the poor little screaming thing was soon left behind, very much to my distress, for his almost human agony of countenance was painful to behold. For this, Jack was punished by being shut up all day in the empty hen-coop, in which he usually passed the night, and which he so hated, that when bed-time came, he generally avoided the clutches of the steward; he, however, committed so much mischief when un-watched, that it had become necessary to confine him at night, and I was often obliged to perform the office of nursemaid. Jack's principal punishment, however, was to be taken in front of the cage in which a panther belonging to me was placed, in the fore part of the deck. His alarm was intense; the panther set up his back and growled, but Jack instantly closed his eyes, and made himself perfectly rigid. I generally held him up by the tail, and if I moved, he cautiously opened one eye; but if he caught sight of even a corner of the cage, he shut it fast, and again pretended to be dead. His drollest trick was practised on a poor little black monkey; taking the opportunity when a calm, similar to that spoken of above, left him nearly the sole possessor of the deck. I do not

'His drollest trick.'

know that he saw me, for I was sitting behind the companion door. The men had been painting the ship outside, and were putting a broad band of white upon her, when they went to dinner below, leaving their paint and brushes on the upper deck. Jack enticed his victim to him, who meekly obeyed the summons; and, seizing him with one hand, he, with the other, took the brush, and covered him with the white fluid from head to foot. The laugh of the man at the helm called my attention to the circumstance, and as soon as Jack perceived he was discovered, he dropped his dripping brother, and rapidly scampered up the rigging, till he gained the main-top, where he stood with his nose between the bars, looking at what was going on below. As the other monkey began to lick himself, I called up the steward, who washed him clean with turpentine, and no harm ensued; but Jack was afraid to come down, and only after three days passed in his elevated place of refuge did hunger compel him to descend. He chose the moment when I was sitting on deck, and, swinging himself by a rope, he dropped suddenly into my lap, looking so imploringly at me for pardon, that I not only forgave him myself, but procured his absolution from others. Jack and I parted a little to the south of the Scilly Islands, after five month's companionship, and never met again; but I was told that he was much distressed at my absence, hunted for me all over the vessel in the most disconsolate manner, even venturing into my cabin; nor was he reconciled to the loss of me when the ship's company parted in the London docks.

MRS R. LEE, *Anecdotes of the Habits and Instincts of Animals*, 1852

On the morning of the auspicious day upon which the Bishop was coming to luncheon, the Vicar went out into the garden where he found Antonia planting primroses on little Jock's grave. It was the anniversary of his death. The camel stood beside her watching the proceeding with an air of interest. The Vicar saw that Antonia had tears in her eyes. What a sweet, tender-hearted woman she was!

Under the stress of her emotion she was digging rather wildly and scattering the earth in all directions.

'Oh! be careful, my dear,' said the Vicar, 'or you will be displacing the body of your little dog. I remember he was not buried very deep.'

Antonia laid down her trowel.

'My poor little Jock,' she said, 'it distresses me to think of his poor little body lying there in the cold earth. I often wish that he had been otherwise disposed of. Lady Bugle had her pug stuffed, and very nice it looks too. I wish now that I had thought of having dear little Jock stuffed. The idea never occurred to me. I should feel far happier if he were in a nice glass case somewhere in the house. But alas! now it is too late. Of course,' she went on, 'I might still dig up his remains and put them in a nice mahogany box. Then I could keep him in the drawing-room, and I shouldn't wake up on cold nights, as I often do, and think of my poor little Jock out here in the freezing soil.'

The Vicar was alarmed at the suggestion.

'My dear,' he said, 'I shouldn't do that. It is better . . .' He was about to say 'to let sleeping dogs lie,' but he thought it might perhaps sound a trifle heartless, and left the sentence unfinished.

He looked at his watch.

'Antonia dear,' he said, 'it is growing late, and the Bishop will soon be arriving. Hadn't you better put the camel away and come into the house? I am sure that there are a lot of little things that need attending to.'

Antonia sighed and gathered up her gardening basket and her trowel.

'Very well,' she said, 'I will finish planting the primroses later.'

She led the camel to the stables. As she was in the act of closing the barn door, she was struck by an extraordinarily sympathetic expression in the eyes of the beast.

'Ah,' she said, 'you seem to understand my feelings so well, even though you are but a dumb animal. Had you known my little Jock, I am sure that you would have loved him too.'

In inviting Lady Bugle to meet the Bishop, Antonia had made a grave mistake. Indeed, had Antonia mentioned in her letter that the Bishop was coming to luncheon, Lady Bugle would have most certainly refused the invitation. She had no friendly feelings for the Bishop; neither did the Bishop care very much for Lady Bugle. He considered her pretentious and vulgar, while Lady Bugle thought that the Bishop was supercilious and inclined to be patronising. In fact, on one occasion, it had come to an actual tiff at a Committee Meeting in Woxham. Lady Bugle was apt to domineer at Committee Meetings and, when there were no important members of County Families present, she would 'throw her weight' so ponderously that all opposition was crushed. The Bishop, however, was not in the least intimidated by Lady Bugle's autocratic methods, and had openly snubbed her in a way that she could neither forgive nor forget.

Thus, when they met in the Vicarage drawing-room, their greetings were anything but cordial. The Vicar and his wife were neither of them gifted with any great subtlety of discernment where social amenities were concerned, and they merely attributed the reciprocated coldness to the reserve peculiar to important people.

As soon as they were seated at the luncheon table, the trouble began. The Bishop proceeded studiously to ignore Lady Bugle who retaliated by treating every remark the Bishop made with contemptuous comments. These, however, were not directed openly at the Bishop but delivered either in the form of a soliloquy, or muttered in an audible sotto voce to the Vicar.

The Husseys, when at last the eccentricity of this conversational technique began to dawn upon them, were very much perplexed. Antonia thought that perhaps Lady Bugle was feeling indisposed. She was known to suffer from bilious attacks which made life rather difficult for people in her immediate vicinity.

The Vicar, realising that something was going wrong, wrestled courageously with the difficulties of the situation. But the air was filled with electricity and it seemed as though, at any moment, the storm might burst. The conversation was tactfully led to a discussion of foreign travel. That, at least, seemed to be a subject that might prove less controversial than many others, and one that the Bishop enjoyed talking about. Antonia supplied a few reminiscences of her early missionary life, and, for a while, Lady Bugle's aggressive policy seemed to have somewhat abated. The Vicar remarked that it was a long time since he had been abroad, and went on to expatiate on the discomforts he and Antonia had been obliged to put up with on a trip to Switzerland they had made during their honeymoon.

Then the Bishop took up the theme. 'My wife and I,' he said, 'had a very curious experience in Italy last year.'

'Oh Italy!' said Lady Bugle, addressing her remarks to no one in particular, 'Nobody goes to Italy nowadays. All the best people are going to Egypt.'

The Bishop flushed and raised his eyebrows in an ominous manner. It looked as though at last he were about to deliver himself of a crushing retort and annihilate Lady Bugle.

Luckily at that moment a diversion was provided by the entry of the leg of mutton. Bessie bore in the dish in her most pontifical manner and set it down before the Vicar. The silver dish cover had been newly polished and filled the little dining-room with effulgence. For a moment Antonia caught a glimpse of the Bishop's angry face reflected in it as though in a distorting mirror.

The Vicar rose to his feet and said to the Bishop with a facetious pomposity, 'Here, My Lord, is your leg of mutton.'

He lifted the cover and, to everyone's amazement, there was disclosed, reposing in a pool of gravy, the skeleton of a small dog.

Antonia realised at once from the mouldering collar that it was the last mortal remains of her darling little Jock and burst into tears. The Vicar gasped and turned to Bessie who stood rooted to the spot, her mouth opening and shutting in a way that reminded one of the spasms of a dying fish.

'What is the meaning of this?' the Vicar enquired.

Bessie's mouth continued to open and shut but no sound came from her lips.

At that moment Annie burst into the room with the missing leg of mutton on a plate. Someone, she declared breathlessly, must have tried to steal it. She had found it on the ground outside the kitchen window.

'I will enquire into this matter afterwards,' said the Vicar, handing her the dish with the skeleton, which she hastily removed.

Although Annie assured the Vicar that the leg of mutton had been well wiped, neither the Bishop nor Lady Bugle fancied that it had been much improved by its strange adventure. The Bishop took a very small helping which he left on his plate, while Lady Bugle refused it altogether.

From this point onwards, as you may well imagine, the luncheon party grew more difficult to cope with than ever. Antonia, tearfully excusing herself, rose from the table and hurried out of the room in order to rescue the remains of her beloved pet and put it in a safe place, leaving her husband alone to deal with the situation. He could think of no apology and no explanation to offer.

'It would appear,' he said, 'that it must have been the skeleton of my wife's little dog, though how it came to appear on the luncheon table I am at a loss to conjecture.'

Neither the Bishop nor Lady Bugle seemed inclined to discuss the matter any further. Nor did either of them seem very conversationally disposed, so that the Vicar was obliged to keep up a sort of desperate monologue.

Antonia returned after a while, but she was so upset that her presence did not contribute very much towards relieving the general gloom.

The Bishop left immediately after luncheon, and Lady Bugle left very soon after the Bishop. 'Please,' she said as she bade them farewell, 'never think of asking me again to meet that detestable man.'

LORD BERNERS (1883–1950),
The Camel, 1936

Here is an account sent to me by a correspondent.

'One of my pets was a frog . . .

'One day I wanted to paint him in a picture, and tried to take a profile view. But he would not let me do it, and whenever I placed him in the right position he would hop round so as to face me, and then get on my paper.

'Then I bethought myself of putting him in a plate with some water, so that he might be comfortable. This plan answered very well as far as keeping him off the paper went; but when I turned the plate round so as to get a side view, he hobbled round also, and *would* face me.

'Then I tried edging round the table myself, but with the same result, so that I was obliged to hold him sideways while I drew him. But whenever I raised my head to look at him he raised his too, and lowered it again when I began to paint, and so we went on nodding at each other like two china mandarins. . . .'

One of these creatures took a great fancy to its mistress. But its affection soon became troublesome. It had a way of hiding itself in some unsuspected spot, and then springing out and clinging to her nose. Removing it was a difficult task, and at last the lady became so used to the creature that she went on with her domestic work not in the least impeded by the frog on her nose.

REV. J. G. WOOD (1827–89),
Petland Revisited, 1890

The first of my succession of five big dogs was one about whom I have often written, the incomparable Marco . . .

. . . he was as great a philosopher, as perfect a comrade as busy woman could desire, walking, skating, doing everything with me. And though in the free Germany of those days a lead was not imperative, he never strolled across the communal flower beds or paid the wrong sort of compliment to projecting flowers. In England he won all hearts, notably my mother's, even when he walked under a wicker table at one of her tea parties and marched about the room as beneath a canopy, bearing tea cups and anchovy sandwiches. . . .

Devoted to music – or at least to noise – for hours he would lie entranced, his head on the pedals of a seldom silent piano. Indeed one day he carried the musical passion to extremes, and instead of staying in the street as he was bidden, dashed up two pairs of stairs and burst into a room where Brahms and four local artists were rehearsing his divine F minor piano quintet, upsetting the 'cellist's desk in his stride. After which, to the horror of his mistress who was seated beside Brahms and turning over, he dived under the piano and pushed in between us. I have sometimes described Brahms as rather a curmudgeon, so let me add that

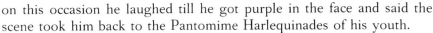

on this occasion he laughed till he got purple in the face and said the scene took him back to the Pantomime Harlequinades of his youth.

ETHEL SMYTH, DBE (1858–1944),
Inordinate(?) Affection, A Story for Dog Lovers, 1936

Of all the dogs that are so sweet,
The spaniel is the most complete;
Of all the spaniels, dearest far
The little loving cockers are.

They're always merry, always hale;
Their eyes are like October ale;
They are so loyal and so black;
So unresentful 'neath the whack....

The Bob-tail is a jolly chap;
The Pekinese commands your lap;
The Dachshound (with Queen Anne her legs)
Your sympathy enchains or begs.

Yet why compare? All dogs on earth
Possess some special charm and worth.
But Cocker spaniels? Every way
They are the kennel's angels, they.

E. V. LUCAS (1868–1938),
'The More I See of Men' ..., 1927

Wake! For the Sun wherein I love to snore
Has like an Eiderdown slid off me to the Floor;
And I a drowsy Step or Two must creep
And flop into its gentle Warmth once more....

Why, if a purring Motor-car outside
Invites us on the Air of Heav'n to ride,
Were't not a Shame – were't not a Shame for us
In this dull Habitation to abide? ...

To lift one's Nose high in the Air,
And feel the grateful Wind trill through one's Hair:
Even on Bus-top or in Taxi-cab,
Of all Adventures, none with this compare! ...

Strange, is it not? of all the wondrous Things
Of Beauty, Joy and Love kind Heaven brings,
There's none so Wonderful as Mistress Mine –
And when I gaze at Her, my heart *just sings*.
SEWELL COLLINS, *The Rubaiyat of a
Scotch Terrier*, 1926

We landed, hauled up the boat, and then feebly sat down on our belongings to review the situation, and Maria came and shook herself over each of us in turn. We had run into a little cove, guided by the philanthropic beam of a candle in the upper window of a house about a hundred yards away. . . .

We trailed up from the cove, laden with emigrants' bundles, stumbling on wet rocks in the half-light, and succeeded in making our way to the house.

It was a small two-storied building, of that hideous breed of architecture usually dedicated to the rectories of the Irish Church; we felt that there was something friendly in the presence of a pair of carpet slippers in the porch, but there was a hint of exclusiveness in the fact that there was no knocker and that the bell was broken. The light still burned in the upper window, and with a faltering hand I flung gravel at the glass. This summons was appallingly responded to by a shriek; there was a flutter of white at the panes, and the candle was extinguished.

'Come away!' exclaimed Miss Shute, 'it's a lunatic asylum!'

We stood our ground, however, and presently heard a footstep within, a blind was poked aside in another window, and we were inspected by an unseen inmate; then some one came downstairs, and the hall door was opened by a small man with a bald head and a long sandy beard. He was attired in a brief dressing-gown, and on his shoulder sat, like an angry ghost, a large white cockatoo. Its crest was up on end, its beak was a good two inches long and curved like a Malay kris; its claws gripped the little man's shoulder. Maria uttered in the background a low and thunderous growl.

'Don't take any notice of the bird, please,' said the little man nervously, seeing our united gaze fixed upon this apparition; 'he's extremely fierce if annoyed.'

The majority of our party here melted away to either side of the hall door, and I was left to do the explaining. The tale of our misfortunes had its due effect, and we were ushered into a small drawing-room, our host holding open the door for us, like a nightmare footman with bare shins, a gnome-like bald head, and an unclean spirit swaying on his shoulder. He opened the shutters, and we sat decorously round the room, as at an afternoon party, while the situation was further expounded on both sides. Our entertainer, indeed, favoured us with the leading items of his family history, amongst them the facts that he was a Dr Fahy from Cork, who had taken somebody's rectory for the summer, and had been prevailed on by some of his patients to permit them to join him as paying guests.

'I said it was a lunatic asylum,' murmured Miss Shute to me.

'In point of fact,' went on our host, 'there isn't an empty room in the house, which is why I can only offer your party the use of this room and the kitchen fire, which I make a point of keeping burning all night.'

He leaned back complacently in his chair, and crossed his legs; then, obviously remembering his costume, sat bolt upright again. We owed the guiding beams of the candle to the owner of the cockatoo, an old Mrs Buck, who was, we gathered, the most paying of all the patients, and also, obviously, the one most feared and cherished by Dr Fahy. 'She has a candle burning all night for the bird, and her door open to let him walk about the house when he likes,' said Dr Fahy; 'indeed, I may say her passion for him amounts to dementia. . . .'

Dr Fahy had evidently a turn for conversation that was unaffected by circumstance; the first beams of the early sun were lighting up the rep chair covers before the door closed upon his brown dressing-gown, and upon the stately white back of the cockatoo, and the demoniac possession of laughter that had wrought in us during the interview burst forth unchecked. It was most painful and exhausting, as such laughter always is; but by far the most serious part of it was that Miss Sally, who was sitting in the window, somehow drove her elbow through a pane of glass, and Bernard, in pulling down the blind to conceal the damage, tore it off the roller.

There followed on this catastrophe a period during which reason tottered and Maria barked furiously. Philippa was the first to pull herself together, and to suggest an adjournment to the kitchen fire ... and, respecting the repose of the household, we proceeded thither with a stealth that convinced Maria we were engaged in a rat hunt. The boots of paying guests littered the floor, the débris of their last repast covered the table; a cat in some unseen fastness crooned a war song to Maria, who feigned unconsciousness and fell to scientific research in the scullery.

We roasted our boots at the range, and Bernard, with all a sailor's gift for exploration and theft, prowled in noisome purlieus and emerged with a jug of milk and a lump of salt butter. No one who has not been a burglar can at all realise what it was to roam through Dr Fahy's basement storey, with the rookery of paying guests asleep above, and to feel that, so far, we had repaid his confidence by breaking a pane of glass and a blind, and putting the scullery tap out of order. I have always maintained that there was something wrong with it before I touched it, but the fact remains that when I had filled Philippa's kettle, no human power could prevail upon it to stop flowing. For all I know to the contrary it is running still.

It was in the course of our furtive return to the drawing-room that we were again confronted by Mrs Buck's cockatoo. It was standing in malign meditation on the stairs, and on seeing us it rose, without a word of warning, upon the wing, and with a long screech flung itself at Miss Sally's golden-red head, which a ray of sunlight had chanced to illumine. There was a moment of stampede, as the selected victim, pursued by the cockatoo, fled into the drawing-room; two chairs were upset (one, I think, broken), Miss Sally enveloped herself in a window curtain, Philippa and Miss Shute effaced themselves beneath a table; the cockatoo, foiled of its prey, skimmed, still screeching, round the ceiling. It was Bernard who, with a well-directed sofa-cushion, drove the enemy from the room. There was only a chink of the door open, but the cockatoo turned on his side as he flew, and swung through it like a woodcock.

We slammed the door behind him, and at the same instant there came a thumping on the floor overhead, muffled, yet peremptory.

'That's Mrs Buck!' said Miss Shute, crawling from under the table; 'the room over this is the one that had the candle in it.'

We sat for a time in awful stillness, but nothing further happened, save a distant shriek overhead, that told the cockatoo had sought and found sanctuary in his owner's room. We had tea *sotto voce*, and then, one by one, despite the amazing discomfort of the drawing-room chairs, we dozed off to sleep.

It was at about five o'clock that I woke with a stiff neck and an uneasy remembrance that I had last seen Maria in the kitchen. The others, looking, each of them, about twenty years older than their age, slept in various attitudes of exhaustion. Bernard opened his eyes as I stole forth to look for Maria, but none of the ladies awoke. I went down the evil-smelling passage that led to the kitchen stairs, and, there on a mat, regarding me with intelligent affection, was Maria; but what – oh what was the white thing that lay between her forepaws?

The situation was too serious to be coped with alone. I fled noiselessly back to the drawing-room and put my head in; Bernard's eyes – blessed be the light sleep of sailors! – opened again, and there was that in mine that summoned him forth. (Blessed also be the light step of sailors!)

We took the corpse from Maria, withholding perforce the language and the slaughtering that our hearts ached to bestow. For a minute or two our eyes communed.

'I'll get the kitchen shovel,' breathed Bernard; 'you open the hall door!'

A moment later we passed like spirits into the open air, and on into a little garden at the end of the house. Maria followed us, licking her lips. There were beds of nasturtiums, and of purple stocks, and of marigolds. We chose a bed of stocks, a plump bed, that looked like easy digging. The windows were all tightly shut and shuttered, and I took the cockatoo from under my coat and hid it, temporarily, behind a box border. Bernard had brought a shovel and a coal scoop. We dug like badgers. At eighteen inches we got down into shale and stones, and the coal scoop struck work.

'Never mind,' said Bernard; 'we'll plant the stocks on top of him.'

It was a lovely morning, with a new-born blue sky and a light northerly breeze. As we returned to the house, we looked across the wavelets of the little cove and saw, above the rocky point round which we had groped last night, a triangular white patch moving slowly along.

'The tide's lifted her!' said Bernard, standing stock-still. He looked at Mrs Buck's window and at me. 'Yeates!' he whispered, 'let's quit!'

It was now barely six o'clock, and not a soul was stirring. We woke the ladies and convinced them of the high importance of catching the tide. Bernard left a note on the hall table for Dr Fahy, a beautiful note of leave-taking and gratitude, and apology for the broken window (for which he begged to enclose half-a-crown). No allusion was made to the other casualties. As we neared the strand he found an occasion to say to me:

'I put in a postscript that I thought it best to mention that I had seen the cockatoo in the garden, and hoped it would get back all right. That's quite true, you know! But look here, whatever you do, you must keep it all dark from the ladies –'

At this juncture Maria overtook us with the cockatoo in her mouth.

SOMERVILLE AND ROSS (1858–1949 and 1862–1915), *Some Experiences of an Irish RM*, 1899

Mr Hawker had a pair of stags, which he called Robin Hood and Maid Marian, given to him by the late Sir Thomas Acland, from his park at Killerton. These he kept in the long open combe in front of the house, through which a stream dashes onwards to the sea. One day ... Mr Knight proceeded too curiously to approach Robin Hood, when the deer ran at him, and butted him down. The clergyman shrieked with fear, and the stag would have struck him with his antlers had not the vicar rushed to the rescue. Being an immensely strong man, he caught Robin by the horns, and drew his head back, and held him fast whilst the frightened man crawled away.

'I was myself in some difficulty,' said Mr Hawker, when telling the story. 'The stag would have turned on me when I let go, and I did not quite see my way to escape; but that wretched man did nothing but yell for his wig and hat, which had come off, and were under the deer's feet; as if my life were of no account beside his foxy old wig and battered beaver.'

Dr Phillpotts, the late Bishop of Exeter, not long after this occurred, came to Morwenstow to visit Mr Hawker. Whilst being shown the landscape from the garden, the bishop's eye rested on Robin Hood.

'Why! that stag which butted and tossed Mr Knight is still suffered to live! It might have killed him.'

'No harm done, my lord,' said Mr Hawker. 'He is a very Low Church parson.'

SABINE BARING–GOULD (1834–1924),
*The Vicar of Morwenstow, A Life
of Robert Stephen Hawker*, 1886

To suddenly become a mother like that, Dodo found, was very demoralizing, and she almost had a nervous breakdown, for she was torn between the desire to stay in one spot with her puppy and the urge to keep as close to Mother as possible. We were, however, unaware of this psychological turmoil. Eventually Dodo decided to compromise, so she followed Mother around and carried the puppy in her mouth. She had spent a whole morning doing this before we discovered what she was up to; the unfortunate baby hung from her mouth by its head, its body swinging to and fro as Dodo waddled along at Mother's heels. Scolding and pleading having no effect, Mother was forced to confine herself to the bedroom with Dodo and her puppy, and we carried their meals up on a tray. Even this was not altogether successful, for if Mother moved out of the chair, Dodo, ever alert, would seize her puppy and sit there regarding Mother with starting eyes, ready to give chase if necessary.

'If this goes on much longer that puppy'll grow into a giraffe,' observed Leslie.

'I know, poor little thing,' said Mother; 'but what can I *do*? She picks it up if she sees me lighting a cigarette.'

'Simplest thing would be to drown it,' said Larry. 'It's going to grow into the most horrifying animal, anyway. Look at its parents.'

'No, indeed you won't drown it!' exclaimed Mother indignantly.

'Don't be *horrible*,' said Margo; 'the poor little thing.'

'Well, I think it's a perfectly ridiculous situation, allowing yourself to be chained to a chair by a dog.'

'It's my dog, and if I want to sit here I *shall*,' said Mother firmly.

'But for how long? This might go on for months.'

'I shall think of something,' said Mother with dignity.

The solution to the problem that Mother eventually thought of was simple. She hired the maid's youngest daughter to carry the puppy for Dodo. This arrangement seemed to satisfy Dodo very well, and once more Mother was able to move about the house. She pottered from room to room like some Eastern potentate, Dodo pattering at her heels, and young Sophia bringing up the end of the line, tongue protruding and eyes squinting with the effort, bearing in her arms a large cushion on which reposed Dodo's strange offspring. When Mother was going to be in one spot for any length of time Sophia would place the cushion reverently on the ground, and Dodo would surge on to it and sigh deeply. As soon as Mother was ready to go to another part of the house, Dodo would get off her cushion, shake herself, and take up her position in the cavalcade, while Sophia lifted the cushion aloft as though it carried a crown. Mother would peer over her spectacles to make sure the column was ready, giving a little nod, and they would wind their way off to the next job.

GERALD DURRELL (b. 1925), *My Family and Other Animals*, 1956

17

Cautionary Tale

'Tis vanity that swells thy mind.
What, heav'n and earth for thee design'd!
For thee! made only for our need;
That more important Fleas might feed

JOHN GAY (1685–1732),
Fables, 1727

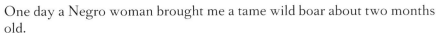

One day a Negro woman brought me a tame wild boar about two months old.

'It is called Josephine, and it will follow you around like a dog,' she said.

We agreed upon five francs as the price. My wife was just then away for a few days. With the help of Joseph and N'Kendju, my hospital assistant, I immediately drove some stakes into the ground and made a pen, with the wire netting rather deep in the earth. Both of my black helpers smiled.

'A wild boar will not remain in a pen; it digs his way out from under it,' said Joseph.

'Well, I should like to see this little wild boar get under this wire netting sunk deep in the earth,' I answered.

'You will see,' said Joseph.

The next morning the animal had already gotten out. I felt almost relieved about it, for I had promised my wife that I would make no new acquisition to our zoo without her consent, and I had a foreboding that a wild boar would not, perhaps, be to her liking.

When I came up from the hospital for the midday meal, however, there was Josephine waiting for me in front of the house, and looking at me as if she wanted to say: 'I will remain ever so faithful to you, but you must not repeat the trick with the pen.' And so it was. . . .

On a Saturday some weeks later, however, Josephine disappeared. In the evening the missionary met me in front of my house and shared my sorrow, since Josephine had also shown some attachment to him.

'I feel sure she has met her end in some Negro's pot,' he said. 'It was inevitable.'

With the blacks a wild boar, even when tamed, does not fall within the category of a domestic animal but remains a wild animal that belongs to him who kills it. While he was still speaking, however, Josephine appeared, behind her a Negro with a gun.

'I was standing,' he said, 'in the clearing, where the ruins of the former American missionary's house are still to be seen, when I saw this wild boar. I was just taking aim, but it came running up to me and rubbed against my legs. An extraordinary wild boar! But imagine what it then did. It trotted away with me after it, and now here we are. So it's your wild boar? How fortunate that this did not happen to a hunter who is not so quick-witted as I.' I understood this hint, complimented him generously and gave him a nice present.

But the thought that my wild boar was in constant danger, as the missionary had told me, troubled me . . .

'Listen, doctor,' the missionary began suddenly, 'tomorrow I have to preach, and as it will soon be necessary to touch upon the sin of theft in every service for our Negroes, I will bring in Josephine right off tomorrow morning as an illustration of the fact that an animal once wild and anybody's property nonetheless may afterwards become private property and inviolable, when it is cherished by someone.'

I thanked him in advance for coming to my assistance with his choice of this significant example.

The next morning, in the second half of the sermon, Josephine was introduced. With rapt attention the Negroes listened as the missionary explained the complicated case and broadened the horizon of their notion of property. At that very moment – it made me almost ill – Josephine took her place beside the preacher! In Lambaréné, you know, we have no chancel. The service of worship takes place in the corrugated-iron barrack in which the school is also held. The preacher stands on the ground. The doors are always left open, so that some air can come in. People are quite accustomed to have hens and sheep come and go during the service. The missionary's dogs regularly take part in it. I always took it for granted that my dog Caramba would go with me to the service. Whenever he heard the bell and the singing he could not be kept in the house. . . .

But that Josephine should also have become religious seemed horrible to me. Moreover, I was soon made forcibly aware that she did not know how to behave. She had come fresh from the marsh, covered with black mire. And now she walked among the benches, where the children sat, while they drew their knees up under their chins! Then she came to the women! Then to the other missionary! Then to the ladies of the mission with their white skirts, trying to rub herself against them! Then to the lady doctor! Then to me! At that moment she received a kick, the first she ever had from me. It was, however, justified.

I was not able to discourage Josephine's delight in the church service. She could not be shut up; neither could she be tied up, for she worked her way out of every harness that I contrived for her. The moment the bell sounded she ran to church. I do not think she missed any of the morning or evening prayers for the children. I proposed to the missionary who was in charge of the station that because of all this I should kill her. But he forbade me to do so: the animal should not lose her life because of such an instinct. In time Josephine began to behave more properly in church. . . .

'When a wild boar is more than six months old, it begins to eat hens,' said N'Kendju.

'Josephine will not go so far as to eat hens,' I replied with an unsteady voice. . . .

One morning, however, as I was examining the blood of some patients under the microscope in the hospital, I heard the cackling of hens and through it all the voices of men calling. Shortly thereafter the boy Akaja appeared with a note from the lady doctor. The writing said: 'Josephine has gotten in among the chickens, has eaten three of them, and has torn off the tail of the clucking hen. I saw it with my own eyes. You know what you have to do.'

I knew it and did it. Josephine was enticed into the hospital, tied up, and expeditiously and artistically slaughtered. Before noon sounded her life came to an end. I estimate that she was about eight months old.

ALBERT SCHWEITZER (1875–1965), *The Animal World of Albert Schweitzer*, TRANS. CHARLES R. JOY

I know not if you have had, as I, the pleasure of seeing a monkey study music; but even now, when I no longer laugh as in those careless days, I never think of my monkey without smiling. The half-man began by gripping the instrument with his palm, and smelling it as if it had been an apple he was about to bite. The breath of his nose was probably enough to draw from the sonorous wood a low harmony, for the orang-outang threw back his head, turned the fiddle over and over, raised it and lowered it, stood it on end, shook it, put it to his ear, threw it down and took it up again, all with a rapid unexpectedness of movement such as belongs only to the animals. He questioned the dumb wood with an unreasoned sagacity, which was somehow the more marvellous for its deficiency. At last, holding it by the finger-board, he made a grotesque attempt to put the violin under his chin; but like a spoilt child, he soon tired of a game demanding long practice, and began to pluck at the strings discordantly. He became angry then, placed the violin on the window-seat and seizing the bow, began to work it backwards and forwards over the strings, like a mason sawing a stone. By this new experiment he only succeeded in offending his fastidious ear; so taking the bow in both hands, he began to rain blows on the innocent fiddle, that might have been a source of harmony and pleasure. I seemed to see a schoolboy holding another face downwards, and treating him to a furious thumping to cure him of some slackness. The violin having been judged and condemned, the monkey sat on the remains, and took a stupid pleasure in tangling the white hairs of the broken bow.

Never since that day have I been able to see the household of a predestinate, without comparing the majority of husbands to that orang-outang trying to play the violin. Love is the most melodious of all music, and a taste for it is inborn in us. Woman is a delightful instrument of pleasure, but it is necessary to know her trembling cords, the attitude in which to approach her, and the difficult changes of fingering needed for so delicate a keyboard. How many orangs – men, I mean – marry without knowing what a woman is! How many predestinates have proceeded with her like the monkey of Cassan with his violin!

HONORÉ DE BALZAC (1799–1850), *The Physiology of Marriage*, TRANS. FRANCIS MACNAMARA

The animals at Cookesborough were treated unusually. . . . Once he was told that a bullock had stumbled into a river and was drowning. After consultation with Dunne and Cruise he gave instructions that all his other cattle should be driven to the bank of the river. 'There they will have an opportunity of seeing their companion drowning, and it will be a warning and a caution to each and everyone during their mortal tenure to shun water.'

PETER SOMERVILLE-LARGE,
Irish Eccentrics, 1975

I met, not long ago, a young man who aspired to become a novelist. Knowing that I was in the profession, he asked me to tell him how he should set to work to realize his ambition. I did my best to explain. 'The first thing,' I said, 'is to buy quite a lot of paper, a bottle of ink, and a pen. After that, you merely have to write.' But this was not enough for my young friend. . . . Disappointed in his hope that I would give him the fictional equivalent of 'One Hundred Ways of Cooking Eggs' or the 'Carnet de la Ménagère', he began to cross-examine me about my methods of 'collecting material'. Did I keep a notebook or a daily journal? Did I systematically frequent the drawing-rooms of the rich and fashionable? Or did I, on the contrary, inhabit the Sussex downs? Or spend my evenings looking for 'copy' in East End gin-palaces? . . . 'My young friend,' I said, 'if you want to be a psychological novelist and write about human beings, the best thing you can do is to keep a pair of cats.' And with that I left him.

Yes, a pair of cats. Siamese by preference; for they are certainly the most 'human' of all the race of cats. Also the strangest, and, if not the most beautiful, certainly the most striking and fantastic. For what disquieting pale blue eyes stare out from the black velvet masks of their faces! Snow-white at birth, their bodies gradually darken to a rich mulatto colour. Their fore-paws are gloved almost to the shoulder like the long black kid arms of Yvette Guilbert; over their hind legs are tightly drawn the black silk stockings with which Félicien Rops so perversely and indecently clothed his pearly nudes. Their tails, when they have tails – and I would always recommend the budding novelist buy the tailed variety; for the tail, in cats, is the principal organ of emotional expression and a Manx cat is the equivalent of a dumb man – their tails are tapering black serpents endowed, even when the body lies in Sphinx-like repose, with a spasmodic and uneasy life of their own. And what strange voices they have! Sometimes like the complaining of small children; sometimes like the noise of lambs; sometimes like the agonized and furious howling of lost souls. Compared with these fantastic creatures, other cats, however beautiful and engaging, are apt to seem a little insipid.

Well, having bought his cats, nothing remains for the would-be novelist but to watch them living from day to day; to mark, learn, and inwardly digest the lessons about human nature which they teach; and finally – for, alas, this arduous and unpleasant necessity always arises – finally write his book about Mayfair, Passy, or Park Avenue, whichever the case may be. . . .

At first there would seem, in . . . feline behaviour, no special 'message' for humanity. But appearances are deceptive; the lids under which civilized people live are so thick and so profusely sculptured with mythological ornaments, that it is difficult to recognize the fact, so much insisted upon by D. H. Lawrence in his novels and stories, that there is almost always a mingling of hate with the passion of love and that young girls very often feel (in spite of their sentiments and even their desires) a real abhorrence of the fact of physical love. Unlidded, the cats make manifest this ordinary obscure mystery of human behaviour. After witnessing a cats' wedding no young novelist can rest content with the falsehood and banalities which pass, in current fiction, for descriptions of love.

ALDOUS HUXLEY (1894–1963),
Music at Night, 1931

Mona Lake is two hundred feet deep, and its sluggish waters are so strong with alkali that if you only dip the most hopelessly soiled garment into them once or twice, and wring it out, it will be found as clean as if it had been through the ablest of washerwomen's hands. While we camped there our laundry work was easy. We tied the week's washing astern of our boat, and sailed a quarter of a mile, and the job was complete, all to the wringing out. If we threw the water on our heads and gave them a rub or so, the white lather would pile up three inches high. This water is not good for bruised places and abrasions of the skin. We had a valuable dog. He had raw places on him. He had more raw places on him than sound ones. He was the rawest dog I almost ever saw. He jumped overboard one day to get away from the flies. But it was bad judgment. In his condition, it would have been just as comfortable to jump into the fire. The alkali water nipped him in all the raw places simultaneously, and he struck out for the shore with considerable interest. He yelped and barked and howled as he went – and by the time he got to the shore there was no bark to him – for he had barked the bark all out of his inside, and the alkali water had cleared the bark all off his outside, and he probably wished he had never embarked in any such enterprise. He ran round and round in a circle, and pawed the earth and clawed the air, and threw double somersaults, sometimes backward and sometimes forward, in the most extraordinary manner. He was not a demonstrative dog, as a general thing, but rather of a grave and serious turn of mind, and I never saw him take so much interest in anything before. He finally struck out over the mountains, at a gait which we estimated at about two hundred and fifty miles an hour, and he is going yet. This was about nine years ago. We look for what is left of him along here every day.

MARK TWAIN (1835–1910),
Roughing It, 1875

Robert Stephen Hawker was . . . ordained priest in 1831, by the Bishop of Bath and Wells. He took his M.A. degree in 1836. He had a favorite rough pony which he rode, and a black pig of Berkshire breed, well cared for, washed, and curry-combed, which ran beside him when he went out for walks, and paid visits. Indeed, the pig followed him into ladies' drawing-rooms, not always to their satisfaction. The pig was called Gyp, and was intelligent and obedient. If Mr Hawker saw that those whom he visited were annoyed at the intrusion of the pig, he would order it out; and the black creature slunk out of the door with its tail out of curl.

SABINE BARING-GOULD (1834–1924), *The Vicar of Morwenstow, A Life of Robert Stephen Hawker*, 1886

THE MAN AND THE FLEA

What dignity 's in human nature,
Says Man, the most conceited creature,
As from a cliff he cast his eye,
And view'd the sea and arched sky!
The sun was sunk beneath the main,
The moon, and all the starry train
Hung the vast vault of heav'n. The Man
His contemplation thus began.
When I behold this glorious show,
And the wide watry world below,
The scaly people of the main,
The beasts that range the wood or plain,
The wing'd inhabitants of air,
The day, the night, the various year,
And know all these by heav'n design'd
As gifts to pleasure human kind,
I cannot raise my worth too high;
Of what vast consequence am I!
Not of th' importance you suppose,
Replies a Flea upon his nose:
Be humble, learn thyself to scan;
Know, pride was never made for man.
'Tis vanity that swells thy mind.
What, heav'n and earth for thee design'd!
For thee! made only for our need;
That more important Fleas might feed.

JOHN GAY (1685–1732), *Fables*, 1727

'You must go to Vienna alone if you are bent on going,' she said; 'I couldn't leave Louis behind, and a dog is always a fearful nuisance in a foreign hotel, besides all the fuss and separation of the quarantine restrictions when one comes back. Louis would die if he was parted from me for even a week. You don't know what that would mean to me.'

Lena stooped down and kissed the nose of the diminutive brown Pomeranian that lay, snug and irresponsive, beneath a shawl on her lap.

'Look here,' said Strudwarden, 'this eternal Louis business is getting to be a ridiculous nuisance. Nothing can be done, no plans can be made, without some veto connected with that animal's whims or convenience being imposed. If you were a priest in attendance on some African fetish you couldn't set up a more elaborate code of restrictions. I believe you'd ask the Government to put off a General Election if you thought it would interfere with Louis's comfort in any way.'

Mrs Strudwarden stooped down again and kissed the irresponsive brown nose. It was the action of a woman with a beautifully meek nature, who would, however, send the whole world to the stake sooner than yield an inch where she knew herself to be in the right.

'It isn't as if you were in the least bit fond of animals,' went on Strudwarden, with growing irritation; 'when we are down at Kerryfield you won't stir a step to take the house dogs out, even if they're dying for a run, and I don't think you've been in the stables twice in your life. You laugh at what you call the fuss that's being made over the extermination of plumage birds, and you are quite indignant with me if I interfere on behalf of an ill-treated, over-driven animal on the road. And yet you insist on every one's plans being made subservient to the convenience of that stupid little morsel of fur and selfishness.'

'You are prejudiced against my little Louis,' said Lena, with a world of tender regret in her voice.

'I've never had the chance of being anything else but prejudiced against him,' said Strudwarden; 'I know what a jolly responsive companion a doggie can be, but I've never been allowed to put a finger near Louis. You say he snaps at any one except you and your maid, and you snatched him away from old Lady Peterby the other day, when she wanted to pet him, for fear he would bury his teeth in her. All that I ever see of him is the tip of his unhealthy-looking little nose, peeping out from his basket or from your muff, and I occasionally hear his wheezy little bark when you take him for a walk up and down the corridor. You can't expect one to get extravagantly fond of a dog of that sort. One might as well work up an affection for the cuckoo in a cuckoo-clock.'

'He loves me,' said Lena, rising from the table, and bearing the shawl-swathed Louis in her arms. 'He loves only me, and perhaps that is why I love him so much in return. I don't care what you say against him, I am not going to be separated from him. If you insist on going to Vienna you must go alone, as far as I am concerned. I think it would be much more sensible if you were to come to Brighton with Louis and me, but of course you must please yourself.'

'You must get rid of that dog,' said Strudwarden's sister when Lena had left the room; 'it must be helped to some sudden and merciful end. Lena is merely making use of it as an instrument for getting her own way on dozens of occasions when she would otherwise be obliged to yield gracefully to your wishes or to the general convenience.'...

'I don't mind admitting,' said Strudwarden, 'that I've dwelt more than once lately on the possibility of some fatal accident putting an end to Louis's existence. It's not very easy, though, to arrange a fatality for a creature that spends most of its time in a muff or asleep in a toy kennel. I don't think poison would be any good; it's obviously horribly over-fed, for I've seen Lena offer it dainties at table sometimes, but it never seems to eat them.'

'Lena will be away at church on Wednesday morning,' said Elsie Strudwarden reflectively; 'she can't take Louis with her there, and she is going on to the Dellings for lunch. That will give you several hours in which to carry out your purpose. The maid will be flirting with the chauffeur most of the time, and, anyhow, I can manage to keep her out of the way on some pretext or other.'

'That leaves the field clear,' said Strudwarden, 'but unfortunately my brain is equally a blank as far as any lethal project is concerned. The little beast is so monstrously inactive; I can't pretend that it leapt into the bath and drowned itself, or that it took on the butcher's mastiff in unequal combat and got chewed up. In what possible guise could death come to a confirmed basket-dweller? It would be too suspicious if we invented a Suffragette raid and pretended that they invaded Lena's boudoir and threw a brick at him. We should have to do a lot of other damage as well, which would be rather a nuisance, and the servants would think it odd that they had seen nothing of the invaders.'

'I have an idea,' said Elsie; 'get a box with an air-tight lid, and bore a small hole in it, just big enough to let in an india-rubber tube. Pop Louis, kennel and all, into the box, shut it down, and put the other end of the tube over the gas-bracket. There you have a perfect lethal chamber.'. . .

Two mornings later the conspirators stood gazing guiltily at a stout square box, connected with the gas-bracket by a length of india-rubber tubing. . . .

Some minutes later, when the fumes had rushed off, he stooped down and lifted out the little kennel with its grim burden. Elsie gave an exclamation of terror. Louis sat at the door of his dwelling, head erect and ears pricked, as coldly and defiantly inert as when they had put him into his execution chamber. Strudwarden dropped the kennel with a jerk, and stared for a long moment at the miracle-dog; then he went into a peal of chattering laughter.

It was certainly a wonderful imitation of a truculent-looking toy Pomeranian, and the apparatus that gave forth a wheezy bark when you pressed it had materially helped the imposition that Lena, and Lena's maid, had foisted on the household. For a woman who disliked animals, but liked getting her own way under a halo of unselfishness, Mrs Strudwarden had managed rather well.

'Louis is dead,' was the curt information that greeted Lena on her return from her luncheon party.

'Louis *dead*!' she exclaimed.

'Yes, he flew at the butcher-boy and bit him, and he bit me too, when I tried to get him off, so I had to have him destroyed. You warned me that he snapped, but you didn't tell me that he was down-right dangerous. I shall have to pay the boy something heavy by way of compensation, so you will have to go without those buckles that you wanted to have for Easter; also I shall have to go to Vienna to consult Dr Schroeder, who is a specialist on dogbites, and you will have to come too. I have sent what remains of Louis to Rowland Ward to be stuffed; that will be my Easter gift to you instead of the buckles. For Heaven's sake, Lena, weep, if you really feel it so much; anything would be better than standing there staring as if you thought I had lost my reason.'

Lena Strudwarden did not weep, but her attempt at laughing was an unmistakable failure.

'*Louis*', SAKI (1870–1916)

It is said of the large-hearted John Wesley, that on one occasion, when a fly pitched on the back of his hand, instead of crushing the life out of it (which is too often done by some people), he gently brushed it off, saying as he did so: 'Go, sir; there is room enough for both of us.'

VERNON S. MORWOOD, *Facts and Phases of Animal Life*, 1882

THE MONKEY-MARTYR

'Tis strange, what awkward figures and odd capers
Folks cut, who seek their doctrine from the papers;
But there are many shallow politicians,
Who take their bias from bewilder'd journals –
    Turn state-physicians,
And make themselves fools'-caps of the diurnals.
One of this kind, not human, but a monkey,
Had read himself at last to this sour creed
That he was nothing but Oppression's flunkey,
And man a tyrant over all his breed.
    He could not read
Of niggers whipt, or over-trampled weavers,
But he applied their wrongs to his own seed,
And nourish'd thoughts that threw him into fevers,
His very dreams were full of martial beavers,
And drilling Pugs, for liberty pugnacious,
    To sever chains vexatious:
In fact, he thought that all his injur'd line
Should take up pikes in hand, and never drop 'em
Till they had clear'd a road to Freedom's shrine, –
Unless perchance the turn-pike men should stop 'em.

    Full of this rancour,
Pacing one day beside St Clement Danes,
    It came into his brains

To give a look in at the Crown and Anchor;
Where certain solemn sages of the nation
Were at that moment in deliberation
How to relieve the wide world of its chains,
  Pluck despots down,
  And thereby crown
Whitee- as well as blackee-man-cipation.
Pug heard the speeches with great approbation,
And gaz'd with pride upon the Liberators;
  To see mere coal-heavers
  Such perfect Bolivars –
Waiters of inns sublim'd to innovators,
And slaters dignified as legislators –
Small publicans demanding (such their high sense
Of liberty) an universal license –
And pattern-makers easing Freedom's clogs –
  The whole thing seem'd
  So fine, he deem'd
The smallest demagogues as great as Gogs!

Pug, with some curious notions in his noddle,
Walk'd out at last, and turn'd into the Strand,
  To the left hand,
Conning some portions of the previous twaddle,
And striding with a step that seem'd design'd
To represent the mighty March of Mind,
  Instead of that slow waddle
Of thought, to which our ancestors inclin'd –
No wonder, then, that he should quickly find
He stood in front of that intrusive pile,
  Where Cross keeps many a kind
  Of bird confin'd,
And free-born animal, in durance vile –
A thought that stirr'd up all the monkey-bile!

  The window stood ajar –
  It was not far,
Nor, like Parnassus, very hard to climb –
The hour was verging on the supper-time,
And many a growl was sent through many a bar.
Meanwhile Pug scrambled upward like a tar,
  And soon crept in,
  Unnotic'd in the din
Of tuneless throats, that made the attics ring
With all the harshest notes that they could bring;

For like the Jews,
Wild beasts refuse,
In midst of their captivity – to sing.

Lord! how it made him chafe,
Full of his new emancipating zeal,
To look around upon this brute-bastille,
And see the king of creatures in – a safe!
The desert's denizen in one small den,
Swallowing slavery's most bitter pills –
A bear in bars unbearable. And then
The fretful porcupine, with all its quills
Imprison'd in a pen!
A tiger limited to four feet ten;

And, still worse lot,
A leopard to one spot!
An elephant enlarg'd,
But not discharg'd;
(It was before the elephant was shot;)
A doleful wanderoo, that wandered not;
An ounce much disproportion'd to his pound.
Pug's wrath wax'd hot
To gaze upon these captive creatures round;
Whose claws – all scratching – gave him full assurance
They found their durance vile of vile endurance.

He went above – a solitary mounter
Up gloomy stairs – and saw a pensive group
Of hapless fowls –
Cranes, vultures, owls,
In fact, it was a sort of Poultry-Compter,
Where feather'd prisoners were doom'd to droop:
Here sat an eagle, forc'd to make a stoop,
Not from the skies, but his impending roof;
And there aloof,
A pining ostrich, moping in a coop;
With other samples of the bird creation,
All cag'd against their powers and their wills,
And cramp'd in such a space, the longest bills
Were plainly bills of least accommodation.
In truth, it was a very ugly scene
To fall to any liberator's share,
To see those winged fowls, that once had been
Free as the wind, no freer than fix'd air.

His temper little mended,
Pug from this Bird-cage Walk at last descended
    Unto the lion and the elephant,
    His bosom in a pant
To see all nature's Free List thus suspended,
And beasts depriv'd of what she had intended.
    They could not even prey
    In their own way;
A hardship always reckon'd quite prodigious.
    Thus he revolv'd
    And soon resolv'd
To give them freedom, civil and religious.

That night there were no country cousins, raw
From Wales, to view the lion and his kin:
The keeper's eyes were fix'd upon a saw;
The saw was fix'd upon a bullock's shin:
    Meanwhile with stealthy paw,
    Pug hastened to withdraw
The bolt that kept the king of brutes within.
Now, monarch of the forest! thou shalt win
Precious enfranchisement – thy bolts are undone;

Thus he resolv'd
And soon resolv'd
To give them freedom,
civil and religious.

Thou art no longer a degraded creature,
But loose to roam with liberty and nature;
And free of all the jungles about London –
All Hampstead's healthy desert lies before thee!
Methinks I see thee bound from Cross's ark,
Full of the native instinct that comes o'er thee,
    And turn a ranger
Of Hounslow Forest, and the Regent's Park –
Thin Rhodes's cows – the mail-coach steeds endanger,
And gobble parish watchman after dark: –
Methinks I see thee, with the early lark,
Stealing to Merlin's cave – (*thy* cave.) – Alas,
That such bright visions should not come to pass!
Alas, for freedom, and for freedom's hero!
    Alas, for liberty of life and limb!
For Pug had only half unbolted Nero,
    When Nero *bolted him*!

THOMAS HOOD (1799–1845)

The mouse is very easily tamed, and a very engaging little pet it becomes. Many a prisoner has had the monotony of his servitude alleviated by the companionship of a mouse. Perhaps some of my readers may remember that in one of our gaols a prisoner had tamed a mouse, and that the warder took it away from him, and, as far as was known, killed it. The man was beside himself with grief at the loss of his little friend, and became so violent that he was taken before the higher authorities. He was simply broken-hearted. He cared nothing for threatened punishment, but only his mouse.

The warder justified himself by saying that prison was intended as a punishment; and, moreover, that if one kind of pet were allowed, others must be permitted, and that the discipline of the prison could not be maintained. Still, as every schoolmaster knows, there are many cases where mercy ought to temper justice, and this was one of them.

REV. J. G. WOOD, Petland Revisited, 1890

THE FAITHFUL BIRD

The greenhouse is my summer seat:
My shrubs displaced from that retreat
Enjoy'd the open air:
Two goldfinches, whose sprightly song
Had been their mutual solace long,
Lived happy prisoners there.

They sang, as blithe as finches sing,
That flutter loose on golden wing,
And frolic where they list;
Strangers to liberty, 'tis true,
But that delight they never knew,
And therefore never miss'd.

But nature works in every breast,
With force not easily suppress'd;
And Dick felt some desires,
That, after many an effort vain,
Instructed him at length to gain
A pass between his wires.

The open windows seem'd t' invite
The freeman to a farewell flight;
But Tom was still confined;
And Dick, although his way was clear,
Was much too gen'rous and sincere,
To leave his friend behind.

So settling on his cage, by play,
And chirp, and kiss, he seem'd to say,
You must not live alone –
Nor would he quit that chosen stand
Till I with slow and cautious hand,
Return'd him to his own.

O ye, who never taste the joys
Of Friendship, satisfied with noise,
Fandango, ball, and rout!
Blush, when I tell you how a bird,
A prison with a friend preferr'd
To liberty without.
Poems of William Cowper (1731–1800)

Upon being asked if he thought dogs would be admitted into heaven, Landor answered: 'And pray, why not? They have all of the good and none of the bad qualities of man.'

KATE FIELD, *The Last Days of*
Walter Savage Landor, 1866

ANECDOTE
Of an extraordinary kind of PAROQUET, its untimely death, and the consequence thereof to two young Ladies of Fashion and Fortune

Most people, at least most people who have honored me with their names to this trifling publication, have heard of my favorite fellow traveller, *Jacko*; but few have heard of Mrs Thicknesse's; this bird, which had the use of his wings as perfect as any bird whatever, travelled from Marseilles to Calais, quite at liberty, in an open chaise, and most part of the day sat upon Mrs Thicknesse's shoulder or bosom; or hung by his bill at her tippet; and he would sit by her for hours at the Inns, *gilding his eyes* with such delight, that it would almost induce one to believe the transmigration of souls, and that the bird was animated by the spirit of a departed parent, or a deceased lover, for to me he was a determined enemy! Upon my return to Calais, where I took a house, some stranger entering the room, while the bird was sitting in the open window, he flew out & was absent a day or two, for the boys had hunted him from tree to tree all round the city, till at length he returned to the very first tree he had alighted upon when he flew from the window, and soon after found his way in again, and perched upon the bosom of his mistress; it is needless, I presume, to say, that this bird was of so inestimable a value to her, that no sum of money could have induced her to part with it.

At this time there passed through Calais, a friend of mine, a gentleman of fashion and fortune . . . he desired I would write to Mrs Thicknesse, and ask her if it would prove agreeable to let the two youngest of his daughters, spend the summer with her at Calais: this being agreed to, I was to conduct the ladies over; the eldest was of the age of fourteen, the youngest between eleven and twelve, both lovely handsome children, but the youngest of uncommon vivacity and beauty. I was a little hurt to find in the arms of the latter, the day we set out, a favorite dog, and hinted to her, that I feared that dog would be attended with great inconvenience to her, and me too; however we all set off in good humour, and to avoid their sleeping at a Inn, I got them lodged with a family at Canterbury for whom I had much esteem, where the dog gnawed the carved clawed feet of the mahogany chairs, and did much injury, the next night however, we were so lucky to be landed at Calais, and at supper, the dog was placed in the charming little girls lap, but I observed that I too had a favorite dog, who had travelled through Spain with me, but that I did not permit him to *sit at table*, and desired she would put hers down, this request was complied with but reluctantly, and I found I had given much offence to one whom I wished to oblige, and with whose animated disposition I was highly delighted. The next morning, the dog was put into the closet where the bird roosted, and there eat for his breakfast, what fifty louidores would not have purchased. I need not say how much I was irritated at this, & how it was aggravated by seeing Mrs Thicknesse in tears, but I leave the reader to imagine, what we both felt, when in the midst of this distress, the little spirited girl, with a single feather flicking in her hair, began to hum Lady Coventry's minuet; I then called for the dog, and threatened to cut his throat, but was told if I did, she would cut hers, and I offered her my pen knife as being better adapted to the purpose than her own, I however sent my servant with the dog to the packet, and returned him to England, and then within the same half hour, I told the young lady that before the expiration of one hour more, she should be in a convent, till the pleasure of her father was known; and turning to her elder sister, desired to know whether she would accompany her sister in the convent, or honour Mrs Thicknesse with her company till we had heard from England? She replied with great propriety, and good sense; that she loved her, and would not part from her, a reply as much to her honour, as it was to my satisfaction, and so giving each a *bras*, I conducted them to the convent door. . . . there I left them, not doubting but that their father's letter would liberate them in a week or ten days at farthest, but circumstanced *as he was*, and knowing that they were in perfect security, he suffered them to remain there, I think near three years, a conduct I could not disapprove. . . .

Memoirs and Anecdotes of Philip Thicknesse, late
Lieutenant Governor of Landguard Fort, 1788

The swine who stole my dog doesn't realise what he did to me.

ADOLF HITLER (1889–1945), 1917

18

Tears and Farewell

Four years! – and didst thou stay above
The ground, which hides thee now, but four?
And all that life, and all that love,
Were crowded, Geist! into no more?

MATTHEW ARNOLD (1828–88)

S.—51. (Revised—April, 1920.)   [587] 3426/D195 5000 6/26 40×6 G & S 125

RETURN OF DEATHS AND DESERTIONS, in the Month

H.M.S. *APHIS*

of *Joyce* 1928, transmitted in accordance with Article 1596 of the King's Regulations. The Service Certificates of the men referred to are to accompany the Return.* If for any reason they are not available, an explanation is to be given.

Date *17th Day of September 1928*

DEATHS.

Official No.	Name in full	Rank or rating	Date and place of burial	Remarks
00000011	*Joyce William Dale*	A CANNARY 1st Class	P.M. 17th Day of September — . YANGTSE PORT SIDE FOc'SLE AMS APHIS	21 Gun Salute was fired in accordance AFO161 KRR's Part II. REST IN PEACE NO COULD SING

Bless the Dumb Creatures
in our care
and listen to
their voiceless prayer.
CRAWSHAWBOOTH PET CEMETERY, LANCASHIRE

IN MEMORY OF AN OLD FISH

Under the soil
The old fish do lie,
20 years he lived
And then did die
He was so tame
You understand
He would come and
Eat out of our hand

Died April the 20th 1855
Aged 20 years
MEMORIAL TO A MURDERED TROUT AT
BLOCKLEY, GLOUCESTERSHIRE

THE NYMPH COMPLAINING FOR THE DEATH
OF HER FAWN

The wanton troopers riding by
Have shot my Fawn, and it will dye....
'Twas on those little Silver Feet;
With what a pretty skipping Grace,
It oft would challange me the Race;
And when't had left me far away
'Twould stay and run again and stay;
For it was nimbler much than Hindes,
And trod as if on the four winds.
I have a garden of my own,
But so with Roses over-grown,
And Lillies, that you would it guess
To be a little wilderness;
And all the Springtime of the Year
It only loved to be there
Among the Beds of Lillies I
Have sought it oft, where it should lye;
Yet could not, till itself would rise
Find it, although before mine Eyes:
For in the flaxen Lillies' shade,
It like a bank of Lillies laid.
Upon the Roses it would feed,
Untill its lips ev'n seemed to bleed;
And then to me 'twould boldly trip,
And print those Roses on my Lip.
But all its chief Delight was still
On roses thus itself to fill,
And its pure Virgin Limbs to fold
In whitest sheets of Lillies cold:
Had it lived long it would have been
Lillies without Roses within...
ANDREW MARVELL (1621–78)

No monument that we are aware of has ever been erected to the memory
of a pig. The town of *Lunenburg*, in *Hanover*, has filled up that blank, and
at the *Hotel de Ville*, in that town, there is to be seen a kind of mausoleum
to the memory of a member of the swinish race. In the interior of that
commemorative structure is to be seen a glass-case, enclosing a ham still
in good preservation. A slab of black marble attracts the eye of visitors,
who find thereon the following inscription in Latin, engraved in letters

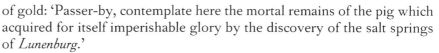

of gold: 'Passer-by, contemplate here the mortal remains of the pig which acquired for itself imperishable glory by the discovery of the salt springs of *Lunenburg*.'

HENRY JAMES LOARING, *Curious Records*, 1872

An obelisk to a pig stands high on a hilltop overlooking Plymouth. Stately, substantial and some thirty feet high, it was put up in the second half of the eighteenth century by the Countess of Mount Edgcumbe to her pet pig Cupid, her most faithful companion. Cupid would follow her about wherever she went, coming in to meals and even going with her on visits to London. When he died, according to a local paper, he was 'buried in a gold casket, the spot being marked by the obelisk on the instruction of the Countess'. Her grief was satirised by Dr John Walcot, who versified under the name of Peter Pinder:

> O dry that tear so round and big,
> Nor waste in sighs your precious wind,
> Death only takes a single pig –
> Your Lord and son are left behind.

The obelisk was moved from the grounds of Mount Edgcumbe in the 1860s to stand on the prominence overlooking the Sound where the River Tamar joins the English Channel.

LUCINDA LAMBTON, *Beastly Buildings*, 1985

Boatswain, Byron's beloved Newfoundland dog, died of rabies at Newstead Abbey in 1808. The poet had a monument erected on what he believed was the site of the high altar of the old abbey church. In his will of 1811 he wrote that he wished to be buried in one of the three vaults beneath this monument, beside his dog and his manservant.
When Boatswain was dying, Byron was unable to comprehend that the dog was going mad before his eyes. According to Thomas Moore, 'so little aware was Lord Byron of the nature of the malady, that he, more than once, with his bare hands, wiped away the saliva from the dog's lips during the paroxysms.'

Near this spot
are deposited the Remains of one,
who possessed Beauty without Vanity,
Strength without Insolence

Courage without Ferocity
and all the virtues of man without his vices,
This Praise, which would be unmeaning flattery
inscribed over human Ashes
is but a just tribute to the Memory of
BOATSWAIN, a DOG,
who was born in Newfoundland May 1803
and died at Newstead Nov.ʳ 18th 1808

When some proud Son of Man returns to Earth,
Unknown to Glory but upheld by Birth,
The sculptor's art exhausts the pomp of woe,
And storied urns record who rests below:
When all is done, upon the Tomb is seen
Not what he was, but what he should have been.
But the poor Dog, in life become the firmest friend,
The first to welcome, foremost to defend,
Whose honest heart is still his Master's own,
Who labours, fights, loves, breathes for him alone.
Unhonour'd falls, unnotic'd all his worth,
Deny'd in heaven the Soul he had on earth:
While man, vain insect! hopes to be forgiven;
And claims himself a sole exclusive heaven,
Oh Man! thou feeble tenant of an hour,
Debas'd by slavery, or corrupt by power,
Who knows thee well, must quit thee with disgust,
Degraded mass of animated dust!
The love is lust thy friendship all a cheat,
Thy tongue hypocrisy, thy heart deceit,
By nature vile, ennobled but by name,
Each kindred brute might bid thee blush for shame,
Ye! who behold perchance this simple urn
Pass on, it honours none you wish to mourn.
To mark a friend's remains these stones arise
I never knew but one – and here he lies.

LORD BYRON (1788–1824),
INSCRIPTION ON THE TOMB OF HIS
DOG BOATSWAIN, NEWSTEAD ABBEY,
NOTTINGHAMSHIRE

Beneath the sculptured form which late you wore
Sleep soundly, Maida, at your master's door.

SIR WALTER SCOTT (1771–1832), ON
THE MONUMENT TO HIS STAGHOUND

Jolyon, who had crossed from Calais by night, arrived at Robin Hill on Sunday morning. He had sent no word beforehand, so walked up from the station, entering his domain by the coppice gate. . . . He passed the pond and mounted the hill slowly. Near the top a hoarse barking greeted him. Up on the lawn above the fernery he could see his old dog Balthasar. The animal, whose dim eyes took his master for a stranger, was warning the world against him. Jolyon gave his special whistle. Even at that distance of a hundred yards and more he could see the dawning recognition in the obese brown-white body. The old dog got off his haunches, and his tail, close-curled over his back, began a feeble, excited fluttering; he came waddling forward, gathered momentum, and disappeared over the edge of the fernery. Jolyon expected to meet him at the wicket gate, but Balthasar was not there, and, rather alarmed, he turned into the fernery. On his fat side, looking up with eyes already glazing, the old dog lay.

'What is it, my poor old man?' cried Jolyon. Balthasar's curled and fluffy tail just moved; his filming eyes seemed saying: 'I can't get up, master, but I'm glad to see you.'

Jolyon knelt down; his eyes, very dimmed, could hardly see the slowly ceasing heave of the dog's side. He raised the head a little – very heavy.

'What is it, dear man? Where are you hurt?' The tail fluttered once; the eyes lost the look of life. Jolyon passed his hands all over the inert warm bulk. There was nothing – the heart had simply failed in that obese body from the emotion of his master's return. Jolyon could feel the muzzle, where a few whitish bristles grew, cooling already against his lips. He stayed for some minutes kneeling, with his hand beneath the stiffening head. The body was very heavy when he bore it to the top of the field; leaves had drifted there, and he strewed it with a covering of them; there was no wind, and they would keep him from curious eyes until the afternoon. 'I'll bury him myself,' he thought. Eighteen years had gone since he first went into the St John's Wood house with that tiny puppy in his pocket. Strange that the old dog should die just now! Was it an omen? He turned at the gate to look back at that russet mound, then went slowly towards the house, very choky in the throat. . . .

In the afternoon he and Jolly took picks and spades and went out to the field. They chose a spot close to the russet mound, so that they need not carry him far, and, carefully cutting off the surface turf, began to dig. . . .

. . . they dug on in silence, till Jolyon said:

'Now, old man, I think it's big enough.' And, resting on their spades, they gazed down into the hole where a few leaves had drifted already on a sunset wind.

'I can't bear this part of it,' said Jolyon suddenly.

'Let me do it, Dad. He never cared much for me.'

Jolyon shook his head.

'We'll lift him very gently, leaves and all. I'd rather not see him again. I'll take his head. Now!'

With extreme care they raised the old dog's body, whose faded tan and white showed here and there under the leaves stirred by the wind. They laid it, heavy, cold, and unresponsive, in the grave, and Jolly spread more leaves over it, while Jolyon, deeply afraid to show emotion before his son, began quickly shovelling the earth on to that still shape. There went the past! If only there were a joyful future to look forward to! It was like stamping down earth on one's own life. They replaced the turf carefully on the smooth little mound and, grateful that they had spared each other's feelings, returned to the house arm-in-arm.

JOHN GALSWORTHY (1867–1933),
The Forsyte Saga

Thistle, elder of our two Jack Russells, has died, notwithstanding the skills of our vet and my wife, who has kept this breed all her life. 'We'll to the woods no more...' I said to myself, when Thistle was being buried under our chestnut trees, and went to another part of the garden in a sad state of mind. Why does the death of these friends affect us so much?

I fancy the years have something to do with it, for after a certain age we know that a long association has ended which cannot be repeated. On the day she died, I glanced at Thistle, and received what struck me as a look of pure innocence. Dogs did not occupy that original garden.

W. F. DEEDES, *Daily Telegraph*, 16 JULY 1990

That was a strange thing, the death of Coco. Not that he should die, for, owing to the unexpected folly of the *concierge* it was inevitable that he should, but his manner of doing it. Even at this distance of time, the remembrance agonises me.

There was a snowstorm the day of my return, alone, to the house I had left that brilliant summer afternoon five years before. A railway had meanwhile been made up the mountain, and when I got out at the little station only the *concierge* was there to meet me – no Coco. For an instant I supposed it was because of the snowstorm, but I immediately remembered that from his birth he was accustomed to every sort of mountain weather, and I asked quickly, having grown during these years very quick to be afraid, 'Coco? Where is he?'

'*Il est un peu souffrant,*' said the *concierge*.

'*Souffrant? Vous avez fait venir le vétérinaire?*'

'*Pas encore. Sachant que madame devait arriver, j'ai attendu sa permission.*'

Attendu. Oh fool, oh fool, I thought, scrambling off homewards, fighting my way through the storm, blinded by the snow, my feet slipping, my heart thumping with fear.

'But you said nothing of this in your last letter,' I paused once to call back to the battling, bent figure behind.

'Three days ago he was well,' came the answer, faint against the wind.

'Where is he?'

'In his kennel. He cannot move from it.'

I struggled on, stumbling and slipping. Oh fool, oh fool, not at once to send for the vet, not to do everything, everything for my dog.

The last bit to the house is up a slope, very slippery in the snow – twice I fell, sprawling – then there are steps on to the terrace, then a porch, and inside the porch the front door. Panting, I climbed that last bit, and from afar began, as well as I could, having but little breath, to whistle and call.

'Coco! Coco!' I called. 'Darling dog – I've come back –'. But what was that I saw inside the porch as I got nearer, what was that dark thing stretched right across the threshold of the door, blocking it up so that I wouldn't be able to go in without stepping over it, so that whatever happened I was bound not to miss it? Coco. Somehow struggled round from his kennel in the yard. He knew I was coming home, knew I was being fetched from the station, and with his last breath made his last effort to be with me again, for a moment, for the last moment of all.

'Coco?' I whispered, standing still, hardly able to believe it. 'Oh *Coco?*'

'It is impossible to imagine,' a voice behind seemed to be saying from a great distance away, 'how the dog could have reached this spot. For three days he has been immovable in his kennel.' I dropped on my knees, and took his paw in my hand. He gave the faintest wag of his tail, and tried to raise his head; but it fell back again, and he could only look at me.

For an instant, for the briefest instant, we looked at each other, and while we looked his eyes glazed.

'Coco – I've come back. Darling – I'll never leave you any more.'

I don't know why I said these things. I knew he was dead, and that no calls, no lamentations, no love could ever reach him again. Sliding down onto the stone flags beside him, I laid my head on his and wept in an agony of bitter grief. Now indeed I was left alone in the world. Even my dog was gone.

ELIZABETH VON ARNIM (1866–1941),
All the Dogs of My Life

A DOG'S GRAVE

My dog lies dead and buried here,
My little Pet for five sweet years.
As I stand here, beside her grave,
With eyes gone dim, and blind with tears –
I see it rising up and down,
As though she lay in a sleeping-gown.

W. H. DAVIES (1871–1940)

Tydeus was tiny in stature, but we find
His little body held a mighty mind
And tiny caskets oft enclose a gem
Of lustre meet for sultans diadem
The stranger thou that visitest this spot
Where tiny's ashes slumber scorn it not
But if the virtues that lie buried here
Love, truth, fidelity to thee are dear
Give all fond memory asks a single tear

MEMORIAL TO A DOG AT
WYNARD PARK, COUNTY DURHAM

Jim Dog is dead – they're saying so,
He's nicely boxed and left behind
A rosebush, suddenly I know
A large contempt for death –

Those gay bones resting calm and shrivven
Ashes of roses? Ten to none
All up and down the hills of heaven
Rabbits are on the run –

I'll wager if I died tonight
And waiting by the river's rim
A bit bewildered at my plight,
Should call, 'Here Jim! Here Jim!' –

Yelping with glory, glad and rough,
He'd hurtle down the further side,
And soon I'd feel a warm wet scruff
Towing me through the tide.

ANON

Flush wandered off into the streets of Florence to enjoy the rapture of smell. He threaded his path through main streets and back streets, through squares and alleys, by smell. He nosed his way from smell to smell; the rough, the smooth, the dark, the golden. He went in and out, up and down, where they beat brass, where they bake bread, where the women sit combing their hair, where the bird-cages are piled high on the causeway, where the wine spills itself in dark red stains on the pavement, where leather smells and harness and garlic, where cloth is beaten, where vine leaves tremble, where men sit and drink and spit and dice – he ran in and out, always with his nose to the ground, drinking in the essence; or with his nose in the air vibrating with the aroma. He slept in this hot patch of sun – how sun made the stone reek! he sought that tunnel of shade – how acid shade made the stone smell! He devoured whole bunches of ripe grapes largely because of their purple smell; he chewed and spat out whatever tough relic of goat or macaroni the Italian housewife had thrown from the balcony – goat and macaroni were raucous smells, crimson smells. He followed the swooning sweetness of incense into the violet intricacies of dark cathedrals; and, sniffing, tried to lap the gold on the window-stained tomb. Nor was his sense of touch much less acute. He knew Florence in its marmoreal smoothness and in its gritty and cobbled roughness. Hoary folds of drapery, smooth fingers and feet of stone received the lick of his tongue, the quiver of his shivering snout. Upon the infinitely sensitive pads of his feet he took the clear stamp of proud Latin inscriptions. In short, he knew Florence as no human being has ever known it; as Ruskin never knew it or George Eliot either. He knew it as only the dumb know. Not a single one of his myriad sensations ever submitted itself to the deformity of words.

But though it would be pleasant for the biographer to infer that Flush's life in late middle age was an orgy of pleasure transcending all description; to maintain that while the baby day by day picked up a new word and thus removed sensation a little further beyond reach, Flush was fated to remain for ever in a Paradise where essences exist in their utmost purity, and the naked soul of things presses on the naked nerve – it would not be true. Flush lived in no such Paradise. The spirit, ranging from star to star, the bird whose furthest flight over polar snows or tropical forests never brings it within sight of human houses and their curling wood-smoke, may, for anything we know, enjoy such immunity, such integrity of bliss. But Flush had lain upon human knees and heard men's voices. His flesh was veined with human passions; he knew all grades of jealousy, anger and despair. Now in summer he was scourged by fleas. With a cruel irony the sun that ripened the grapes brought also the fleas. . . .

The sun seemed droning in the sky. Keeping to the shady side of the street, Flush trotted along the well-known ways to the market-place. The whole square was brilliant with awnings and stalls and bright umbrellas. The market women were sitting beside baskets of fruit; pigeons were fluttering, bells were pealing, whips were cracking. The many-coloured mongrels of Florence were running in and out sniffing and pawing. All was as brisk as a bee-hive and as hot as an oven. Flush sought the shade. He flung himself down beside his friend Catterina, under the shadow of her great basket. A brown jar of red and yellow flowers cast a shadow beside it. Above them a statue, holding his right arm outstretched, deepened the shade to violet. Flush lay there in the cool, watching the young dogs busy with their own affairs. They were snarling and biting, stretching and tumbling, in all the abandonment of youthful joy. They were chasing each other in and out, round and round, as he had once chased the spotted spaniel in the alley. His thoughts turned to Reading for a moment – to Mr Partridge's spaniel, to his first love, to the ecstasies, the innocences of youth. Well, he had had his day. He did not grudge them theirs. He had found the world a pleasant place to live in. He had no quarrel with it now. The market woman scratched him behind the ear. She had often cuffed him for stealing a grape, or for some other misdemeanour; but he was old now; and she was old. He guarded her melons and she scratched his ear. So she knitted and he dozed. The flies buzzed on the great pink melon that had been sliced open to show its flesh.

The sun burnt deliciously through the lily leaves, and through the green and white umbrella. The marble statue tempered its heat to a champagne freshness. Flush lay and let it burn through his fur to the naked skin. And when he was roasted on one side he turned over and let the sun roast the other. All the time the market people were chattering and bargaining; market women were passing; they were stopping and fingering the vegetables and the fruit. There was a perpetual buzz and hum of human voices such as Flush loved to listen to. After a time he drowsed off under the shadow of the lilies. . . .

Suddenly every muscle in his body twitched. He woke with a violent start. Where did he think he was? . . .

Whatever it was, he woke from his dream in a state of terror. He made off as if he were flying to safety, as if he were seeking refuge. The market women laughed and pelted him with rotten grapes and called him back. He took no notice. Cartwheels almost crushed him as he darted through the streets – the men standing up to drive cursed him and flicked him with their whips. Half-naked children threw pebbles at him and shouted '*Matta! Matta!*' as he fled past. Their mothers ran to the door and caught them back in alarm. Had he then gone mad? Had the sun turned his brain? Or had he once more heard the hunting horn of Venus? Or had one of the American rapping spirits, one of the spirits that live in table legs, got possession of him at last? Whatever it was, he went in a bee-line up one street and down another until he reached the door of Casa Guidi. He made his way straight upstairs and went straight into the drawing-room.

Mrs Browning was lying, reading, on the sofa. She looked up, startled, as he came in. No, it was not a spirit – it was only Flush. She laughed. Then, as he leapt on to the sofa and thrust his face into hers, the words of her own poem came into her mind:

> You see this dog. It was but yesterday
> I mused forgetful of his presence here
> Till thought on thought drew downward tear on tear,
> When from the pillow, where wet-cheeked I lay,
> A head as hairy as Faunus, thrust its way
> Right sudden against my face, – two golden-clear
> Great eyes astonished mine, – a drooping ear
> Did flap me on either cheek to dry the spray!
> I started first, as some Arcadian,
> Amazed by goatly god in twilight grove;
> But, as the bearded vision closelier ran
> My tears off, I knew Flush, and rose above
> Surprise and sadness, – thanking the true Pan,
> Who, by low creatures, leads to heights of love.

She had written that poem one day years ago in Wimpole Street when she was very unhappy. Now she was happy. She was growing old now and so was Flush. She bent down over him for a moment. Her face with its wide mouth and its great eyes and its heavy curls was still oddly like his. Broken asunder, yet made in the same mould, each, perhaps, completed what was dormant in the other. But she was woman; he was dog. Mrs Browning went on reading. Then she looked at Flush again. But he did not look at her. An extraordinary change had come over him. 'Flush!'

she cried. But he was silent. He had been alive; he was now dead. That was all. The drawing-room table, strangely enough, stood perfectly still.

VIRGINIA WOOLF (1882–1941),
Flush: A Biography, 1933

I have sometimes thought of the final cause of dogs having such short lives, and I am quite satisfied that it is in compassion to the human race; for if we suffer so much in losing a dog after an acquaintance of ten or twelve years, what should we do if they were to live double that time?

LETTER TO MARIA EDGEWORTH
(1767–1849), APRIL 1822, FROM
SIR WALTER SCOTT (1771–1832)

To the memory of
Signor FIDO,
An Italian of good Extraction,
Who came into England,
Not to *bite* us, like most of his Countrymen,
But to gain an honest Livelyhood.
He *hunted* not after Fame,
Yet acquired it.
Regardless of the Praise of his Friends,
But most sensible of their Love.
Tho' he lived among the Great,
He neither learnt nor flatter'd any Vice.
He was no Bigot,
Tho' he doubted of none of the Thirty-nine
Articles:
And if to follow Nature,
And to respect the Laws of Society,
Be Philosophy;
He was a perfect Philosopher,
A faithful Friend,
An agreeable Companion,
A loving Husband;
And, tho' an Italian,
Was distinguished by a numerous Offspring:
All which he liv'd to see take good Courses.
In his old Age he retir'd
To the house of a Clergyman in the Country,
Where he finish'd his *earthly Race*,
And died an Honour and Example to the
whole species.

Reader,
This Stone is guiltless of Flattery;
For he to whom it was inscrib'd
Was not a Man,
But a Grey Hound.

INSCRIPTION ON A STONE IN THE MAUSOLEUM
AT STOWE, BUCKINGHAMSHIRE, 1744

'Poor Nero's last good days were with us at Aberdour, in 1859. Twice or thrice I flung him into the sea there, which he didn't at all like; and in consequence of which he even ceased to follow me at bathing time, the very strongest measure he could take – or pretend to take. For two or three mornings accordingly I have seen nothing of Nero, but the third or fourth morning, on striking out to swim a few yards, I heard gradually a kind of swashing behind me; looking back, it was Nero out on voluntary humble partnership – ready to swim with me to Edinburgh, or to the world's end, if I liked. . . .

'Nero had done his mistress, and still more him, a great deal of good. But, alas, in Cook's grounds here, within a month or two, a butcher's cart (in her very sight) ran over him, neck and lungs: all winter he wheezed and suffered; Feb. 1st, 1860, he died (prussic acid, and the doctor obliged at last!). I could not have believed my grief then and since would have been the twentieth part of what it was – nay, that the want of him would have been to me other than a riddance. Our last midnight walk together (for he insisted on trying to come), Jan. 31st, is still painful to my thought. "Little dim, white speck, of Life, of Love, Fidelity, and Feeling, girdled by the Darkness as of Night Eternal!" Her tears were passionate and bitter, but repressed themselves, as was fit, I think, the first day. Top of the garden, by her direction, Nero was put underground. A small stone tablet with date she also got, which, broken by careless servants, is still there – a little protected now.'

The stone is there still, in 1927, but few visitors to the gloomy Chelsea house, where two geniuses, a man and woman, failed sufficiently to subdue and blend their individualities for so many years, ever walk down the garden to see it. Underneath are the remains of one who could neither read nor write nor frame systems nor eternally scold, but who lived the only successful life of the three.

LETTER TO JOHN FORSTER FROM THOMAS
CARLYLE (1795–1881), QUOTED IN
E. V. LUCAS (1868–1938),
'The More I See of Men . . .', 1927

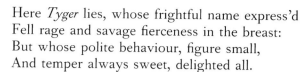
Here *Tyger* lies, whose frightful name express'd
Fell rage and savage fierceness in the breast:
But whose polite behaviour, figure small,
And temper always sweet, delighted all.

1750

At Rousham [in Oxfordshire], a stone slab was put up to commemorate
Faustina Gwynne, a cow. She and her sister, Goody Gwynne, were
short-horns brought from Northamptonshire in 1873. Very brilliant chestnut with white speckles, Faustina was the terror of the village and would
chase people down the street at every opportunity. There is a watercolour
of her in the house – handsome and huge with a table-flat back. The
Cottrell Dormer family still have her horns.

LUCINDA LAMBTON, *Beastly Buildings*, 1985

JIMMY
A TINY MARMOSET
AUGUST 16TH 1937

There isn't enough
Darkness in the world
To quench the light
Of one small candle

GRAVESTONE ON THE OXFORD ROAD,
HENLEY-ON-THAMES

GEIST'S GRAVE

Four years! – and didst thou stay above
The ground, which hides thee now, but four?
And all that life, and all that love,
Were crowded, Geist! into no more?

Only four years those winning ways,
Which make me for thy presence yearn,
Call'd us to pet thee or to praise,
Dear little friend! at every turn?

That liquid, melancholy eye
From whose pathetic, soul fed springs
Seemed surging the Virgilian cry,*
The sense of tears in mortal things – . . .

Stern law of every mortal lot!
Which man, proud man, finds hard to bear,
And builds himself I know not what
Of second life I knot not where.

But thou, when struck thine hour to go,
On us, who stood despondent by,
A meek last glance of love didst throw
And humbly lay thee down to die.

Yet would we keep thee in our heart –
Would fix our favorite on the scene,
Nor let thee utterly depart
And be as if thou ne'er hadst been.

And so there rise these lines of verse
On lips that rarely form them now;
While to each other we rehearse:
Such ways, such arts, such looks hadst thou!

We stroke thy broad brown paws again,
We bid thee to thy vacant chair,
We greet thee by the window-pane
We hear thy scuffle on the stair;

We see the flaps of thy large ears
Quick raised to ask which way to go;
Crossing the frozen lake, appears
Thy small figure on the snow....

We lay thee, close within our reach,
Here, where the grass is smooth and warm,
Between the holly and the beech,
Where oft we watch'd thy couchant form,

Asleep, yet lending half an ear
To travellers on the Portsmouth road; –
There choose we thee, O guardian dear,
Mark'd with a stone, thy last abode!

Then some, who through this garden pass,
When we too, like thyself, are clay,
Shall see thy grave upon the grass,
And stop before the stone, and say;

People who lived here long ago
Did by this stone, it seems, intend
To name for future times to know
The dachs-hound, Geist, their little friend.

*Sunt lacrimae rerum!

MATTHEW ARNOLD (1828–88)

PHILIP SPARROW

It had a velvet cap,
And would sit upon my lap,
And seek after small wormès,
And sometime white bread-crummès;
And many times and oft
Between my brestès soft
It would lie and rest;
It was proper and prest.

Sometime he would gasp
When he saw a wasp;
A fly or a gnat,
He would fly at that;
And prettily he would pant
When he saw an ant,
Lord, how he would pry
After the butterfly!

'Accepting all the comforts that Providence has sent.'

Lord, how he would hop
After the gressop!
And when I said, 'Phip, Phip!'
Then he would leap and skip,
And take me by the lip.
Alas, it will me slo
That Philip is gone me fro!
Si in i qui ta tes . . .
Alas, I was evil at ease!
De pro fun dis cla ma vi,
When I saw my sparrow die! . . .

            JOHN SKELTON (1460–1529)

For many years I have had a habit of fox-terriers. I am aware that this way of putting it would seem to imply a considerable number, as one speaks of a Covey of Partridge, or a Stand of Plover, or (with some effort of culture) a Pride of Lions. I do not deny that I find the phrase a Pride of Fox-terriers suitable to so arrogant and self-assertive a company of small dogs as, in varying numbers, and for many years, I have habitually possessed. Three is perhaps the perfect number for what I may speak of as a Habit of dogs – a habit that has for long been mine – and for some years I owned a small family party of three very small fox-terriers, a mother, her daughter, and her sister. I know not why, but, for some obscure psychological reason, while two little dogs excite no enthusiasm, the sight of three little white creatures, yoked in Russian Troika fashion three abreast, moving in sweet accord on a single lead, can fill the most insensitive observer with an admiration that has sometimes surprised me. I have once taken these three on the long journey from Ireland to England, and although I, personally, found the effort something of a strain, their social successes made it feel worth while. The Ticket Office beamed upon them, and nearly forgot to charge me for their tickets. The porters failed to exact their fees and stood at gaze. The very Guard hesitated to interfere as I steered my Troika-harnessed trio to the entrance of what appeared to be an empty carriage, but he warned me in a conspiratorial whisper.

'There's a lady in it – ye must ask her –'

But she was a right-thinking lady. She looked at the three and at me, and said with *empressement,*

'I should be honoured!'

So, in all love, we proceeded to Dublin. On English railways the Troika was treated with a similar indulgence. On London pavements people stood aside, as if for Royalty. In Twickenham one summer's day, a butcher's boy, trundling his grim barrow-load, dropped the handles and went down on his knees – even as young Paris may have knelt, worshipping, when '*Goddesses three to Ida came*' – and murmured in adoration "Naice little fices!'

That was the feeling they invariably evoked. Now they have preceded me, to join the throng of small white ladies that, I know, awaits me on the farther side of the Border. I have loved and lost four and a half couple of them, and as each little life ended its meagre span of something less than fifteen years, I have felt how much charm, affection, and faithfulness would be wasted if such lives as these knew no future. For me the Fields of Paradise would be lonesome and empty, if I found myself there without a little white dog or two at my heels.

SOMERVILLE AND ROSS (1858–1949 and
1862–1915), '*Happy Days!*', 1946

He asks no angel's wing, no seraph's fire
But thinks, admitted to that equal sky,
His faithful dog shall bear him company.
ALEXANDER POPE (1688–1744),
Essay on Man

Bob came to me, a gift from my mother, when he was two years old. He had belonged to a butcher and had been 'thoroughly well trained'; that is to say he had been systematically ill treated, beaten and abused, was as hard as nails and quite without illusions. He feared no one, for fear was not in him, but he mistrusted everyone – with good cause – and neither gave nor expected affection. It was sad to meet the hard, undog-like glance of his yellow eyes, to see the involuntary flinch before he stiffened himself to bear the punishment which he felt sure must be coming.

It took about a year to thaw his icebound but unbroken spirit, and indeed all through his life the memory of evil days remained with him; he was stern and grave, unable to play or to take a joke, chary of his friendship and difficult to approach. But all the stunted and unwanted love hidden in his deep heart he gradually bestowed upon me; at first grudgingly and tentatively, but at last with a fierce concentration almost tragic in its intensity. . . .

Where I went he did not care so long as he might go too; London or the country were the same to him; he would thread dense crowds at my heels like my shadow, he would wait patiently outside galleries or the residences of those strange people who 'like dogs in their proper place'; nothing came amiss to him save only separation from me.

Anything might happen if he were not there to look after me. He was sternly upright, a stickler for law and order; in fact, the only being capable of turning him from the path of rectitude was a railway porter. These officials had decided that he was too large a dog to travel in a carriage with me; he himself thought otherwise and would take infinite trouble to set their authority at naught.

It was an understood thing between us that at a station we did not belong to each other. I chose my carriage and took my seat innocently dogless. Bob apparently had 'found the receipt of fern seed and walked invisible,' for, though he never got into that carriage, he always got out of it at the end of the journey. When and how he entered, where he disposed his large body I never knew, but when I alighted, so did he.

Once, on a very tumultuous Boxing Day, I was wedged into a much overcrowded carriage into which, as the train was moving off, shot a little pink man, finding a temporary resting-place on my lap. As there was nowhere else for him to sit, he stayed there amicably conversing, and as we neared Witley I said to him, 'Now watch – you'll see something in a minute.' Sure enough, as the train slowed down, the apparition of a large grey sheep dog materialised in our midst and was received with an incredulous gasp, then with shouts of wonder. On the following Easter Monday, I, in my turn, nearly missed my train, and hurling myself into a crammed carriage, found myself on the lap of – the little pink man. But this time he was preoccupied, evidently filled with but one idea, and, after we passed Milford, he whispered in tones of awe, 'Is it going to happen again?' I nodded, and soon the usual sensation announced the mystic burgeoning of Bob.

'Well,' cried the little man, bouncing me on his knee in his excitement, 'I'm glad I've seen that again! I didn't believe it last time, but I suppose seeing it *twice* –' He said it with such fervour that it suggested a 'Nunc Dimittis.' We have never met since. . . .

Towards the end of his long life – he lived to be nearly sixteen – Bob became almost blind, and the son of his old age, Benny, tended him like a nurse, waiting for him, going back for him if he were not in sight, pushing him gently round corners just as a professional plough horse will turn a novice at the end of a furrow. Despite his rigorous upbringing – for Bob was still a stern parent – Benny loved his father deeply and had the greatest reverence and respect for him. . . .

In his last year I could hardly leave him. He was well but feeble and seemed to live only in my life. If I were away he would lie by the door waiting; he did not want food nor companionship and would hardly stir from his post lest he should miss the first sound of my footstep. As I came round the garden gate, I would see the grand old figure at strained attention, the sightless eyes striving to pierce their veil of dimness, then – the sudden shock of joy – the transfiguration, as, with a rush, a young dog – young again for one wild moment – hurled himself towards me across the grass. Then – his body pressed against me, his grey paw in my hand, and there would come a deep quiet. His vigil was over, we were together again, all was very, very well, but – 'don't let us make a fuss about it.'

That was Bob, the grandest, most gallant creature I have ever known. . . .

. . . evidently there had been long and serious talks between them as to Benny's future duties, for . . . Benny immediately took over various little intimate services and attentions which his father had rendered me and in which he had never been allowed to take part. 'This and this you must do for the Old Man when I am gone,' Bob had evidently said, and even on the day after his death, Benny fell into line. . . .

Beloved by all, Benny stayed with us till he was fourteen, and, with his passing, there came to me a strange experience. Tranquilly as he had lived, Benny lay a-dying. Death was coming softly to him as if loath to disquiet his gentle spirit. I sat by him holding his paw. Suddenly, and entirely unprompted by any previous train of thought, my mind became dominated and possessed by the grim, strong personality of Bob. If I had seen and touched him I could not have been more keenly aware of his presence – in fact, I looked hastily round for him, forgetting that he was long dead.

And Benny's quiet breathing ceased – he was gone.

Benny had always waited for his blind father, had gone back for him and guided him gently round corners, and I shall always feel that Bob, the strong-hearted, came back in his turn for the timid Benny to help him across the great barrier.

W. GRAHAM ROBERTSON (1866–1948),
Time Was, 1931

TO THE MEMORY OF A PUG WHO DEPARTED THIS LIFE JUNE 24 1754 IN THE THIRD YEAR OF HER AGE

No Blazon'd Coat or Sculptor'd Bone
(Honours we scarcely deem our own)
Adorn this simple rustic stone

But Love and Friendship without Blame
With Gratitude we justly claim:
Where will Faith ever find the same?

Not unlamented now she dies:
Besprinkled here this tribute lies:
With heavenly tears from Angels' eyes.

MEMORIAL AT EASTON NESTON,
NORTHAMPTONSHIRE

FIDELE'S GRASSY TOMB

The Squire sat propped in a pillowed chair,
His eyes were alive and clear of care,
But well he knew that the hour was come
To bid good-bye to his ancient home.

He looked on garden, wood, and hill,
He looked on the lake, sunny and still:
The last of earth that his eyes could see
Was the island church of Orchardleigh.

The last that his heart could understand
Was the touch of the tongue that licked his hand:
'Bury the dog at my feet,' he said,
And his voice dropped, and the Squire was dead.

Now the dog was a hound of the Danish breed,
Staunch to love and strong at need:
He had dragged his master safe to shore
When the tide was ebbing at Elsinore.

From that day forth, as reason would,
He was named 'Fidele,' and made it good:
When the last of the mourners left the door
Fidele was dead on the chantry floor.

They buried him there at his master's feet,
And all that heard of it deemed it meet:
The story went the round for years,
Till it came at last to the Bishop's ears.

Bishop of Bath and Wells was he,
Lord of the lords of Orchardleigh;
And he wrote to the Parson the strongest screed
That Bishop may write or Parson read.

The sum of it was that a soulless hound
Was known to be buried in hallowed ground:
From scandal sore the Church to save
They must take the dog from his master's grave.

The heir was far in a foreign land,
The Parson was wax to my Lord's command:
He sent for the Sexton and bade him make
A lonely grave by the shore of the lake.

The Sexton sat by the water's brink
Where he used to sit when he used to think:
He reasoned slow, but he reasoned it out,
And his argument left him free from doubt.

'A Bishop,' he said, 'is the top of his trade;
But there's others can give him a start with the spade:
Yon dog, he carried the Squire ashore,
And a Christian couldn't ha' done no more.'

The grave was dug; the mason came
And carved on stone Fidele's name;
But the dog that the Sexton laid inside
Was a dog that never had lived or died.

So the Parson was praised, and the scandal stayed,
Till, a long time after, the church decayed,
And, laying the floor anew, they found
In the tomb of the Squire the bones of a hound.

As for the Bishop of Bath and Wells
No more of him the story tells;
Doubtless he lived as a Prelate and Prince,
And died and was buried a century since.

And whether his view was right or wrong
Has little to do with this my song;
Something we owe him, you must allow;
And perhaps he has changed his mind by now.

The Squire in the family chantry sleeps,
The marble still his memory keeps:
Remember, when the name you spell,
There rest Fidele's bones as well.

For the Sexton's grave you need not search,
'Tis a nameless mound by the island church:
An ignorant fellow, of humble lot –
But he knew one thing that a Bishop did not.

SIR HENRY NEWBOLT (1862–1938)

DINAH IN HEAVEN

She did not know that she was dead,
But, when the pang was o'er,
Sat down to wait her Master's tread
Upon the Golden Floor

With ears full-cock and anxious eyes,
Impatiently resigned;
But ignorant that Paradise
Did not admit her kind.

Persons with Haloes, Harps, and Wings
Assembled and reproved,
Or talked to her of Heavenly things,
But Dinah never moved.

There was one step along the Stair
That led to Heaven's Gate;
And, till she heard it, her affair
Was – she explained – to wait.

And she explained with flattened ear,
Bared lip and milky tooth –
Storming against Ithuriel's Spear
That only proved her truth!

Sudden – far down the Bridge of Ghosts
That anxious spirits clomb –
She caught that step in all the hosts,
And knew that he had come.

She left them wondering what to do,
But not a doubt had she.
Swifter than her own squeals she flew
Across the Glassy Sea;

Flushing the Cherubs everywhere,
And skidding as she ran,
She refuged under Peter's Chair
And waited for her man.

There spoke a Spirit out of the press,
'Said: – 'Have you any here
That saved a fool from drunkenness,
And a coward from his fear?

'That turned a soul from dark to day
When other help was vain?
That snatched it from wan hope and made
A cur a man again?'

'Enter and look,' said Peter then,
And set The Gate ajar.
'If I know aught of women and men
I trow she is not far.'

'Neither by virtue, speech nor art
Nor hope of grace to win;
But godless innocence of heart
That never heard of sin:

'Neither by beauty nor belief
Nor white example shown.
Something a wanton – more a thief;
But – most of all – mine own.'

'Enter and look,' said Peter then,
'And send you well to speed;
But, for all that I know of women and men,
Your riddle is hard to read.'

Then flew Dinah from under the Chair,
Into his arms she flew –
And licked his face from chin to hair
And Peter passed them through!
                    RUDYARD KIPLING (1865–1936)

The beasts upon which Dante's affections were prodigalized were the
first wombat and his successor the woodchuck. The second wombat,
having died immediately, counts for little. No more engagingly lumpish
quadruped than the first wombat could be found, and none more obese
and comfortable than the woodchuck. They were both tame, especially
the woodchuck; and Dante would sit with either in his arms by the
half-hour together, dandling them paunch upward, scratching gently at
their cheeks or noses, or making the woodchuck's head and hind-paws

meet. With the wombat no such operation was possible. Each of them was his house-mate for some time, and each expired without premonition. I do not assume that my brother wept over them, but certainly 'his heart was sair.' For the wombat (not having yet seen it) he wrote from Penkill Castle the following quatrain:–

Oh how the family affections combat
Within this heart, and each hour flings a bomb at
My burning soul! Neither from owl nor from bat
Can peace be gained until I clasp my wombat....
WILLIAM MICHAEL ROSSETTI, *Dante Gabriel
Rossetti: His Family-Letters, with a Memoir*, 1895

❧ ILLUSTRATIONS ❧

page

3 Florence Nightingale and Athena (*Florence Nightingale 1820–1856*, I. B. O'Malley, 1931)

7 Two maiden ladies and an alligator (*Alligators and Old Lace*, 1948)

14 The lions at home (*Queer Pets and Their Doings*, Olive Thorne Miller, 1880)

18 'Articles useful to mankind' (*St Nicholas Magazine*, January 1895)

24 'A fierce little fellow' (*Queer Pets and Their Doings*)

29 Divers Exercices des Éléphants (*Animaux Savants*, 1816)

34 Steeds of metal and muscle, and Phantoms foot it (*Episodes of Insect Life*, Acheta Domestica, 1851)

38 Deer on tightrope (*Animaux Savants*)

42 Railway Jack (Gerald L. Wood)

47 'I can turn round ten times . . .' (*Sold for a Farthing*, Clare Kipps, 1953)

54 The three Apollos (*Queer Pets and Their Doings*)

60 'A Turkey carpet was their lawn . . .' (*Cowper's Poetical Works*, 1840)

65 'Comfortably tucked up in bed . . .' (*Drawn from Memory*, Ernest H. Shepard, 1957)

71 Greyhounds (*Animal Life*, Rev. J. G. Wood, 1854)

79 Hens' procession (*A Wonder Book of Beasts*, Ed. F. J. Harvey, 1909)

85 A new musical instrument (*La Nature*, c. 1885)

91 'Assassin'd by a thief' (*Grandville – Des Gesamte Werk*, 1972)

94 Kangaroo pie (*Mrs Beeton's Book of Household Management*)

96 'Mustapha seizes the still lit fuse . . .' (*Chiens Célèbres*, A. Antoine, 1796)

100 A noble sacrifice (Imperial War Museum)

107 'Peace at any Price' (*Researches into the History of the British Dog*, George R. Jesse, 1866)

111 Help the Horse (Imperial War Museum)

119 Her Master's Breath (Royal Naval Museum, Portsmouth)

126 Peter Spots (*St Nicholas Magazine*, August 1897)

135 'The threads of insect life . . .' (*Episodes of Insect Life*)

139 'It is positively affecting . . .' (Royal Naval Museum, Portsmouth)

144–7 Crabs, swan, good cow, bald mice (*Animaux Célèbres*, A. Antoine, 1813)

150 The rat conducting his blind companion (*Anecdotes of Animals*, Mrs R. Lee, 1852)

155 Thomas Henry and Mary Ann (*Petland*, Rev. J. G. Wood, 1890)

159 'I wonder whether . . .' (*The Life of John Mytton Esq.*, C. J. Appleby, 1870)

168 Sherry (painting by Frances Broomfield)

172 Fairground tricks (*Sports and Pastimes of the People of England*, Joseph Strutt, 1805)

177 Elephants (*St Nicholas Magazine*, January 1891)

184 Les Serins Savants (*Animaux Savants*)

187 Toby the Sapient Pig (*Learned Pigs and Fireproof Women*, Ricky Jay, 1987)

192 'The bow and the curtsey . . .' (*Animaux Savants*)

200 'The little animal would always find . . .' (*Petland*)

207 Le Cheval Aréonaute (*Animaux Savants*)

209 'A pretty little fellow' (*Notes and Jottings from Animal Life*, Frank Buckland, 1882)

214 'Something between a mole and a pig' (*Queer Pets and Their Doings*)

220 'He bid his horse to supper' (*The Wonders of the Little World*, Nathaniel Wanley, 1678)

224 'Miss Reynard, *la belle* . . .' (*Fox's Frolic*, Sir Francis Burnand)

227 The pig-faced lady (*Learned Pigs and Fireproof Women*)

231 Zoo (*Scènes des Animaux*, 1842)

239 'He drinks the air . . .' (*Animaux Savants*)

246 The fishing cat (*An Introduction to the Study of Backboned Animals, Especially Mammals*, St. George Mivart, 1881)

252 The wasps in their aesthetic home (*St Nicholas Magazine*, June 1889)

258 Mary Tofts (*The Works of William Hogarth*, 1872)

261 'Her heart beats grew slower...' (*Dante's Inferno, Purgatory and Paradise*, Illustrated by Gustave Doré, 1904)

265 'I am still hopeful...' (*Grandville – Des Gesamte Werk*)

269 'The handsome horse...' (*St Nicholas Magazine*, March 1891)

277 'His drollest trick' (*Anecdotes of Animals*)

283 'To lift one's Nose high in the Air' (*The Rubaiyat of a Scotch Terrier*, Sewell Collins, 1926)

288 Gnasher (*Dennis the Menace Annual*, 1983)

294 Grandville's cats (*Scènes des Animaux*)

302 'Thus he resolv'd...' (*Scènes des Animaux*)

308 Death certificate (British Naval Museum, Portsmouth)

315 'My dog lies dead' (*True Dog Stories*, Rudyard Kipling, 1934)

323 'Accepting all the comforts...' (*Queer Pets and Their Doings*)

332 Rossetti's wombat (Rossetti/Burn Jones Sketchbook)

⌐ INDEX OF SUBJECTS ⌐

Acland, Sir T. 203, 288
Addams, Charles 157
Aelian 39
Aldrovandus 23
Alexander, King of Yugoslavia
    123
Allgood, Mr Hunter 153–4
Alligators
    attacks fox with its tail 157
    Beelzebub 16–18
    cayman as beast of burden 162–3
    friendship with cat 157
    Uncle Tom 6–8
Ant-bear, served as food 151
Apes
    car-driving 99–100
    as railway signalman 41
    see also Monkeys
Aristotle 39
Armadillos, Rossetti's pets 166
Ascanio, Cardinal 38
Asses, milk of 259
Astley, Philip 133
Athens, Panayoti the seal 12
Aylesbury, horse's visit to hotel
    dining room 104–5

Baboon, Jack the railway worker
    41–2
Badgers, Ernest Thompson Seton's
    story about 85–9
Baigne, the Abbot of 84–5
Balzac, Honoré de 21
Barker, Betsy 148
Barlowe, Captain 105
Barnes, John 38
Barraud, Enid 120
Barraud, Francis 119
Barraud, Mark 120

Bears
    Edward James's pet 202
    Kola 210–12
    London barber's 'bear-killings'
        114–15
    Tiglath Pileser 202–4
Bees
    idiot-boy's propensity to 149–50
    performing 185
Beetles, Primo Levi on 33–6
Bell, Mrs M. E. 154
Belli, Giuseppe Gioacchino 20
Bergeret, M. 66–8
Birds
    performing 184–5
    at Woburn Abbey Aviary 245
    see also individual references, e.g.
        Dove, Sparrow, etc.
Blenheim Palace, emus at 15–16
Brahms, Johann 282
Bridgewater, Eighth Earl of 155
Brighton Aquarium 28
Browning, Robert 166
Buckland, Dean William 221
Buckland, Frank 79–80, 81–2, 165,
    169, 174, 176
Buffon, Comte de 22, 23
Bugle, Lady 278–80
Bug Mee, The 231
Bullfinch, Lady Throckmorton's
    90–2
Bullock, drowning of 293
Bull, ridden for shooting and
    hunting 154
Burney, Fanny
    164–5
Burton, Sir Richard 231–2
Butterflies as pets 133–6
Byron, Lord 176

Caligula 221
Camels
    coat-eating 164
    tame 204–6, 278
Campbell, Dr 224–6
Canaries
    André Gide on 234–5
    death certificate of 308
    in Dickens' *Bleak House* 163
Canino, Prince (Charles
    Buonaparte) 203
Carlton, Lady Maria 165
Cats
    Aldous Huxley on 293–4
    alligator's friendship with 157
    attachment to a mouse 147–8
    Babou, strawberry-eater 151
    Bajazet, Persian pet of M.
        Bergeret 66–7
    Candide, Elinor Glyn's Persian 165
    character of 29–30
    church-attending, at Morwenstow
        173
    Don Pierrot 245–6
    Jeoffrey 269
    Lunardi's ballooning pet 207–8
    Maltese, Rudyard Kipling's 97–9
    petty officer rescues drowning cat
        in Malta 131
    Seraphita 246–7
    ship-board pet 130
    Sir Henry Wyatt's prison pet 118
    Stré-ici, Max Beerbohm's pet 169
    Swift on 264
    Tabby, and Beelzebub the
        alligator 18
    Tessimond on 242
    Thomas Henry's friendship with
        tortoise Mary Ann 155–6

Cats – *cont.*
  Tom, Benjamin West uses fur of
    to make pencils 173
  Verlaine on 245
  William Cowper's pet 140–2
Chameleons
  Florida chameleon 24
  John Bargrave on 196–7
  mode of progression 25
  taming of 24–5
Charles II, King, spaniels of 16
Charlotte, Queen 164
Chestertown, Maryland 176–8
Chickens *see* Hens
Chimpanzees
  Apollo of Berlin Zoological
    Garden 53–4
  fire-using habits 41
  see also Monkeys
Clifton, Yorkshire 62
Coaitimundi, Kiko the tame
    199–201
Cockatoo, Mrs Buck's 284–6
Coleoptera 33–6
Colvin, Sir Sidney 54
Cook, Captain 93
Cooke, Adolphus 222–3, 230–1
Cooke, Robert 230
Cookesborough 222, 223, 230, 231,
    293
Cottrell Dormer family 321
Cows
  Burritt on 248–9
  Cherry, plays like a dog 156
  Faustina Gwynne, memorial to 321
  flannel-wearing Alderney 148
Cranes, as farmyard guardians
    123–4
Cromwell, Sir Henry 5
Cromwell, Oliver 5
Cruise, Tom 230
Currey, Fanny 265–6

Darwin, Charles, on worms 28
Davis, William 133
Deer
  Lulu, Karen Blixen's fawn 243–4
  Marvell on 309
  as a mount like a horse 38
Des Esseintes 240–2
Devonshire, Duke of 196
Dimnik, huntsman to Yugoslav
    king 132

Dingley, Mrs, verse on collar of her
    lapdog 68
Dogs
  Alice D. Cur, attends master's
    marriage in pet cemetery
    157
  Balthasar, in *The Forsyte Saga*
    312–13
  Benny, sheepdog 326, 327
  Billy, rat-killing terrier 103–4
  Bingo, black poodle 82–4
  Boatswain, Byron's dog 310–11
  Bob, London fireman's dog 113,
    121
  Bob, sheepdog, train travel 325
  Bob-tail, E. V. Lucas on 282
  bull terriers, 27–8, 40, 43, 99
  Camp, bull terrier 40, 99
  Caramba, Albert Schweitzer's pet
    291
  character of the dog 13
  church-attending, Morwenstow
    173
  cocker spaniels, E. V. Lucas on
    282
  Coco, Elizabeth von Arnim's pet
    313–14
  Dachshunds
    Geist 321–3
    Lucas on 282
  Diamond, Isaac Newton's pet 68
  Dodo and her puppy 287
  Duchess of York's pets 175–6
  Dusk, Karen Blixen's pet 174,
    175
  Eskimaux dog 117
  Feng Hou 236
  Fidele, Dane 328–9
  Flush, Elizabeth Barrett
    Browning's pet 316–19
  fox terriers, travels with 324–5
  Geist, dachshund 321–3
  General Norman Schwarzkopf on
    107
  George Eliot on 32
  Gusty, Adolphus Cooke's pet
    230–1
  harrier pack lives with tame fox
    157
  Her Imperial Majesty Tzu Hsi on
    236–7
  Hippy, Sir Nevile Henderson's
    pet 122–3

Hitler's pet 306
hounds
  guinea fowl and 153
  Shakespeare on 242
  Somerville on 115–16
human affection for 55–6
Jack Russell, origin of 208
James, Max Beerbohm's pet 168
Jim, memorial verse to 315–16
Jock, bizarre fate of remains of
    278–80
King Charles II and his spaniels
    16
Landor and 54, 304
lap-seeking habits 20
life as (Dean Spanley) 221–2
Louis the Pomeranian 296–9
maddened by swim in alkali
    water 295
Maida, Sir Walter Scott's
    staghound 70–2
Marco, musical passion of 281–2
Maria, and the cockatoo 283–6
Mary Queen of Scots' pet 62
mastiffs 121
measles cure myth 268
Nero, Thomas Carlyle on 320
Nipper, the 'His Master's Voice'
    dog 118–20
Owd Bob of Kenmuir 68–70
Paris shoe-black's poodle 114
Pekinese, E. V. Lucas on 282
Percy, Sir Walter Scott's dog 104
performing 188–94
Peter Spots 125–8
Pomeranians
  Louis 296–9
  on *On Forsyte 'Change* 237–8
poodles 82–4, 114, 188–9, 193–4
Pomero, Walter Savage Landor's
    pet 54–5
Rex the bull terrier 43
Riquet, M. Bergeret's pet 66–8
Roger, and Ulysses the owl
    169–70
Rough, shares master's breakfast
    with hen 154
Scotch Terrier, Sewell Collins on
    282–3
served up as 'leg of mutton'
    279–80
service to man 114, 124–5
sheepdogs 325, 326, 327

Signor Fido, greyhound, memorial
    to 319–20
Sir Walter Scott and 70–2, 319
spaniel called Poodle, musical
    appreciation 193–4
Sylvanus Urban on 128
Thistle, W. F. Deedes' Jack
    Russell 313
'Tich', Desert Rat 100
Toby, bull-terrier, moral
    excellence of 27–8
Tray, pointer, scents live
    partridges inside shark 130
Trump, Jack Russell terrier 208
Tulip, Alsatian bitch 250–1
Twain on 74–8
Tydeus, memorial to at Wynard
    Park 315
Wessex, wire-haired terrier, fan of
    the wireless 156
white poodle, performs on hearing
    Italian language 188–9
Whuff, Sir William Beach
    Thomas's dog 117–18
wild, in Masai Reserve 174–5
Donkey, Modestine 25–7
Doves 1, 5
Droll Stories (Balzac) 21
Ducks
    Doctor Hobson, Charles
        Waterton's pet 212–14
    Humphrey, Indian runner drake
        52–3
Dunne, Billy 230

Eagle, Frank Buckland's pet 81–2
Edwards, Milne 203
Elephants
    affection between 36
    empty railroad water-tank 176–8
    learning abilities 38–9
Ellenborough, Lady 196
Emin Pasha 40–1
Emus, at Blenheim Palace 15–16

Farah, Karen Blixen's native helper
    174–5
Faust 21
Ferret, Sredni Vashtar 255–8
Fish
    gilding of 162
    trout, memorial to 308
Fitzsimons, F. W. 42

Fleas
    Balzac on 21
    feeding of 183
    Gay on 296
    mention in Faust 21
    performing 180–3
    power in legs of 19, 21
    taming of 21
    training 181–3
    traps for 20
    true gypsy spirit of 21–2
Flower, Mr 105
Flies
    Cresta run 179
    Freddy the Fearless 54
    John Wesley and 299
Food, animals served up as 151,
    152, 169, 174, 279–80
Fotheringay 62
Foxes
    alligator attacks 157
    Burnand on 224
    clerical fox-hunter's tame fox
        159–60
    dog-riding 224
    Masefield on 129, 131–3
    Mr Tebrick's vixen 216–19
    tame fox lives with harrier pack
        157
Frogs
    green treefrog, morning song of
        272–3
    as pets 165, 281

Galba 39
Game-birds
    Hardy on 79
    Saki on 136
Garner, Mr 232
Geese, Mr Staveley's gander 62
Germanicus Caesar 39
Giraffes, character of 251
Glyn, Elinor 165
Goats
    Montagu on 272
    saint's bones declared to be those
        of goat 221
Goldfinches, Cowper on 303–4
Goldsworthy, Colonel 165
Gozzano 34
Guinea fowl, accompanies hounds
    153–4
Guinness, Mrs Benjamin 202

Hadfield, Lady 202
Haldane, J. B. 33
Hall, Basil 139
Hares, William Cowper's pets
    58–61
Hatton, Sir Christopher 118
Hawker, Robert Stephen, pets of
    173, 288, 295
Hayes, headquarters of 'Her
    Master's Voice' 120
Hedgehogs, as domestic animals
    120
Hens
    dog shares food with 154
    friendship with horse 158
    Houdan hen 255–6
    Shepard's tame black hen
        65–6
Herne Bay 109–10
Herrings, at Brighton Aquarium
    28
Hippopotamuses
    Hippo, zoo animal 200
    Obaysch, Sir Charles Murray's
        pet 204
Hirst, Mr Jemmy 154
History of the Westmeath Hunt 222
HMS Ariadne, tame lion on board
    52
HMS Leander, crew member rescues
    drowning cat 131
Hobson, Dr 219
Hogs
    Abbot of Baigne and 84–5
    gardener and the hog 267–8
    hog-faced gentlewoman 226–9
    see also Pigs
Horses
    African giant mare 269–70
    Arab proverb 252
    Baronet, Squire Mytton's pet 158
    Bataille, pet pony 80, 81
    biblical reference to 105
    Billy, little circus animal 133
    Caligula entertains horse to
        supper 221
    Copenhagen, Wellington's horse
        110
    cruelties against 92
    dancing upon a rope 180
    Fanney Currey's charm 265–6
    friendship between pig and pony
        153

Horses – *cont.*
  friendship with hen 158
  Frou Frou in *Anna Karenina*
    247–8
  the Genuine Mexican Plug 30–2
  how to encourage or restrain
    staling 185–6
  King Pippin 262–3
  Lady Ellenborough's ponies 196
  Mazeppa tied to back of wild
    horse 167–8
  memorial to a 93
  military 101, 112, 273
  Mrs Knox's stolen colt 197–9
  racehorse, Right Royal 101–3
  Shakespeare on 249–50
  Squire Mytton's 158–9
  Sybarian dancing 189
  Trompette, pet pony 80–1
  Valiant, soldier's horse 101
  Virgil on 239–40
  visits dining room 104–5
Houghton, Lord 203
Houseflies
  Freddy the Fearless 54
  John Wesley and the fly 299
Howe, Rev. George 42

Insects *see* Beetles, Flies, Spiders
Islip 203–4

Jackal, Frank Buckland's pet 79–80
Jamrach's 166
Jocelyn, Lord 104
Jollyboy, Captain 206

Kafka, Franz 35
Kangaroos
  Captain Cook and 93–4
  Rossetti's pets 166, 167
Kearney, Mr 154
Knox, Mrs 197–9
Kremlin, Museum 20

Laistre, Mademoiselle de 22
Landor, Walter Savage 54–5, 304
Lawrence, D. H. 294
Leather-mouthed Jemmy 115
Lincoln, Mrs 13–14
Lion-monkey 202
Lions
  as domestic animals 13–14
  Prince, HMS *Ariadne* pet 52

Lizards
  Carolina lizards 165
  see also Chameleons
Lobsters as pets 153
Long, Mr 15–16
Lunardi, Mr 206–8
Lyell, Sir Charles 203

Malta 131
Manatee as pet 214
Manning, Mr 105
Marlborough, Duke of 15–16
Marmoset
  Jimmy, memorial to 321
  see also Monkeys
Mary, Queen of Scots 62
Masai Reserve 174
Maximilian the Second, Emperor 38
Meercat, Frank Buckland's pet
    208–10
Merriman, Dr 169
Mice
  prison mouse trained by prisoner
    303
  reaction to violin-playing 39
  singing 37, 148–9
Milnes, Monckton 203
Mitchell, Mr 170
Molande, Countess de 55
Monastier 25–6
Monkeys
  Billy, whisky-drinking pet 224–6
  Carroty Jane and Little Jack 210
  dressed as postillion 128
  'The Hag' 107–10
  Jack, Cercopithecus monkey
    276–7
  Jacko 82
  language of 232
  Master Jacko, the horse-rider 144
  served as food 151
  steals Oliver Cromwell from
    cradle 5
  Tiny 107–10
  violin-playing 292
Moore, James 68, 70
Moore, Thomas 310
Morris, Janey 166
Morwenstow, vicar's
    church-attending cats and
    dogs 173
Mount Edgecumbe, Countess of
    310

Mouse *see* Mice
Msongwa, forest of 41
Mswa Station 41
Mule
  kicking power 19, 21
  kiss of, as cold cure 204
Murray, Sir Charles 204
Museum of Natural History, Paris
    36
Mytton, Squire John 158–9

Nerval, Gerard der 153
Newton, Isaac 68

Orang-outang, Charles Waterton
    and 170–1
Otter, Gavin Maxwell's pet, Mijbil
    62–5
Owen, Mrs Mostyn 110
Owls
  Athena 2–5
  as pets 273–4
  screech owl 222–3
  Ulysses and Roger the dog
    169–70
  Wol 273–4
Oxen, horse-riding 180
Oxford 202–3
Oysters, 'Lord' George Sanger's
    tame 171–2

Pakenham, Hon. Edward 223
Panther, Frank Buckland eats 169
Paris 114
Parke, Surgeon 41
Parrakeet, Mrs Thicknesse's pet
    305–6
Parrots
  Greek-speaking 254
  language-learning abilities 43
  Loulou 259–61
  speaking and singing abilities 38
Parthenon 2–5
Peaston, George 28
Philpotts, Dr 288
Pigeon, Gerald Durrell's pet,
    Quasimodo 57
Pigs
  capital punishment of 266–7
  character of 32–3
  Cupid, memorial to 310
  domestication 33
  friendship with pony 153

Gyp, pet 295
Latimer Springfield as host to
    136–9
monument to a, in Lunenburg
    309–10
performing 186–8
pig-faced lady 229–30
as ship's pet 139–40
Slut, the pointer-pig 43–4
Toby 186–8
vocal powers 33
see also Hogs *and* Wild Boar
Pliny 39
Prairie dog, Frank Buckland's pet,
    Jemmy 208–10
Purdon, Dr Wellington 222, 223

Rabbits, Blue Ridge 221
Raccoon, Rossetti's pet 166, 167
Raikes, Thomas 176
Rats
    blind rat led by companion 150–1
    found in lady's bustle 275
    as pets 40, 81
    served as food 152
Raven, Pliny on talking raven 78–9
Rawlinson, Canon and Mrs 203
Reinham, Mons. Auguste 180
Ricardo, Mr 104
Romanes, Professor 41
Rossetti, Dante Gabriel 66, 166–7,
    330
Ryall, Arthur 201–2

Saint-Pierre, Bernardin de 20
Saunders, Abraham 133

Sclater, Phillip Lutlet 204
Schwellenberg, Mrs 165
Scott, Sir Walter 70–2
Seal, Panayoti 8–12
Sheep as pets 57, 74
Slark, Captain 109
Slugs, red, Frank Buckland's pets
    176
Snakes
    Coulacanara 105–6
    Frank Buckland's pets 81
Sodre, Luisina de 55
Southport 8
Sparrows
    Clarence 44–50
    Philip Sparrow 323–4
Spiders, music-loving 151
Stags, Robert Hawker's pets 288
Staveley, Mr 62
Stevens, Madame 229–30
Stolberg, Count of 38
Strand Magazine 119
Strozza 23
Suetonius 39
Swinburne, Algernon Charles
    166
Sybaris, Calabria 189
Symonds, Mr Chas 104–5

Tebrick, Mr 216–19
Thicknesse, Mrs 305–6
Tiberius 39
Toad, Bahia toad 249
Tortoises
    Achilles, Gerald Durrell's pet 57
    encased in jewels 240–2

Mary Ann, cat's friendship with
    155–6
Turkey, Robert Cooke believed
    reincarnated as 230
Tzu Hsi, Her Imperial Majesty
    236–7

Vandermeerscht, Emilie 184
Vanno, the Bear Man 211–12

Walcot, Dr John 310
Walker, Mrs 109–10
Wallaby, roasted 94
Walpole, Edward 272
Wasps, and curtain tassel 251–2
Waterton, Charles 170–1, 219, 249
Watson, Mr 223
Weasel, taming of 22–3
Wellington, Duke of 110
Wesley, John 299
West, Benjamin 173
Westmeath Hunt 222
Wide, Mr 41–2
Wild boar, as pet 290–2
Wild dogs in Masai Reserve 174–5
Wildman, Daniel 185
Williams, Dr William 223
Wirkham 226–9
Wombats, Rossetti's pets 66, 330
Woodchuck, Rossetti's pet 330
Worms, mental capacity of 28
Wyatt, Sir Henry 118

Yeates, Major 197, 198
York, Duchess of 175–6
Yugoslavia 122–3

❧ INDEX OF AUTHORS ❧
AND SOURCES

Abbeys and Castles (John Timbs) 118
Achievement of the Cat, The (Saki)
    29–30
Ackerley, J. R. 250–1
Adam Bede (George Eliot) 32
Aeschylus 72
All the Dogs of My Life (Elizabeth von
    Arnim) 313–14
American Red Star Animal Relief
    poster 111
Amethyst Ring, The (Anatole France)
    68
*Anecdotes of the Habits and Instincts of
    Animals* (Mrs R. Lee) 39, 151,
    276–7
'Animal Alphabet, An' (Neville
    Braybrooke) 28
Animal World of Albert Schweitzer
    (Albert Schweitzer) 290–2
Anna Karenina (Leo Tolstoy) 247–8
Anstey, F. 82–4
Arnim, Elizabeth von 313–14
Arab proverb 252
À Rebours (J.-K. Huysmans) 240–2
Arnold, Matthew 321–3

Balzac, Honoré de 292
Bargrave, John 196–7
Baring-Gould, Sabine 173, 288, 295
Barrie, J. M. 156
Beastly Buildings (Lucinda Lambton)
    310, 321
Beerbohm, Max 168–9
Berners, Lord 204–6, 278–80
Bingley's Animal Biography 23, 36,
    120
Biot, Jean-Baptiste 68
Black Poodle and Other Stories, The (F.
    Anstey) 82–4

Bleak House (Charles Dickens)
    163
Blixen, Karen 174–5, 243–4, 254
Blockley, Gloucestershire, memorial
    to murdered trout 308
Boardman, Eleanor Page 1, 5
Bompas, George C. 79–80, 81–2,
    161, 169, 173–4, 176, 202–4,
    272–3
Book of Marco Polo, The 121
*Book of Quadrupeds for the Instruction of
    Young People, The* (W. R.
    Macdonald) 156
Boy and the Badger, The (Ernest
    Thompson Seton), 85–9
Braybrooke, Neville 28
Brown, John 27–8
Browning, Elizabeth Barrett 1, 5
Buckland, Frank 6–8, 15–16, 28,
    93–4, 107–10, 114–15, 180–3,
    208–10, 214, 243
Bucks Chronicle 104–5
Burghley, Lord 62
Burnand, Sir Francis 224
Burritt, Elihu 249
Burton, Isabel 231–2
Byron, Lord 167–8, 310–11

Camel, The (Lord Berners) 204–6,
    278–80
Canetti, Elias 79, 232
Carlyle, Thomas 320
Cavelarice, or the English Horseman
    (Gervase Markham) 92, 185–6
*Certaine Relation of the Hog-faced
    Gentlewoman called Mistris Tannakin
    Skinker . . .* 226–9
'Chace, The' (William Somerville)
    115–16

*Chamber's Miscellany of Useful and
    Entertaining Tracts* 189–94
*Charles Waterton: His Home, Habits
    and Handiwork* (Richard Hobson)
    170–1, 212–14, 249
Coeur Simple, Un (Gustave Flaubert)
    253
Colette 151
Collins, Sewell 282–3
*Constable's Miscellany Vol XLVII, Life
    of Oliver Cromwell* 5
Courtiers: 900 Years of Court Life
    (Thomas Hinde) 165
Cowper, William 57–61, 90–2,
    140–2, 303–4
Cranford (Mrs Gaskell) 148
Crawshaw Booth Pet Cemetery 308
*Criminal Prosecution and Capital
    Punishment of Animals, The* (E. P.
    Evans) 266–7
Croxton-Smith, A. 154
Curiosities of Natural History (Frank
    Buckland) 114–15, 243
*Cyclopaedia, The; or, Universal
    Dictionary of Arts, Sciences, and
    Literature* (Abraham Rees) 162
Daily Telegraph 54, 157, 171, 313
*Dante Gabriel Rossetti: His Family
    Letters with a Memoir* (William
    Michael Rossetti) 330

Darwin, Charles 239
Davies, Edward W. L. 208
Davies, W. H. 74
Deedes, W. F. 313
Diary, John Evelyn's 16
Dickens, Charles 163
Dog Lover's Week-End Book, The (E.
    Parker and E. Croxton-Smith) 154

Dog Who Wouldn't Be, The (Farley Mowat) 273–4

Dogs, Birds and Others (Hugh Massingham) 188–9

Dogs' Cemetery, Oatlands, The (J. W. Lindus Forge) 175–6

Drawn from Memory (Ernest Shepard) 65–6

Duncan, Ronald 112

Dunn, Harry 166–7

Dunsany, Lord 221–2

Durrell, Gerald 57, 169–70, 287

Edgeworth, Maria 319

Elinor Glyn: A Biography (Anthony Glyn) 165

Eliot, George 32

Ellis's Letters (Lord Burghley) 62

English Eccentrics (Edith Sitwell) 154, 158–9, 219

English Eccentrics and Eccentricities (John Timbs) 185

Epitaphs: Quaint, Curious and Elegant (Henry James Loaring) 309–10

Essay on Man (Alexander Pope) 325

Evans, E. P. 266–7

Evelyn, John 16

Fables (John Gay) 267–8, 296

Facts and Phases of Animal Life (Vernon S. Morwood) 299

Fidele's Grassy Tomb (Sir Henry Newbolt) 328–9

Field, Kate 304

Flaubert, Gustave 253

Flush: A Biography (Virginia Woolf) 316–19

Forge, J. W. Lindus 175–6

Forster, John 320

Forsyte Saga, The (John Galsworthy) 312–13

'Fox's Frolic' (Sir Frances Burnand) 224

France, Anatole 66–8

Galsworthy, John 237–8, 312–13

Garnett, David 216–19

Gaskell, Mrs 148

Gaunt, William 166

Gautier, Théophile 153, 245–7

Gay, John 267–8, 296

Geist's Grave (Matthew Arnold) 321–3

Gentleman's Magazine, The 57–60, 128, 155

Georgics (Virgil) 239–40

Germinal (Émile Zola) 80–1

Gide, André 234–5

Gleanings in Natural History (Edward Jesse) 114, 157

Glyn, Anthony 156

G. M. 321

Goldsmith, Oliver 114

Gordon, Mrs E. O. 221

Gosson, Stephen 221

'*Happy Days!*' (Somerville and Ross) 20, 265–6, 271, 275, 324–5

Hardy, Thomas 79, 112

Harper's Magazine 8–12

Haswell House, Goathurst, Somerset 93

Henderson, Sir Nevile 122–3

Henley-on-Thames, gravestone on Oxford Road 321

Henry V (William Shakespeare) 97, 233, 249–50

Henty, G. A. 19, 21–2, 32–3

Herbert, A. P., a Biography (Reginald Pound) 273

Hinde, Thomas 165

Hippy, The Story of a Dog (Sir Nevile Henderson) 122–3

His Master's Voice Picture, The Story of Nipper and the (Leonard Petts) 118–20

Hitler, Adolf 306

Hobson, Richard 170–1, 212–14, 249

Hon. Sir Charles Murray KCB: A Memoir (Sir Herbert Maxwell) 204

Hood, Thomas 299–303

Horae Subsecivae (John Brown) 27–8

'Horse, The' (Ronald Duncan) 112

Horse, The (William Youatt) 262–3

'Horses Abroad' (Thomas Hardy) 112

Huxley, Aldous 293–4

Huysmans, J.-K. 240–2

If I Die (André Gide) 234–5

Ilford Pet Cemetery 100

Illustrated London News 184

In My Mother's House (Colette) 151

Inordinate Affection, A Story for Dog Lovers (Ethel Smyth) 55–6, 281–2

Irish Eccentrics (Peter Somerville-Large) 222–3, 230–1, 293

Irving, Washington, 70–2

Jack-in-the-Pulpit 176–8, 251–2

Jackson, Rev. T. 101, 113, 121, 147–8, 157

James, Edward 201–2

Jay, Ricky 133

Jesse, Edward 114, 157

Job 39: 19–25 105

Journals of Canetti (Elias Canetti) 79, 232

'Jubilate Agno' (Christopher Smart) 269

Kane, Elisha Kent 117

Kenny, Muriel 52–3

Kipling, Rudyard 97–9

Kipps, Clare 44–50

Kola the Bear (Goerge and Helen Papashvily) 210–12

Ladas, Alexis 8–12

Lady into Fox (David Garnett) 216–19

Lambton, Lucinda 310, 321

Land and Water 183

Last Days of Walter Savage Landor, The (Kate Field) 304

Learned Pigs and Fireproof Women (Ricky Jay) 133

Lee, Mrs R. 39, 151, 276–7

Leicester Square: Its Associations and Its Worthies (Tom Taylor) 180

Levi, Primo 20–1, 32–6

Life of Captain Sir Richard Burton by his Wife Isabel 231–2

Life of Dean Buckland (Mrs E. O Gordon) 221

Life of Frank Buckland (George C. Bompas) 79–80, 81–2, 161, 169, 173–4, 176, 202–4, 272–3

Life of Newton (Jean-Baptiste Biot) 68

Life with Rossetti (Gale Pedrick) 66, 166–7

Lincoln, Abraham 221

Loaring, Henry James 309–10

Log Book of a Fisherman and Zoologist (Frank Buckland) 16, 28, 93–4, 107–10

Longford, Elizabeth 110

Lucas, E. V. 13, 54–5, 236–7, 282, 320

Lunardi's Grand Aerostatic Voyage through the Air . . . 206–8

Macaulay, Rose 99–100

Macdonald, W. R. 156

Maëterlinck, Maurice 124–5

'Maltese Cat, The' (Rudyard Kipling) 97–9

Marco Polo, The Book of 121

Markham, Gervase 92, 185–6

Marvell, Andrew 309

Masefield, John 101–3, 129, 131–3

Massingham, Hugh 53, 188–9

Maxwell, Gavin 62–5

Maxwell, Sir Herbert, Bart, MP, FRS 204

'Mazeppa' (Lord Byron) 167–8

Memoir of the Rev. John Russell (Edward W. L. Davies) 208

Midsummer Night's Dream, A (William Shakespeare) 242

Miller, Olive Thorne 13–14, 24, 53–4

Montagu, Elizabeth 272

Monuments of Genii of St Paul's and Westminster Abbey (George Lewis Smyth) 173

'More I see of Men . . . , The' (E. V. Lucas) 13, 54–5, 236–7, 282, 320

Morwood, Vernon S. 299

Mowat, Farley 273–4

Munthe, Axel 224–6

Monro, H. H. *see* Saki

My Dog (Maurice Maëterlinck) 124–5

My Dog Tulip (J. R. Ackerley) 250–1

My Family and Other Animals (Gerald Durrell) 57, 169–70, 287

My Talks with Dean Spanley (Lord Dunsany) 221–2

Natural History (Pliny the Elder) 78–9, 259, 268

Natural History and Antiquities of Selborne, The (Gilbert White) 149–50, 158

Nature 40–2

Newbolt, Sir Henry 328–9

Nightingale, Florence 2–5

Nipper and the His Master's Voice Picture, The Story of (Leonard Petts) 118–20

Nisbet, E. G. 41–2

Notes and Jottings from Animal Life (Frank Buckland) 6–8, 114, 208–10, 214

Notes and Queries 268

Ollivant, Alfred 68–70

On Forsyte 'Change (John Galsworthy) 237–8

Original Travels and Surprising Adventures of Baron Munchausen (Rudolf Eric Raspe) 130

Other People's Trades (Primo Levi) 20–1, 32–6

Our Dumb Companions (Rev. T. Jackson) 101, 113, 121, 147–8, 157

Out of Africa (Karen Blixen) 174–5, 243–4, 254

Owd Bob: The Grey Dog of Kenmuir (Alfred Ollivant) 68–70

Pantagruel (François Rabelais) 269–70

Papashvily, George and Helen 210–12

Parker, E. 154

Parlour Menagerie, The 43, 52, 62, 123–4

Pearson, Albert Carlton 16–18

Pedrick Gale 66, 166–7

Petland Revisited (Rev. J. G. Wood) 25, 37, 40, 43–4, 133–6, 139–40, 148–9, 152, 155–6, 199–201, 281, 303

Petts, Leonard 118–20

Philip Sparrow (Matthew Arnold) 323–4

Physiology of Marriage, The (Honoré de Balzac) 292

Pliny the Elder 78–9, 259, 268

Pope, Alexander 325

Pope Alexander the Seventh and the College of Cardinals . . . (John Bargrave) 196–7

Portland, Duchess of 272

Portraits et Souvenirs Littéraires (Théophile Gautier) 153

Pound, Reginald 273

Praeterita (John Ruskin) 165

Pre-Raphaelite Tragedy, The (William Gaunt) 166

Pückler-Muskau, Prince M. L. H. von 103–4, 196, 202, 235–6, 245, 251

Queer Pets and Their Doings (Olive Thorne Miller) 13–14, 24, 53–4

Rabelais, François xiv, 269–70

Raspe, Rudolf Erich 130

Rees, Abraham 162

Retired Cat, The (William Cowper) 140–2

Reynard the Fox (John Masefield) 129, 131–3

'Right Royal' (John Masefield) 101–3

Ring of Bright Water (Gavin Maxwell) 62–5

Robertson, W. Graham 325–7

Romanes' Animal Intelligence 157

Ross *see* Somerville and Ross

Rossetti, William Michael 330

Roughing It (Mark Twain) 30–2, 164, 295

Rubaiyat of a Scotch Terrier, The (Sewell Collins) 282–3

Ruskin, John 165

St Nicholas Magazine 125–8

Saki (H. H. Munro) 29–30, 136–9, 254–8, 296–9

Sanger, 'Lord' George 171–2, 186–8, 229–30

Saunders, A. D. 131

Schoole of Abuse, The (Stephen Gosson) 221

Schwarzkopf, General Norman 107

Schweitzer, Albert 290–2

Scott, Sir Walter 40, 99, 104, 311, 319

Second Grimmel Expedition in Search of Sir John Franklin, The (Elisha Kent Kane) 117

Sermons on Cats (Aldous Huxley) 294

Seton, Ernest Thompson 85–9

Seventy Years a Showman ('Lord' George Sanger) 171–2, 186–8, 229–30

Shakespeare, William 97, 233, 242, 249–50

Shelley, Percy Bysshe 176
Shepard, Ernest 65–6
Si le grain ne meurt (André Gide)
    234–5
Sitwell, Edith 154, 158–9, 219
Skelton, John 323–4
Sketches and Anecdotes of Animal Life
    (Rev. J. G. Wood) 144
Smart, Christopher 269
Smyth, Ethel 55–6, 281–2
Smyth, George Lewis 173
Sold for a Farthing (Clare Kipps)
    44–50
Some Experiences of an Irish R.M.
    (Somerville and Ross) 197–9,
    283–6
Somerville-Large, Peter 222–3,
    230–1, 293
Somerville and Ross 20, 197–9,
    265–6, 271, 275, 283–6, 324–5
Somerville, William 115–16
Spectator, the 153
*Sports and Pastimes of the People of
    England* (Joseph Strutt) 180, 189
Stevenson, Robert Louis 25–7
Story of San Michele, The (Axel
    Munthe) 224–6
Stowe, Buckinghamshire, inscription
    on a stone in mausoleum at
    319–20
Strutt, Joseph 180, 189

Swans Reflecting Elephants (Edward
    James) 201–2
Swift, Jonathan 68, 264

Taylor, Tom 180
Tessimond, A. S. J. 242
Thomas, Sir William Beach 117–18
Thorne Miller, Olive 13–14, 24, 53–4
Those Other Animals (G. A. Henty)
    19, 21–2, 32–3
Thurber, James 43
Timbs, John 118, 185
Time Was (W. Graham Robertson)
    325–7
Tolstoy, Leo 247–8
*Tour in Germany, Holland and
    England . . .* (Prince M. L. H. von
    Pückler-Muskau) 202, 235–6,
    245, 251
Towers of Trebizond, The (Rose
    Macaulay) 99–100
Travels with a Donkey in the Cevennes
    (R. L. Stevenson) 25–7
Twain, Mark 30–2, 74–8, 164, 295

Urban, Sylvanus 128, 155

Verlaine, Paul 245
Vicar of Morwenstow, The (Sabine
    Baring-Gould) 173, 288, 295
Virgil 239–40

*Walk from London to Land's End and
    Back* (Elihu Burritt) 249
Walton and Weybridge Local
    History Society 175–6
Wanderings in South America (Charles
    Waterton) 105–6, 151, 163
Wanley, Nathaniel 38–9, 84–5,
    185
Waterton, Charles 105–6, 151,
    163
Way of a Dog, The (Sir William
    Beach Thomas) 117–18
Wellington: The Years of the Sword
    (Elizabeth Longford) 110
White and Black Dynasties, The
    (Théophile Gautier) 245–7
White, Gilbert 149–50, 158
Wonders of the Little World, The
    (Nathaniel Wanley) 38–9, 84–5,
    185
Wood, Rev. J. G. 25, 37, 40, 43–4,
    133–6, 139–40, 144, 148–9, 152,
    155–6, 199–201, 281, 303
Woolf, Virginia 316–19
Worsthorne, Peregrine 171
Wynard Park, County Durham,
    memorial to a dog 315

Youatt, William 262–3

Zola, Émile 80–1